Histopatholog
Acute Radiation S
and Delayed E
Rhesus M

Kidney, Lung, Heart, Int
Mesenteric Ly

Histopathology Atlas of Acute Radiation Syndrome and Delayed Effects in Rhesus Macaques

Kidney, Lung, Heart, Intestine, and Mesenteric Lymph Node

George A. Parker

Global Pathology, Charles River Laboratories, Durham, NC, United States

ELSEVIER

ACADEMIC PRESS

An imprint of Elsevier

Academic Press is an imprint of Elsevier
125 London Wall, London EC2Y 5AS, United Kingdom
525 B Street, Suite 1650, San Diego, CA 92101, United States
50 Hampshire Street, 5th Floor, Cambridge, MA 02139, United States
The Boulevard, Langford Lane, Kidlington, Oxford OX5 1GB, United Kingdom

Notices
Knowledge and best practice in this field are constantly changing. As new research and experience broaden our understanding, changes in research methods, professional practices, or medical treatment may become necessary.

Practitioners and researchers must always rely on their own experience and knowledge in evaluating and using any information, methods, compounds, or experiments described herein. In using such information or methods they should be mindful of their own safety and the safety of others, including parties for whom they have a professional responsibility.

To the fullest extent of the law, neither the Publisher nor the authors, contributors, or editors, assume any liability for any injury and/or damage to persons or property as a matter of products liability, negligence or otherwise, or from any use or operation of any methods, products, instructions, or ideas contained in the material herein.

ISBN: 978-0-323-91393-5

For Information on all Academic Press publications
visit our website at https://www.elsevier.com/books-and-journals

Printed in the United States of America
Last digit is the print number: 9 8 7 6 5 4 3

Publisher: Charlotte Cockle
Acquisitions Editor: Anna Valutkevich
Editorial Project Manager: Lena Sparks
Production Project Manager: Niranjan Bhaskaran
Cover Designer: Miles Hitchen

Typeset by MPS Limited, Chennai, India

Contents

Foreword

Histopathology Atlas of Acute Radiation Syndrome and Delayed Effects in Rhesus Macaques: Kidney, Lung, Heart, Intestine and Mesenteric lymph Node

The Consortium, Medical Countermeasures Against Radiological Threats (MCART), funded by the National Institutes of Allergy and Infectious Diseases was formed with a focus on assessment of medical countermeasure (MCM) efficacy against acute, radiation-induced multiple organ injury (MOI). This required the development of validated animal models in small and large species, mouse, rat, and nonhuman primate (rhesus macaque) focused on obtaining MCM approval under the criteria of the US Food and Drug Administration (FDA) "animal rule" and guidance criteria. The mouse, rat, and rhesus macaque (NHP) models anchored the MCART "research platform" required for definitive experimental design to assess MCM efficacy and the duration of acute and delayed radiation-induced MOI within the context of a radiation terrorist event or nuclear disaster.

One of the critical components of study design is the effort to establish the natural history of the multiorgan sequelae within the concomitant acute radiation syndrome (ARS), gastrointestinal (GI-) and hematopoietic (H-ARS), and the delayed effects of acute radiation exposure (DEARE) focused on prolonged GI injury, and delayed lung, heart, and kidney injury. Establishing the natural history of the organ-specific sequelae over the duration of latency, incidence, severity, progression, and duration of effects is difficult when considering the range of radiation dose- and time-dependent, clinical, radiographic, cellular, humoral mediators and histopathology. The gaps in knowledge are significant, especially relative to the potential link between the ARS and the time course and duration of the DEARE, in all animal models. It is of significant import relative to the NHP (rhesus macaque). To date, the NHP is considered the pivotal animal model that most closely mimics the human response to acute radiation and treatment. Therefore the NHP is widely considered the best model to assess MCM efficacy for FDA approval under the animal rule. The NHP has been used in adequate and well-controlled efficacy studies to gain FDA approval for four MCM under the animal rule; these are Neupogen, Neulasta, Sargramostim, and Romiplostim.

The radiation exposure in the animal model was designed to mimic that experienced in a nuclear terrorist event. The exposure protocol used partial-body irradiation with marginal bone marrow sparing. The exposure protocol allowed analysis of acute radiation-induced organ-specific injury in young, adult male rhesus macaques that had survived lethal doses to the GI- and H-ARS, due to marginal bone marrow sparing. The higher threshold exposure resulted in continued development of the delayed effects of irradiation and allowed analysis of dose- and time-dependent lethal injury to the lung and overt sublethal injury to the kidney and heart. An added value is the time-dependent analysis of the concomitant MOI within both the ARS and DEARE to support definition of the latency, incidence, severity, progression, and duration of organ-specific injury through the 180-day study duration.

The objective herein was to close a major gap in the documentation of acute radiation exposure in the rhesus macaque. The analysis defines the histopathological time course of acute radiation-induced injury to the GI, jejunum and proximal colon, mesenteric lymph node, lung, kidney, and heart via definitive histochemical staining of radiation-induced cellular and mediator-based organ pathology.

The invaluable analysis provided herein presents a database in support of future natural history studies, using the 3Rs—reduce, refine, and replace, to focus on selected gaps in knowledge that will define the time course of critical trigger points for efficient definition of acute radiation effects in the nonhuman primate as well as clinically relevant administration and efficacy of MCM against lethal MOI.

Thomas J. MacVittie
Department of Radiation Oncology, School of Medicine,
University of Maryland, Baltimore, MD, United States

Preface

OVERVIEW OF STUDIES

The histopathological specimens serving as a basis for this Atlas were derived from a series of studies performed at the University of Maryland School of Medicine, Baltimore, MD, USA. The objectives of these studies included an assessment of the natural history of delayed effects of acute radiation exposure in a nonhuman primate (NHP) model of acute high-dose, partial-body irradiation with 5% or 2.5% bone marrow sparing (PBI/BM5 or PBI/BM2.5, respectively) to include the clinical, radiographic, and histopathological changes in the kidney, lung, heart, jejunum, colon, and mesenteric lymph node.

IRRADIATION PROTOCOL

Male rhesus macaques were exposed to 10.0, 11.0, or 12.0 Gy or 11.0 Gy utilizing 6 MV LINAC-derived photons (2 MV average) at a dose rate of approximately 0.80 Gy min^{-1} (TrueBeam, Varian Medical Systems, Palo Alto, CA, USA). Radiation dose calculations were performed by a board-certified medical physicist, based on a sagittal measurement taken at the xiphoid process. The dose was calculated for delivery to midline tissue, including correction factors to account for the inverse-square distance, tissue depth, and sources of photon scatter as per a published protocol (MacVittie et al., 2012). The NHPs were positioned with their tibiae, ankles, and feet outside of the beam field. Half of the prescribed dose was delivered with an anteroposterior beam, and half with a posteroanterior beam.

ANIMAL CAGE-SIDE OBSERVATIONS

Cage-side observations including nonsedated respiratory rate (NSRR) were performed twice daily by trained staff at least 6 hours apart. Clinical parameters were assessed two to three times per week on sedated animals as described previously (MacVittie et al., 2012). The NSRR was obtained while the animal was in its home cage to reduce NSRR alterations associated with anxiety. Respiration was measured by observing the rise and fall of the chest cavity. A cycle of one inspiration and exhalation was counted as one breath. Respiratory effects were also characterized by chest involvement and open-mouth breathing.

MEDICAL MANAGEMENT

All animals were administered medical management in accordance with an Institutional Animal Care and Use Committee-approved protocol that specified the clinical triggers to start and stop treatments. Medical management included hydration fluids, antibiotics, analgesics, antidiarrheal agents, antipyretics, antiemetics, antiulcerative agents, nutritional support, and blood transfusions. Specific details as to initiation, dosages, and duration of treatments were described (Farese et al., 2012).

DEXAMETHASONE TREATMENT, NONSEDATED RESPIRATORY RATE

Dexamethasone (Butler Schein, Dublin, OH, USA) was administered to animals noted to be in respiratory distress (> 80 bpm) during daily cage-side observations. The animals were treated with intramuscular injections using a planned taper at the end of dosing: 1 mg kg^{-1}, twice daily (BID) on the first day of treatment, 0.5 mg kg^{-1} BID for 3 days, 0.5 mg kg^{-1} once daily for 3 days, and 0.5 mg kg^{-1} every other day for three doses (Garofalo et al., 2014).

EUTHANASIA

Most animals were euthanized based on predetermined clinical parameters that were largely focused on respiratory effects. A subset of animals in one study was euthanized by a predetermined schedule: 6–7 ($N=2$), 34 ($N=3$), 44–45 ($N=3$), 61–62 ($N=4$), 99–102 ($N=3$), or 119 ($N=1$) days postirradiation. Animals irradiated at 10 Gy were necropsied from postirradiation days 6 through 191. Animals irradiated at 11 Gy were necropsied from postirradiation days 9 through 185. Animals irradiated at 12 Gy were necropsied from postirradiation days 4 through 102. Of the animals irradiated at 12 Gy, only two survived beyond 30 days postirradiation. The survival of those two animals was 81 and 102 days.

NECROPSY AND HISTOLOGY PROCEDURES

Animals were necropsied immediately after death and selected specimens of the kidney, lung, heart, jejunum, ascending colon, and mesenteric lymph node were immersion fixed in neutral buffered formalin. After a variable period of fixation, the tissue specimens were trimmed and subjected to routine histological processing for embedding in paraffin. The preceding procedures were conducted at the University of Maryland School of Medicine, Baltimore, MD, USA. Subsequently, paraffin blocks were shipped to Charles River Laboratories, Durham, NC, USA for histologic sectioning, staining, and microscopic examination. Immunohistochemical and in situ hybridization staining were performed using Ventana immunostainers. Microscopic examinations were performed by the author on the original glass microslides and observations were entered into Provantis histopathology data system. Selected microscopic observations were transferred into Microsoft Excel spreadsheets for further analysis. Stained sections were scanned using a Hamamatsu Nanozoomer scanner, and illustrative microscopic images were extracted using the Hamamatsu NDP Viewer 2 software.

A summary of the histologic, histochemical, immunohistochemical, and in situ hybridization stains performed on the various tissue specimens is presented in the following table:

Stain/marker	Jejunum	Colon	MesLN	Kidney	Lung	Heart
Routine morphologic stain						
Hematoxylin & eosin	X	X	X	X	X	X
Histochemical stains						
Masson's trichrome	X	X	X	X	X	X
Toluidine blue	X	X	X	X	X	X
Periodic acid-Schiff/hematoxylin				X		
Perls' iron stain				X		
Periodic acid-Schiff/Alcian blue	X	X				
Phosphotungstic acid/hematoxylin				X		
Chromotrope aniline blue				X		
Lendrum's stain	X					
Immunohistochemical stains						
Alpha smooth muscle actin	X	X	X	X	X	X
Bmi-1	X	X		X		
Caspase 3	X	X		X	X	X
CD3	X	X		X		
CD13	X	X				
CD20				X		
CD31	X	X		X	X	
CD38	X	X	X	X	X	
CD68				X		
CD163	X	X	X	X	X	
CD206					X	
Collagen-1	X	X	X	X	X	X
Connective tissue growth factor	X	X	X	X	X	X
IL-22	X	X		X		
IL-22R	X	X		X		
Ki-67	X	X				
Lipopolysaccharide	X	X	X			
Myeloperoxidase	X	X	X	X	X	
Pro-renin				X		
Transforming growth factor-beta	X	X	X	X	X	X
Tryptase	X	X	X	X	X	X
von Willebrand factor				X		
In situ hybridization						
Lgr5	X					

MesLN, *Mesenteric lymph node.*

ACKNOWLEDGMENTS

The contributions of the following individuals are gratefully acknowledged: Thomas J. MacVittie, MS, PhD; Ann M. Farese, MS; Eric P. Cohen, MD; Maureen A. Kane, PhD; Catherine Booth, PhD; Gregory L. Tudor; and Michael C. Garofalo, MD.

FUNDING

This work was supported by NIAID contract HHSN272201000046C and SRI/NIAID Contract HHSN272201500013I

REFERENCES

Farese AM, Cohen MV, Katz BP, Smith CP, Jackson W 3rd, Cohen DM, MacVittie TJ: A nonhuman primate model of the hematopoietic acute radiation syndrome plus medical management, *Health Phys* 103 (4):367–382, 2012.

Garofalo M, Bennett A, Farese AM, Ward A, Taylor-Howell C, Cui W, Gibbs A, Lasio G, Jackson WI, MacVittie TJ: The delayed pulmonary syndrome following acute high-dose irradiation: a Rhesus macaque model, *Health Phys* 106(1):56–72, 2014.

MacVittie TJ, Bennett A, Booth C, Garofalo M, Tudor G, Ward A, Shea-Donohue T, Gelfond D, Mcfarland E, Jackson W III, Lu W, Farese AM: The prolonged gastrointestinal syndrome in rhesus macaques: the relationship between gastrointestinal, hematopoietic, and delayed multi-organ sequelae following acute, potentially lethal, partial-body irradiation, *Health Phys* 103(4):427–453, 2012.

FURTHER READING

Carter CL, Hankey KG, Booth C, Tudor GL, Parker GA, Jones JW, Farese AM, MacVittie TJ, Kane MA: Characterizing the natural history of acute radiation syndrome of the gastrointestinal tract: combining high mass and spatial resolution using MALDI-FTICR-MSI, *Health Phys* 116:454–472, 2019.

Cohen EP, Hankey KG, Bennett AW, Farese AM, Parker GA, MacVittie TJ: Acute and chronic kidney injury in a non-human primate model of partial-body irradiation with bone marrow sparing, *Radiat Res* 188 (6):661–671, 2017.

Cohen EP, Hankey KG, Farese AM, Parker GA, Jones JW, Kane MA, Bennett A, MacVittie TJ: Radiation nephropathy in a nonhuman primate model of partial-body irradiation with minimal bone marrow sparing-part 1: acute and chronic kidney injury and the influence of Neupogen, *Health Phys* 116(3):401–408, 2019.

MacVittie TJ, Farese AM, Parker GA, Jackson W 3rd: The time course of radiation-induced lung injury in a nonhuman primate model of partial-body irradiation with minimal bone marrow sparing: clinical and radiographic evidence and the effect of Neupogen administration, *Health Phys* 116(3):366–382, 2019.

MacVittie TJ, Farese AM, Parker GA, Jackson W 3rd, Booth C, Tudor GL, Hankey KG, Potten CS: The gastrointestinal subsyndrome of the acute radiation syndrome in rhesus macaques: a systematic review of the lethal dose-response relationship with and without medical management, *Health Phys* 116(3):305–338, 2019.

Parker GA, Cohen EP, Li N, Takayama K, Farese AM, MacVittie TJ: Radiation nephropathy in a nonhuman primate model of partial-body irradiation with minimal bone marrow sparing-part 2: histopathology, mediators, and mechanisms, *Health Phys* 116(3):409–425, 2019.

Parker GA, Li N, Takayama K, Booth C, Tudor GL, Farese AM, MacVittie TJ: Histopathological features of the development of intestine and mesenteric lymph node injury in a nonhuman primate model of partial-body irradiation with minimal bone marrow sparing, *Health Phys* 116(3):426–446, 2019.

Parker GA, Li N, Takayama K, Farese AM, MacVittie TJ: Lung and heart injury in a nonhuman primate model of partial-body irradiation with minimal bone marrow sparing: histopathological evidence of lung and heart injury, *Health Phys* 116(3):383–400, 2019.

Prado C, MacVittie TJ, Bennett AW, Kazi A, Farese AM, Prado K: Organ doses associated with partial-body irradiation with 2.5% bone marrow sparing of the non-human primate: a retrospective study, *Radiat Res* 188(6):615–625, 2017.

CHAPTER

Radiation-associated changes in the kidney of rhesus macaques

1

Overview

Animals irradiated at 12 Gy, the majority of which were necropsied within 30 days of irradiation, had an increased amount of eosinophilic, presumably proteinic, material in urinary spaces of glomeruli. The glomerular changes were consistent with radiation-associated injury to glomerular capillaries, with an increase in mesangial matrix and a spectrum of microvascular changes. Animals surviving beyond 100 days postirradiation commonly had proteinaceous casts in renal tubules, consistent with increased passage of plasma proteins into the glomerular filtrate. The proteinaceous casts commonly had histologic features suggesting intra-glomerular or intratubular microhemorrhage contributed to the cast formation.

The kidneys of animals necropsied approximately 100 days postirradiation and thereafter had a high incidence of glomerular changes as well as interstitial and glomerular fibrosis. Majority of radiation-associated renal changes were diffusely distributed throughout the renal cortex, with no pattern of distribution relative to vascular structures or the overall topography of the kidney, but the delayed fibrotic changes were accentuated around medullary rays in the cortex.

Interstitial and glomerular fibrosis were apparent with Masson's trichrome staining and collagen 1 immunohistochemical (IHC) staining. The fibrosis was associated with increased IHC staining for α-smooth muscle actin (αSMA) and connective tissue growth factor (CTGF). In addition to the fibrosis noted on trichrome staining, glomeruli had increased mesangial matrix noted on routine H&E-stained sections and sections stained with periodic acid-Schiff/hematoxylin (PASH). Fibrotic glomeruli of some animals had reduced expression of CD31, a marker for endothelial cells, suggesting the radiation-associated glomerular injury progressed to loss of endothelial cells. Loss of microvascular integrity could lead to localized tissue hypoxia, which could promote fibroplasia.

Glomeruli had multiple manifestations of vascular injury, including capillary congestion, thrombosis, and segmental dilatation. The formation of hyaline casts in the lumen of cortical and medullary tubules indicated the loss of proteins into the glomerular filtrate, further indicating a deficit in glomerular function. Less common glomerular changes included mesangial aggregates of $PTAH^+$ particles that were consistent with fragmented erythrocytes, mesangial deposits of PAS^+ material that was presumed to be of basement membrane origin, and increased quantity and atypical dispersion of von Willebrand factor (vWF).

Glomeruli of animals irradiated at 12 Gy commonly had globular accumulations of eosinophilic material, presumably proteinic in nature. The exact origin and nature of the proteinic material was not apparent from microscopic examination, but it was consistent with the passage of plasma proteins through the glomerular filtration apparatus. The proteinic material was observed to extend through the glomerulotubular neck into the proximal tubules. Proximal tubules had increased content of proteinic material, some of which had vague globular outlines as seen in the glomerular

Histopathology Atlas of Acute Radiation Syndrome and Delayed Effects in Rhesus Macaques. DOI: https://doi.org/10.1016/B978-0-323-91393-5.00001-3

urinary spaces. The globular material in urinary spaces was not PAS^+. Accumulations of proteinic material in distal tubules commonly were PAS^+, presumably due to the contribution of uromodulin as glomerular filtrate passed the ascending straight limb of the nephron.

Inflammatory cell infiltration was not a prominent feature of the renal lesion complex, as based on examination of routinely stained histologic sections, but IHC staining indicated an increase in interstitial B cells, T cells, and macrophages. IHC staining indicated a particularly pronounced increase in the population of $CD163^+$ macrophages. A phenotypical switch to a pro-fibrotic macrophage phenotype could be significant with regard to the subsequent development of radiation-associated renal fibrosis.

Toluidine blue staining revealed no definitive alterations of mast cell populations relative to irradiation level or postirradiation interval therapy, but IHC staining for tryptase revealed a slight increase in mast cell populations in irradiated animals. Mast cell populations were more intense in areas of interstitial fibrosis, but it was unclear whether the mast cell populations were a cause or result of the interstitial fibrosis.

Perls' iron stains revealed iron-positive material in glomeruli of only one animal. IHC staining for transforming growth factor β (TGFβ), IL22, IL22R, and pro-renin receptor revealed no alterations associated with radiation dose or postirradiation interval.

Kidney specimens from animals necropsied soon after irradiation had equivocal alterations in cytoplasmic structure and contents of renal tubular epithelial cells. These observations were considered inconclusive, particularly in view of the suboptimal immersion fixation. The animals were euthanized due to clinical evidence of debilitation; therefore it was not possible to distinguish between direct radiation effects and the early manifestations of multisystemic organ failure.

Staining and figures list

Routine histology stain-

Hematoxylin and eosin (H&E)—Figs. 1.1, 1.2, 1.3, 1.5, 1.7, 1.8, 1.10, 1.12A, 1.12B, 1.15A, 1.15B, 1.16, 1.18, 1.21, 1.22A, 1.22B, 1.22C, 1.26A, 1.33

Histochemical stains-

Chromotrope-aniline blue—Fig. 1.9
Masson's trichrome—Figs. 1.25, 1.26B, 1.26C, 1.27A, 1.27B, 1.28A, 1.28B
Periodic acid-Schiff/hematoxylin stain—Figs. 1.4, 1.17A, 1.17B, 1.19, 1.23, 1.2 (1.24)
Phosphotungstic acid-hematoxylin stain—Figs. 1.6, 1.11, 1.20

Immunohistochemical stains-

Alpha smooth muscle actin—Figs. 1.31A, 1.31B, 1.31C, 1.31D
CD20—Figs. 1.36A, 1.36B
CD3—Figs. 1.37A, 1.37B
CD31—Figs. 1.29A, 1.29B
CD38—Figs. 1.34A, 1.34B

CD68—Figs. 1.35A, 1.35B
CD163—Figs. 1.39A, 1.39B, 1.39C, 1.39D
Collagen 1—Figs. 1.30A, 1.30B
Connective tissue growth factor—Figs. 1.32A, 1.32B, 1.32C, 1.32D
Myeloperoxidase—Figs. 1.38A, 1.38B, 1.38C
Tryptase—Figs. 1.40A, 1.40B
von Willebrand factor (vWF) —Figs. 1.13, 1.14

Graphic presentations (scatter plots) of individual histologic observations in the kidney are presented in the following figures:

Fig. 1.41. Kidney—congestion, glomerular capillaries.
Fig. 1.42. Kidney—thrombosis, glomerular.
Fig. 1.43. Kidney—segmental dilation, glomerular capillaries.
Fig. 1.44. Kidney—decreased CD31 IHC staining, glomeruli.
Fig. 1.45. Kidney—increased proteinic material, urinary space.
Fig. 1.46. Kidney—intratubular proteinaceous casts.
Fig. 1.47. Kidney—$PTAH^+$ intratubular casts.
Fig. 1.48. Kidney—increased mesangial matrix, glomeruli.
Fig. 1.49. Kidney—PAS^+ mesangial deposits, glomeruli.
Fig. 1.50. Kidney—increased proteinic material, tubules.
Fig. 1.51. Kidney—tubular proteinic globules, CAB stain.
Fig. 1.52. Kidney—glomerular proteinic globules, CAB stain.
Fig. 1.53. Kidney—increased vWF staining, glomerular.
Fig. 1.54. Kidney—degeneration, tubular epithelial cells.
Fig. 1.55. Kidney—dilatation, tubular.
Fig. 1.56. Kidney—fibrosis, glomerular, trichrome stain.
Fig. 1.57. Kidney—fibrosis, interstitial, trichrome stain.
Fig. 1.58. Kidney—alpha smooth muscle actin staining, glomerular.
Fig. 1.59 Kidney—alpha smooth muscle actin (αSMA) staining, interstitial.
Fig. 1.60. Kidney—increased collage 1 staining, glomerular.
Fig. 1.61. Kidney—increased collage 1 staining, interstitial.
Fig. 1.62. Kidney—connective tissue growth factor (CTGF) staining, glomeruli.
Fig. 1.63. Kidney—connective tissue growth factor (CTGF) staining, tubular epithelial cells.
Fig. 1.64. Kidney—IL-22 staining, collecting ducts.
Fig. 1.65. Kidney—IL-22 staining, tubular epithelial cells.
Fig. 1.66. Kidney—IL-22R staining, collecting ducts.
Fig. 1.67. Kidney—IL-22R staining, tubular epithelial cells.
Fig. 1.68. Kidney—Pro-renin receptor staining, collecting ducts.
Fig. 1.69. Kidney—Pro-renin receptor staining, glomerular.
Fig. 1.70. Kidney—Pro-renin receptor staining, interstitial.
Fig. 1.71. Kidney—infiltration, mononuclear cell.
Fig. 1.72. Kidney—$CD163^+$ interstitial cells.
Fig. 1.73. Kidney—myeloperoxidase$^+$ interstitial cells.
Fig. 1.74. Kidney—$CD3^+$ interstitial cells.

Fig. 1.75. Kidney—CD20$^+$ interstitial cells.
Fig. 1.76. Kidney—CD38$^+$ interstitial cells.
Fig. 1.77. Kidney—CD68$^+$ interstitial cells.

Detailed microscopic observations

Hyaline casts

Kidneys of animals in multiple-dose groups had intratubular accumulations of homogeneous proteinic material, which were recorded as hyaline casts. Changes of this type commonly indicate glomerular injury, with passage of serum proteins into the glomerular filtrate.

Glomerular congestion

Kidneys of irradiated animals had a low to moderate incidence of glomerular changes that included glomerular capillary congestion, thrombosis, and segmental capillary dilation. Glomerular congestion consisted of glomerular capillaries that were engorged with intact erythrocytes, while capillaries in the surrounding renal parenchyma were microscopically normal. Limitation of the capillary congestion to glomerular capillaries suggests a local vascular alteration in the glomeruli, as opposed to generalized congestion of the kidney.

Glomerular thrombosis

Glomeruli of a moderate number of animals had accumulations of eosinophilic material that was consistent with degenerated erythrocytes, which was recorded as glomerular thrombosis. Thrombotic capillaries were filled with degenerated erythrocytes and tended to be slightly dilated. Glomerular thrombosis was considered to be a result of persistent glomerular congestion.

Segmental glomerular capillary dilatation

Segmental capillary dilation consisted of pronounced dilation of capillary segments, resulting in bulbous cavities filled with degenerated erythrocytes. The histologic appearance suggested that the segmental glomerular capillary dilation was an accentuation of glomerular thrombosis.

Increased mesangial matrix

Glomeruli of irradiated animals commonly had an increase in amorphous mesangial material that was recorded as increased mesangial matrix. The increased mesangial matrix often was coincident with increased α-SMA staining, positive PASH staining, and/or increased trichome-positive staining in glomeruli.

Inflammatory cell infiltration

Low-level infiltrations of mononuclear inflammatory cells (lymphocytes, macrophages, and plasma cells) were present in the renal interstitium of irradiated animals necropsied throughout the observation period. Neutrophilic infiltration was not a prominent feature of the cellular infiltration. The overall inflammatory cell infiltration was modest relative to the degree of interstitial, glomerular, and tubular injury. IHC staining revealed the infiltrating cells were most commonly $CD163^+$ and myeloperoxidase$^+$, with lower populations of $CD3^+$, $CD20^+$, $CD38^+$, and $CD68^+$ cells.

Miscellaneous changes observed in H&E-stained sections

There was a low incidence of multifocal vacuolation or hyaline droplet accumulation in cortical tubules. These changes were consistent with early stages of degenerative changes in tubular epithelium, perhaps coupled with an excessive load of proteinic material received due to damaged glomerular structures.

Areas of fibrosis in kidneys (see below under Interstitital fibrosis- *trichrome stain*) occasionally had tubules lined by basophilic epithelial cells, consistent with regenerating tubular epithelium. However, tubular regeneration was not a prominent histologic feature of the overall renal lesion complex.

Dilatation of renal tubular lumina was present in a number of animals, most commonly in animals that were irradiated at 12 Gy. This change was suspected to reflect the overall poor clinical condition of the animals, perhaps coupled with hemodynamic alterations associated with administration of therapeutic fluids.

Occasional animals had glomerular accumulations of faintly fibrillar material that was consistent with fibrin, which was recorded as fibrin thrombosis. Glomeruli of a few animals contained brightly eosinophilic globules of various sizes that were consistent with fragments of degenerated erythrocytes, possibly representing a form of eryptosis. These changes were suspected to be associated with radiation-associated nephropathy but appeared to be randomly scattered among the radiation dose groups.

Interstitial fibrosis—trichrome stain

Masson's trichrome stains revealed renal interstitial fibrosis in the majority of animals irradiated at 10 or 11 Gy. Interstitial fibrosis was present in the kidney of the single animal irradiated at 12 Gy that survived to day 102. Only one additional animal irradiated at 12 Gy survived beyond 30 days postirradiation, and that animal had no histologic evidence of renal interstitial fibrosis when necropsied 81 days after irradiation.

Glomerular fibrosis—trichrome stain

Glomerular fibrosis typically occurred in conjunction with interstitial fibrosis, though different forms of fibrosis were not necessarily topographically colocated within the kidney of an individual animal. Glomerular fibrosis was not as pronounced as interstitial fibrosis.

PAS⁺ mesangial deposits

Deposits of periodic acid/Schiff (PAS)-positive material were noted in the glomerular mesangium of a moderate number of animals. These observations in the PASH-stained sections were related to the increased mesangial matrix noted in the H&E-stained sections, glomerular fibrosis in the trichrome-stained sections, and positive α-SMA staining.

Phosphotungstic acid-hematoxylin stain

Glomerular fibrin deposition was noted in animals from the most irradiated groups, but the fibrin deposition was not of a degree that it would be considered a major component of the pathological process. It was interesting to note that phosphotungstic acid-hematoxylin (PTAH)-positive glomerular fibrin deposition was noted in only one animal from the 12 Gy radiation exposure group, that being the longest surviving animal (day 102). This suggests that the glomerular fibrin deposition was not an acute effect of irradiation.

Casts within the lumen of renal tubules were a common manifestation of radiation-associated injury. The PTAH stains revealed that the casts commonly had low-level PTAH positivity, similar to the degree of PTAH staining exhibited by erythrocytes. It is likely the PTAH positivity of tubular casts reflects an element of glomerular hemorrhage that released damaged erythrocytes or erythrocyte components into the urinary filtrate.

The PTAH stain revealed positively stained droplets of various sizes in the glomerular mesangium and within intratubular casts of irradiated animals. The droplets did not have the physical outline of fibrin deposits and were not as intensely PTAH-positive as fibrin deposits. The degree of PTAH positivity and general configuration of the droplets suggest that they originated from degenerated erythrocytes. These droplets were also visible in the H&E-stained sections as brightly eosinophilic droplets, also consistent with erythrocyte fragments.

Perls' iron stain

Perls' iron stain revealed a minimal number of iron-positive deposits in glomeruli of only one animal in this series of studies.

Toluidine blue stain

Mast cell populations as identified by the toluidine blue stain were sparse in the kidney. There was no apparent radiation-associated effect on mast cell populations, but mast cell populations tended to be more prominent in areas of renal fibrosis. Mast cells in nonmucosal tissues are technically known as "connective tissue mast cells," which reflects the tendency for accumulation of these mast cells near the collagenous components of various tissues. As such, it was not possible to determine whether the increase in localized mast cell prominence was a cause or a result of the increased connective tissue in localized areas of the irradiated kidneys.

Toluidine blue staining of mast cells, as well as the IHC staining presented below, is based on the content of mast cell granules. As a result, the observations on mast cell populations with both staining procedures indicate the population of granulated mast cells and would not reveal populations of degranulated mast cells.

Tryptase immunohistochemical stain

Mast cell populations as identified by the tryptase IHC stain were sparse in the kidney, but somewhat more plentiful than the mast cell population observed in the toluidine blue-stained sections. For this reason, the tryptase IHC stain was considered to be a more sensitive indicator of mast cell populations in the kidney. There was no apparent radiation-associated effect on mast cell populations other than the increased prominence of mast cells in areas of interstitial fibrosis.

Alpha smooth muscle actin immunohistochemical stain

Histologic scoring of α-SMA staining in the kidney was performed in compliance with the scoring procedure described by Cohen et al. (1996). In this scoring regimen, a grade of 1 indicates the background vascular staining by α-SMA. Grades of 2 or above indicate presumed pathological α-SMA staining.

α-SMA staining was present to a minimal degree in all kidneys, reflecting the expected positive staining in blood vessels in the kidney. More pronounced α-SMA staining was observed in association with interstitial fibrosis. Glomerular α-SMA staining was most commonly observed in glomeruli that had increased mesangial matrix visible on routine H&E staining.

von Willebrand factor immunohistochemical stain

The expected low level of vWF staining was present in the naive control animals. The vWF staining typically presented as a crisp delineation of endothelial cells, which was particularly visible in glomerular capillaries. Increased glomerular vWF staining was commonly observed in irradiated animals that were necropsied more than 105 days following irradiation. The increased vWF staining in glomeruli of irradiated animals sometimes had a smudged, indistinct presentation, suggesting a breakdown of the glomerular capillary endothelium.

CD31 immunohistochemical stain

Positive staining for CD31 was present in vascular endothelial cells throughout the kidney of all animals, including the naive controls, and was most noteworthy in glomerular capillaries. Decreased CD31 staining in the interstitium was invariably associated with interstitial fibrosis, but not all areas of interstitial fibrosis were accompanied by decreased CD31 staining. Decreased glomerular CD31 staining occurred in glomeruli that had other histologic indications of radiation-associated injury, particularly fibrosis, increased mesangial matrix, and increased αSMA staining.

Collagen 1 immunohistochemical stain

A low level of positive staining for collagen was noted in the interstitium throughout the renal cortex and outer medulla of all animals, including the naive control animals, but the renal papilla was largely devoid of collagen 1 staining. Increased interstitial collagen 1 staining was noted in all except one of the irradiated animals that were necropsied after postexposure day 91. As expected, increased collagen 1 staining paralleled observations of interstitial fibrosis based on Masson's trichrome-stained sections. The collagen 1 IHC stains tended to reveal more collagen deposition than was discernible in the trichrome-stained sections. This was particularly noticeable in areas of very slight collagen deposition. In some instances, the interstitial collagen deposition was visible in the collagen 1-stained sections but not visible in the trichrome-stained sections. Increased collagen 1 staining in glomeruli was less commonly noted and of lower intensity than the collagen 1 staining in the interstitium. As with interstitial collagen 1 staining, IHC staining for collagen 1 was a more sensitive indicator of glomerular fibrosis than Masson's trichrome staining.

Connective tissue growth factor immunohistochemical stain

Positive CTGF staining in tubules and collecting ducts was related to interstitial fibrosis and interstitial αSMA staining, and glomerular staining was related to glomerular αSMA staining.

Transforming growth factor β immunohistochemical stain

Positive staining for TGF-β was topographically subcategorized as interstitial, tubular, collecting duct, or glomerular. There was a very low incidence of TGF-β staining in each subcategory location of irradiated animals. The sparse TGF-β staining was unexpected, given the evidence of renal fibrosis that was apparent in the trichrome-stained sections and evidence of CTGF activity in the kidney of irradiated animals. This incongruity raised suspicion that the TGF-β primary antibody used in the staining procedure was not optimal for this animal species, or the specimens were not optimally preserved for demonstration of TGF-β. The specimens originated from animals that were euthanized for cause, rather than through a planned sequential sacrifice, thus it is also possible that TGF-β activity was expressed at some time prior to necropsy. However, in the context of this study, the available data suggest that TGF-β activity did not have a major role in the development of the observed renal lesions.

IL-22 immunohistochemical stain

Low-level IL-22 immunostaining was present in collecting ducts of animals from all treatment groups, including the naive control animals. Low-level IL-22 immunostaining was present in renal cortical tubules of animals from all irradiated groups, but not the naive control animals.

IL-22 receptor (IL-22R) immunohistochemical stain

Positive IL-22R staining was more abundantly present than IL-22 staining in the kidneys of animals from all groups, including the naive controls. There was no apparent radiation-associated effect on expression of IL-22R in the kidney.

Pro-renin receptor immunohistochemical stain

Positive staining for pro-renin receptor was present in the collecting ducts, but not present in the glomeruli or interstitium of the naive control animals. There was no apparent radiation-associated alteration in renal pro-renin receptor IHC staining in collecting ducts. Glomerular pro-renin receptor staining was present in a small number of animals necropsied after day 105 postirradiation. Interstitial pro-renin receptor staining was present in three animals necropsied at the end of the observation period.

CD38 immunohistochemical stain

Staining for CD38, a marker for many leukocytes, including lymphocytes and natural killer (NK) cells, was virtually absent in the kidneys of naive control rhesus macaques. A population of $CD38^+$ cells was present in the renal interstitium of irradiated animals.

CD68 immunohistochemical stain

Staining for CD68, a marker for cells of the monocyte/macrophage lineage, was virtually absent in the kidneys of naive control rhesus macaques. A population of $CD68^+$ cells was present in the renal interstitium of irradiated animals.

CD20 immunohistochemical stain

Staining for CD20, a marker for B lymphocytes, revealed a sparse population of $CD20^+$ cells in the kidneys of naive control rhesus macaques. An increased population of $CD20^+$ cells was present in the renal interstitium of irradiated animals.

CD3 immunohistochemical stain

Staining for CD3, a marker for T lymphocytes, revealed a sparse population of $CD3^+$ cells in the kidneys of naive control rhesus macaques. An increased population of $CD3^+$ cells was present in the renal interstitium of irradiated animals.

Myeloperoxidase immunohistochemical stain

Staining for myeloperoxidase (MPO), a marker for neutrophils and macrophages of the monocytic line, revealed a sparse population of MPO^+ cells in the kidneys of naive control rhesus macaques. An increased population of MPO^+ cells in the renal interstitium of irradiated animals consisted primarily of macrophages, with few neutrophils.

CD163 immunohistochemical stain

Staining for CD163, a marker for a pro-fibrotic subset of macrophages, revealed few $CD163^+$ cells in the kidneys of naive control rhesus macaques. Irradiated animals had a marked increase in the population of $CD163^+$ cells in the renal interstitium.

Tryptase immunohistochemical stain

Staining for tryptase, a mast cell marker, revealed the virtual absence of $tryptase^+$ mast cells in the kidneys of naive control rhesus macaques. Irradiated animals had an increased population of $tryptase^+$ cells in the renal interstitium.

Additional details of kidney observations have been previously presented (Cohen et al., 2019; Parker et al., 2019).

References

Cohen EP, Molteni A, Hill P, Fish BL, Ward WF, Moulder JE, Carone FA: Captopril preserves function and ultrastructure in experimental radiation nephropathy, *Lab Invest* 75(3):349–360, 1996.

Cohen EP, Hankey KG, Farese AM, Parker GA, Jones JW, Kane MA, Bennett A, MacVittie TJ: Radiation nephropathy in a nonhuman primate model of partial-body irradiation with minimal bone marrow sparing-part 1: acute and chronic kidney injury and the influence of Neupogen, *Health Phys* 116(3):401–408, 2019.

Parker GA, Cohen EP, Li N, Takayama K, Farese AM, MacVittie TJ: Radiation nephropathy in a nonhuman primate model of partial-body irradiation with minimal bone marrow sparing-part 2: histopathology, mediators, and mechanisms, *Health Phys* 116(3):409–425, 2019.

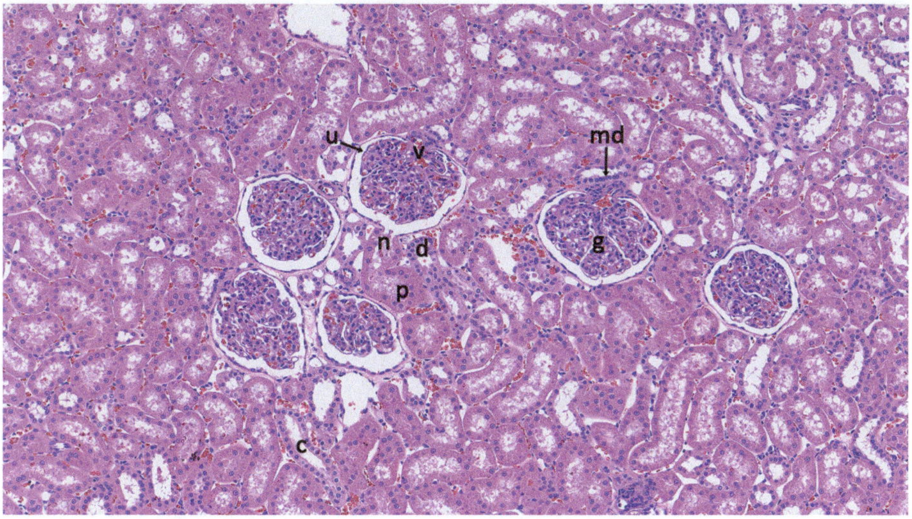

FIGURE 1.1

This image shows the normal renal microarchitecture of a male rhesus macaque, with glomeruli (g) composed of multiple capillary tufts that enter the glomerulus at the vascular pole (v). Filtrate passing the glomerulus enters the urinary space (u) surrounding the glomerulus, which is continuous with the neck (n) of the proximal tubule (p) that exits the urinary space at the equatorial pole opposite the vascular pole. The proximal tubule continues through descending and ascending loops of Henle (not identified in the figure) to form distal tubules (d). A segment of the distal tubule lies adjacent to the vascular pole of the glomerulus, where the distal tubular epithelial cells form a dense cellular cluster known as the macula densa (md) that is involved in regulation of blood flow into the glomerulus. The highly altered filtrate exists the nephron into collecting ducts (c) that serve multiple nephrons. Hematoxylin and eosin stain, 12.7 × objective magnification.

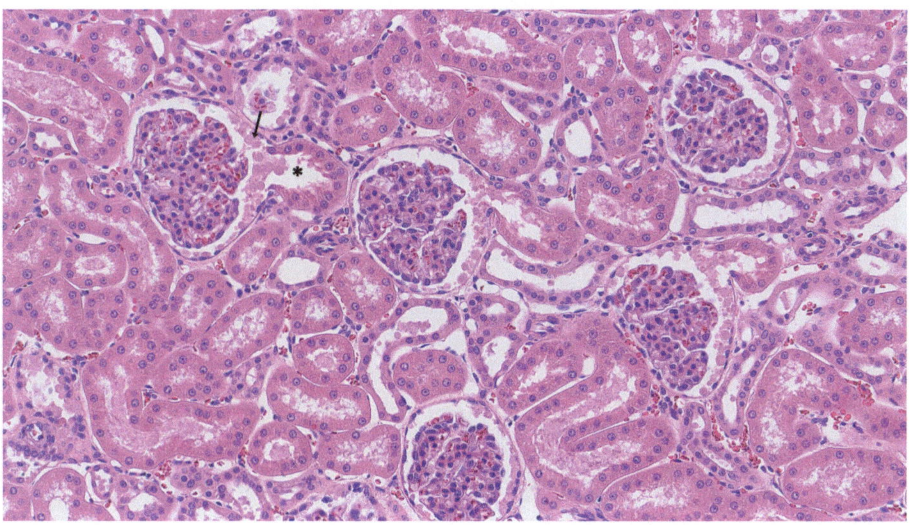

FIGURE 1.2

Renal cortex of a male rhesus macaque necropsied 8 days following irradiation at 12 Gy. Note the accumulations of eosinophilic material (arrow) in urinary spaces of multiple glomeruli. There is direct evidence of transition of the eosinophilic (presumed proteinic) material through glomerular necks into proximal tubules (*). Hematoxylin and eosin stain, 20 × objective magnification.

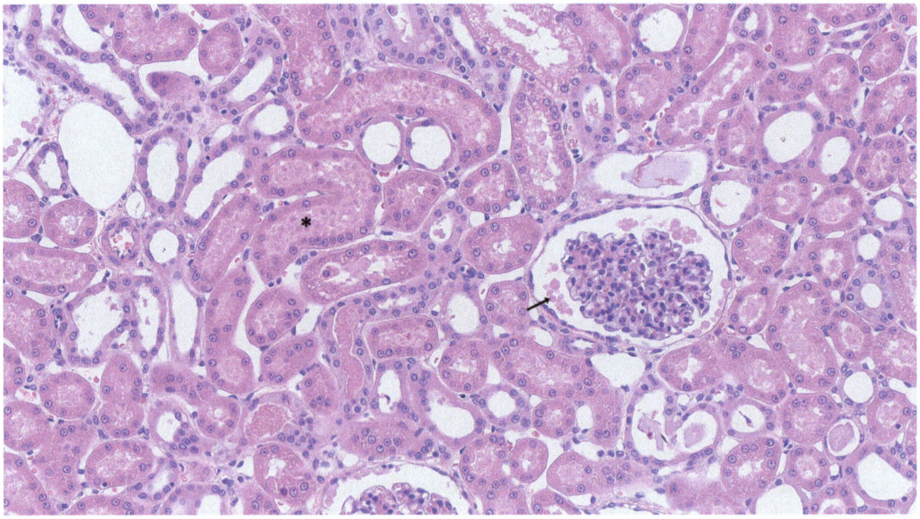

FIGURE 1.3

Renal cortex of a male rhesus macaque necropsied 9 days after irradiation at 12 Gy. Note the accumulations of eosinophilic material (arrow) in the glomerular urinary space (arrow) and in the lumen of proximal tubules (*). Hematoxylin and eosin stain, 20 × objective magnification.

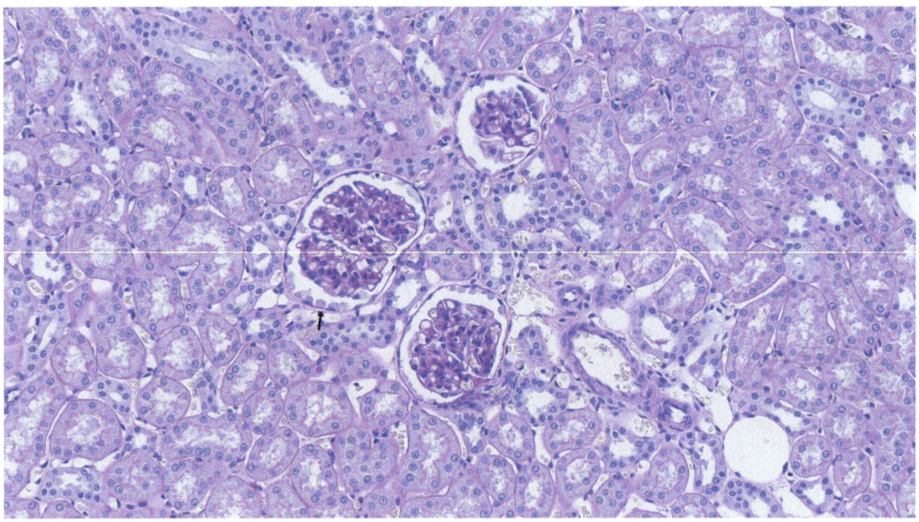

FIGURE 1.4

Periodic acid-Schiff/hematoxylin (PASH) stain on renal cortex of a male rhesus macaque necropsied 8 days following irradiation at 12 Gy. Note the accumulations of (presumed proteinic) material in the glomerular urinary space (arrow) are not selectively stained, indicating absence of significant carbohydrate content in the material. 20 × objective magnification.

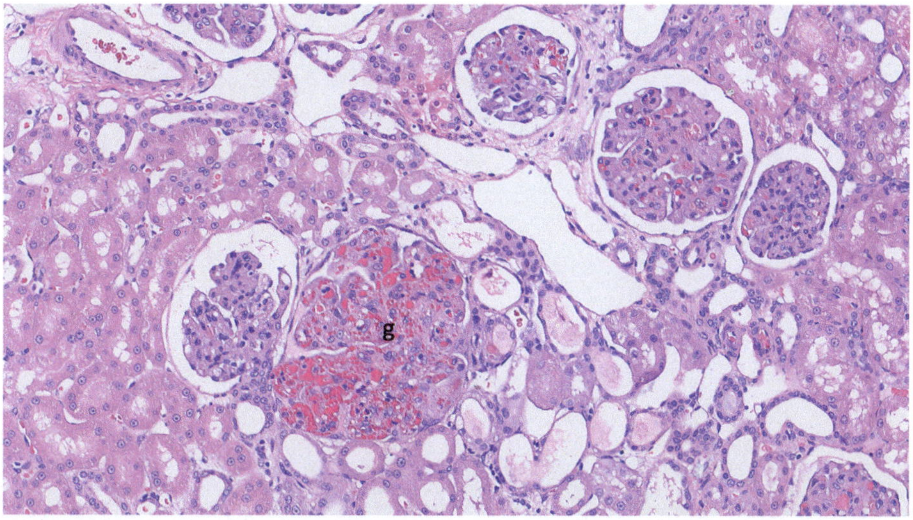

FIGURE 1.5

Renal cortex of a male rhesus macaque necropsied 182 days following irradiation at 10 Gy. Note the accumulations of brightly eosinophilic erythrocytes and erythrocytic fragments in the glomerulus (g). Hematoxylin and eosin stain, 20 × objective magnification.

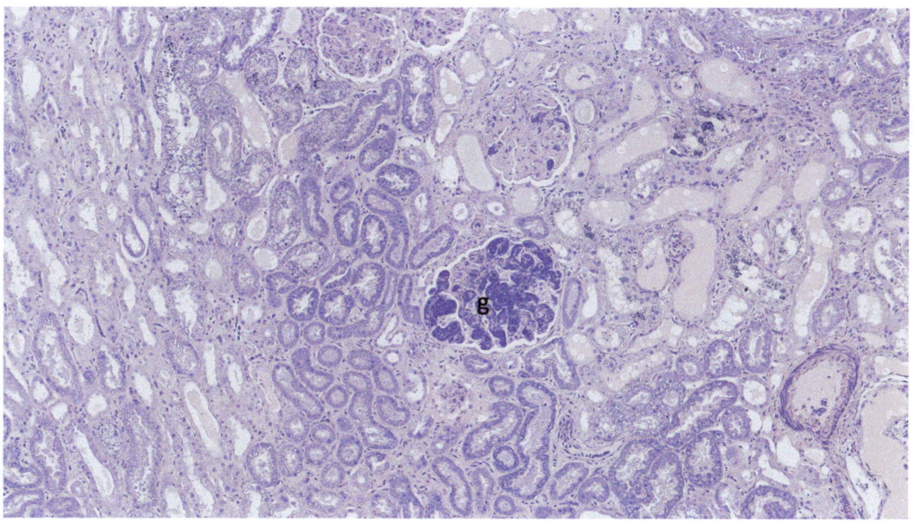

FIGURE 1.6

Phosphotungstic acid/hematoxylin (PTAH) stain on the kidney of a male rhesus macaque necropsied 107 days after irradiation at 11 Gy. Note the accumulation of flocculent to punctate $PTAH^+$ material in the glomerulus (g). This material was consistent with degenerated erythrocytes, which presented as glomerular capillary congestion and/or thrombosis in the H&E-stained sections. 10 × objective magnification.

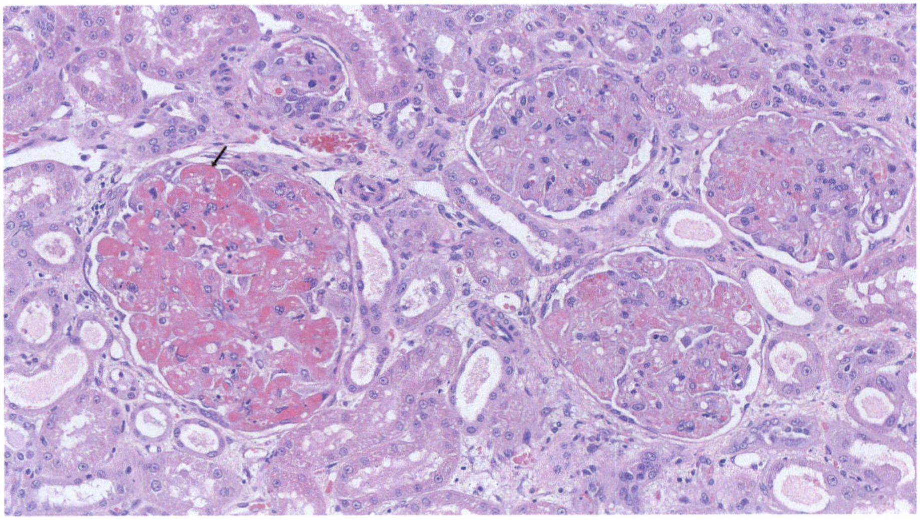

FIGURE 1.7

Renal glomerular thrombosis in the kidney of a male rhesus macaque necropsied 105 days following irradiation at 11 Gy. Note the accumulations of brightly eosinophilic erythrocyte debris within distended glomerular capillaries (arrow). Hematoxylin and eosin stain, 20 × objective magnification.

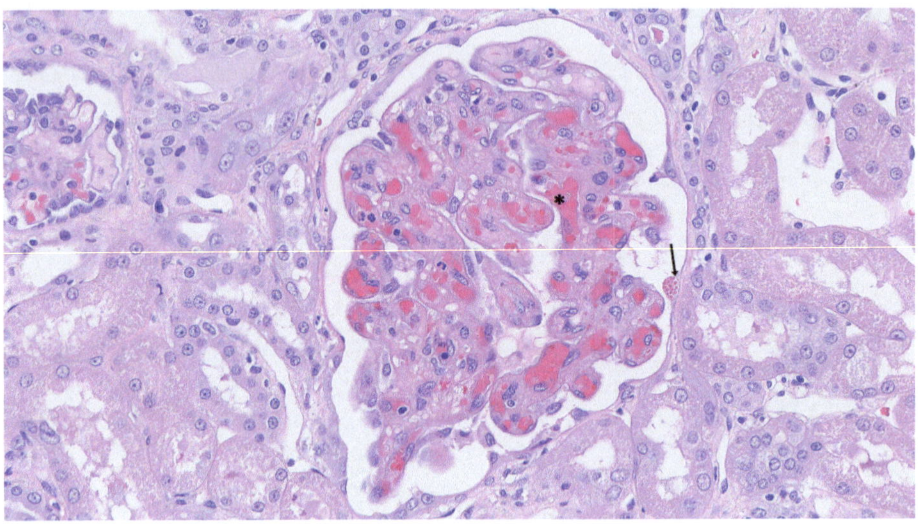

FIGURE 1.8

Glomerular thrombosis (*) in the kidney of a male rhesus macaque necropsied 180 days following irradiation at 10 Gy. Note the accumulation of brightly eosinophilic erythrocyte debris (arrow) in a cell lining the glomerular capsule. Hematoxylin and eosin stain, 30.3 × objective magnification.

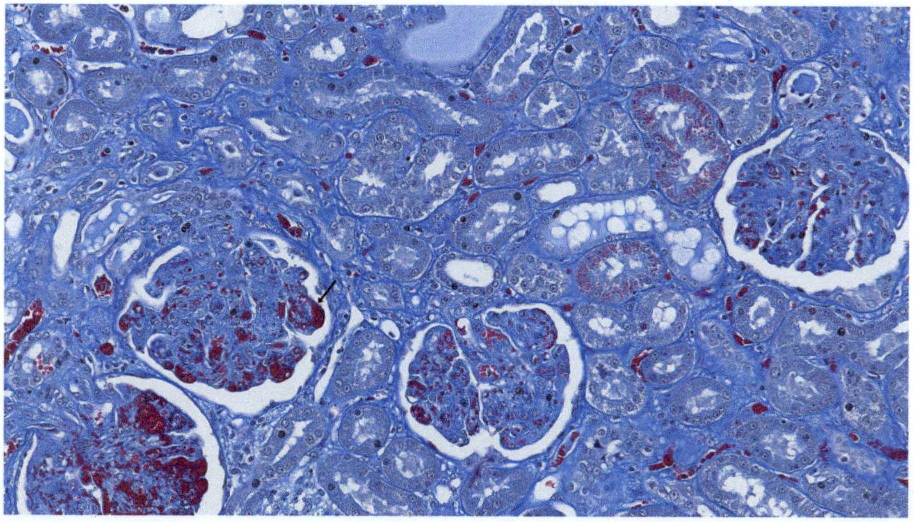

FIGURE 1.9

Glomerular thrombosis in the kidney of a male rhesus macaque necropsied 115 days following irradiation at 11 Gy. Note the accumulations of bright red erythrocytic debris within distended glomerular capillaries (arrow), and similarly stained punctate bodies within the cytoplasm of proximal tubules. Chromotrope-aniline-blue (CBA) stain, 20 × objective magnification.

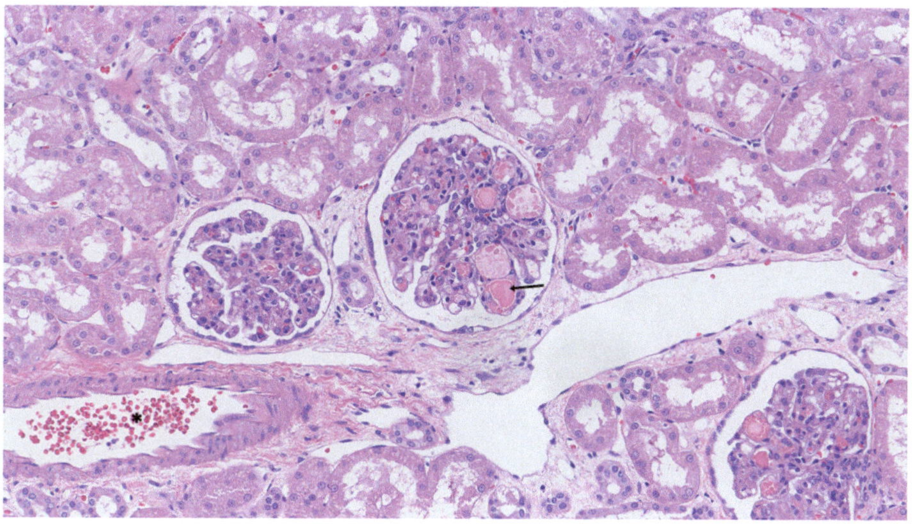

FIGURE 1.10

Glomerular dilation in the kidney of a male rhesus macaque necropsied 180 days following irradiation at 10 Gy. Note the markedly distended glomerular capillaries (arrow) filled with amorphous eosinophilic debris, consistent with degenerated erythrocytes. Hematoxylin and eosin stain, 20 × objective magnification.

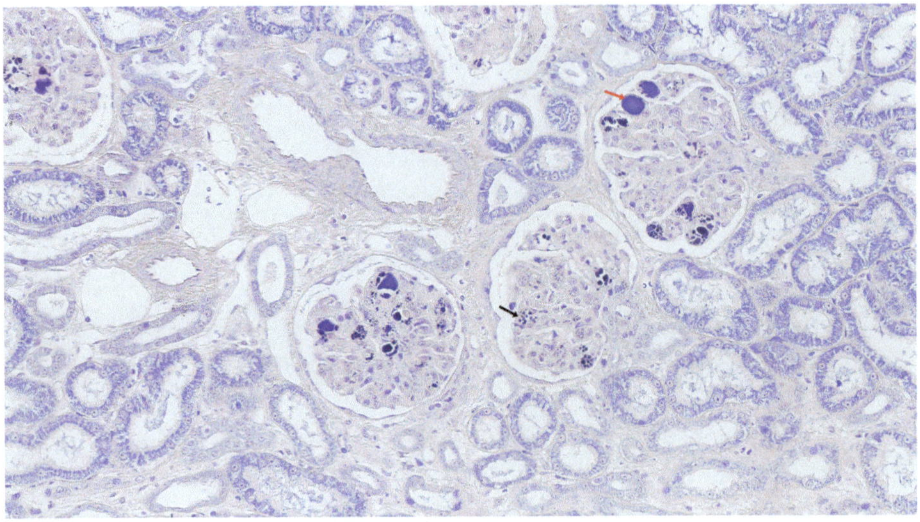

FIGURE 1.11

The phosphotungstic acid-hematoxylin (PTAH) stain on the renal cortex of a male rhesus macaque necropsied 180 days following irradiation at 10 Gy reveals homogeneous and punctate accumulations of $PTAH^+$ material within capillaries. The smaller, punctate structures (black arrow) are consistent with degenerated, fragmented erythrocytes, possibly indicating eryptosis of stagnant erythrocytes in congested or thrombotic capillaries. The larger, homogeneous accumulations of $PTAH^+$ material (red arrow) correspond to the dilatation of glomerular capillaries noted in the H&E-stained sections. 20 × objective magnification.

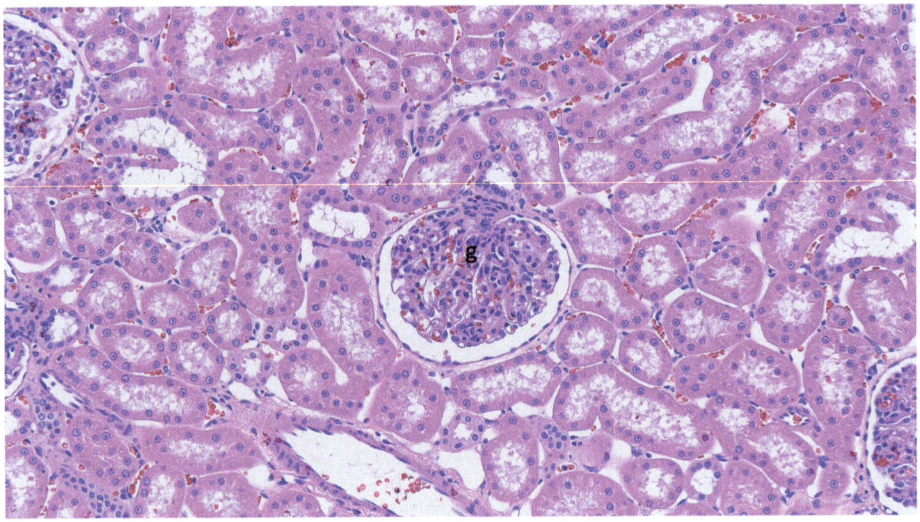

FIGURE 1.12A

Renal cortex of naïve male rhesus macaque. Note the overall size and cellularity of the glomerulus (g). Compare to the following image that has multiple radiation-associated histologic changes. Hematoxylin and eosin stain, 20 × objective magnification.

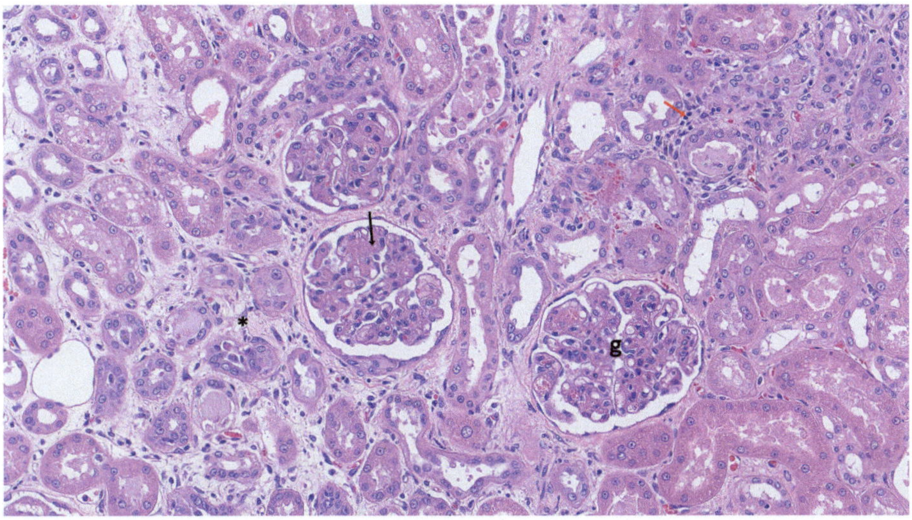

FIGURE 1.12B

Renal cortex of male rhesus macaque necropsied 141 days after irradiation at 11 Gy. Compare the overall size and cellularity of the glomeruli (g) to those seen in the previous image. Glomeruli have expanded mesangial matrix (black arrow). The interstitium (*) has an increased amount of fibrillar connective tissue and an increased population of small, densely stained inflammatory cells (red arrow) that consist primarily of lymphocytes and macrophages. Hematoxylin and eosin stain, 20 × objective magnification.

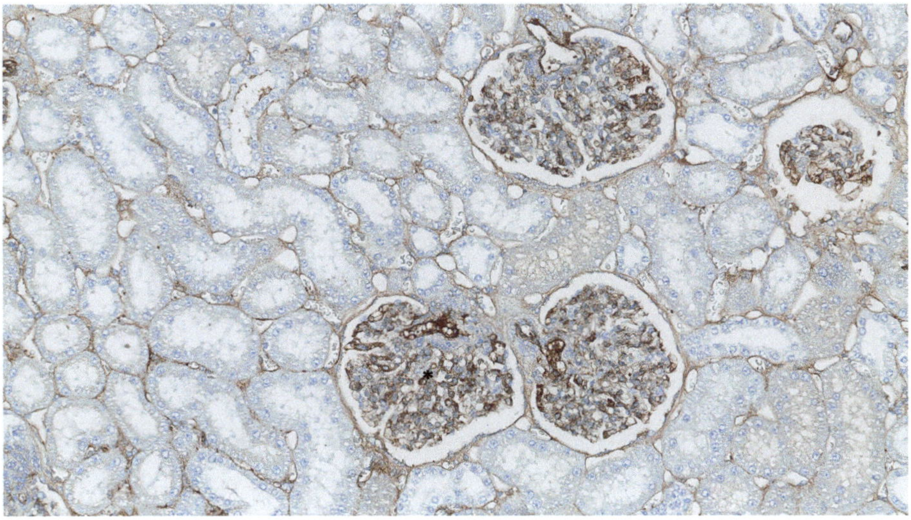

FIGURE 1.13

Immunohistochemical staining for von Willebrand factor (vWF) performed on the renal cortex of a naive control male rhesus macaque. Note the orderly dispersion of low-level vWF staining in glomerular capillaries (*). Stain performed with 3,3′-diaminobenzidine chromogen and hematoxylin counterstain, 20 × objective magnification.

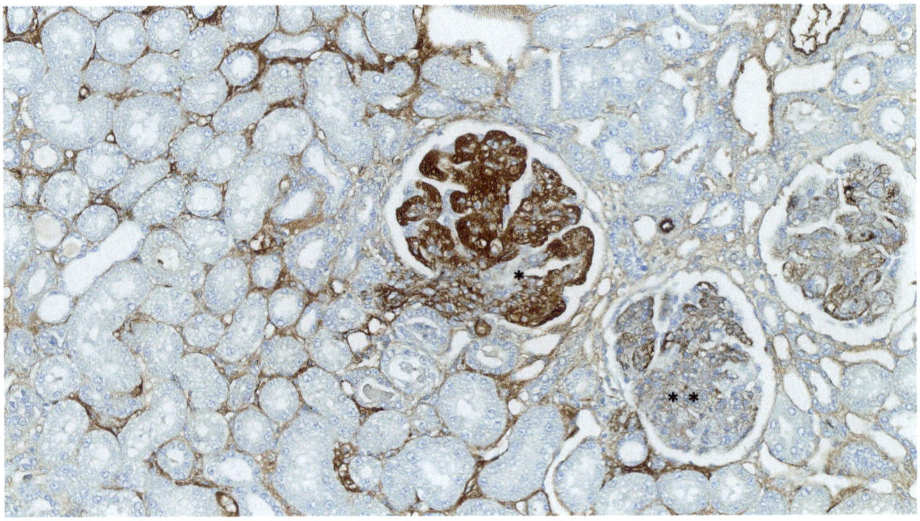

FIGURE 1.14

Immunohistochemical staining for von Willebrand factor (vWF) on the renal cortex of a male rhesus macaque necropsied 93 days following irradiation at 10 Gy. Note the increased vWF staining in one glomerulus (*) and reduced vWF staining in segments of other glomeruli (**). Stain performed with 3,3′-diaminobenzidine chromogen and hematoxylin counterstain, 20 × objective magnification.

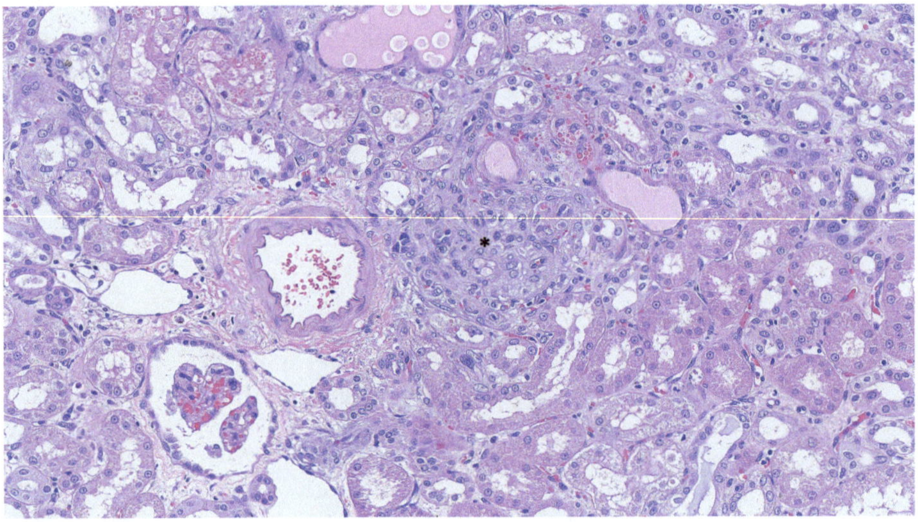

FIGURE 1.15A

Renal cortex of a male rhesus macaque necropsied 183 days after irradiation at 11 Gy. Large clusters of unidentified cells and capillaries (*) represent the remnants of severely altered glomeruli. Hematoxylin and eosin stain, 20 × objective magnification.

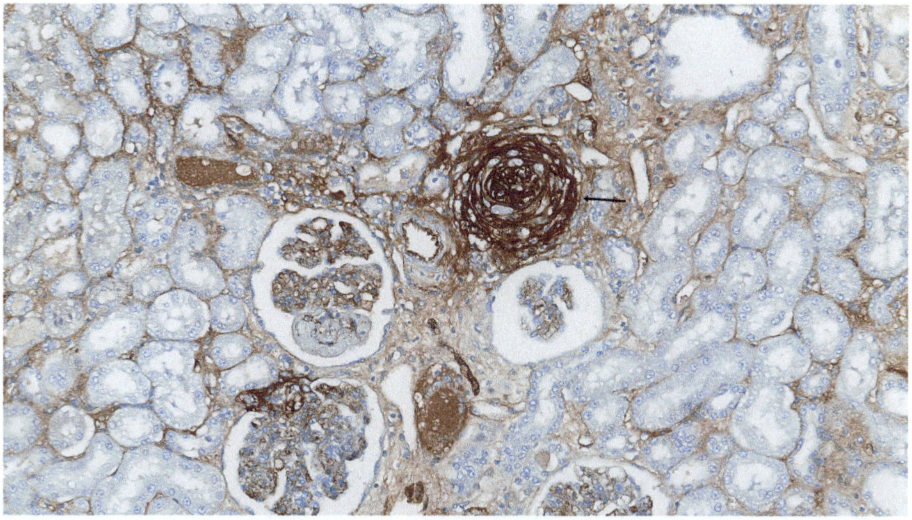

FIGURE 1.15B

Immunohistochemical staining for von Willebrand factor (vWF) on the same specimen shows one glomerulus (arrow) replaced by a tangled mass of vWF^+ vascular remnants. Stain performed with 3,3′-diaminobenzidine chromogen and hematoxylin counterstain, 20 × objective magnification.

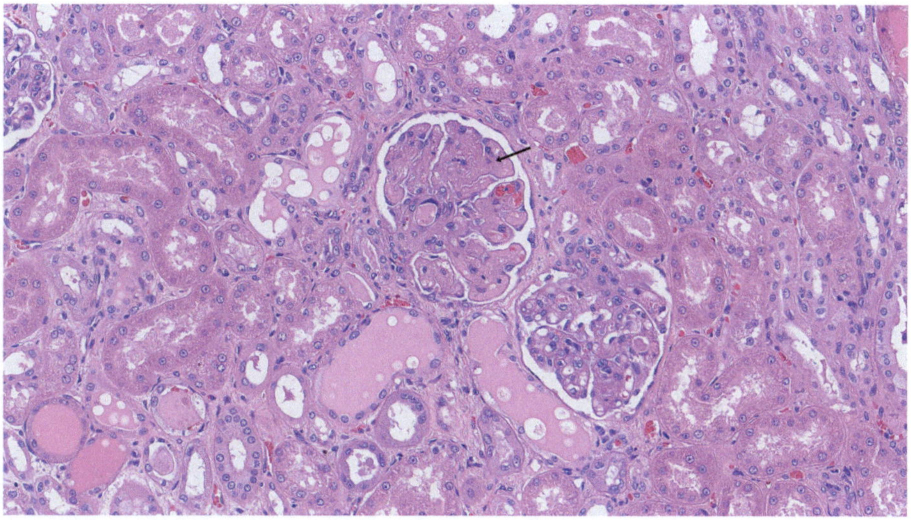

FIGURE 1.16

Note the homogeneous eosinophilic material (arrow) representing increased glomerular mesangial matrix and thickened glomerular capillary basement membranes in a male rhesus macaque necropsied 141 days after irradiation at 11 Gy. Hematoxylin and eosin stain, 20 × objective magnification.

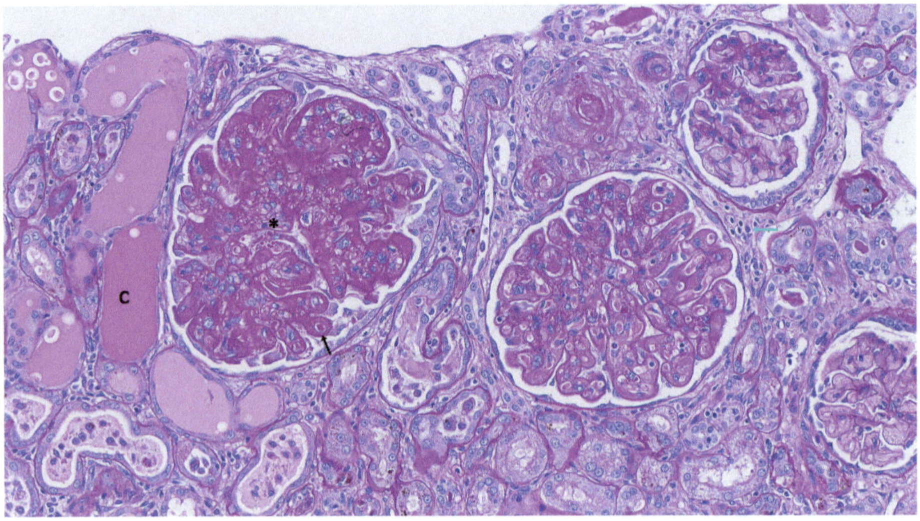

FIGURE 1.17A

Periodic acid-Schiff/hematoxylin (PASH) stain on kidney of male rhesus macaque necropsied 138 after irradiation at irradiation at 10 Gy. One glomerulus (*) has a marked increase in magenta-colored material, indicating carbohydrate components of mesangial matrix and glomerular capillary walls (black arrow). Desquamated epithelial cells and homogeneous casts (c) are present in distal tubules. A minor population of infiltrating inflammatory cells (blue arrow) is present. 20 × objective magnification.

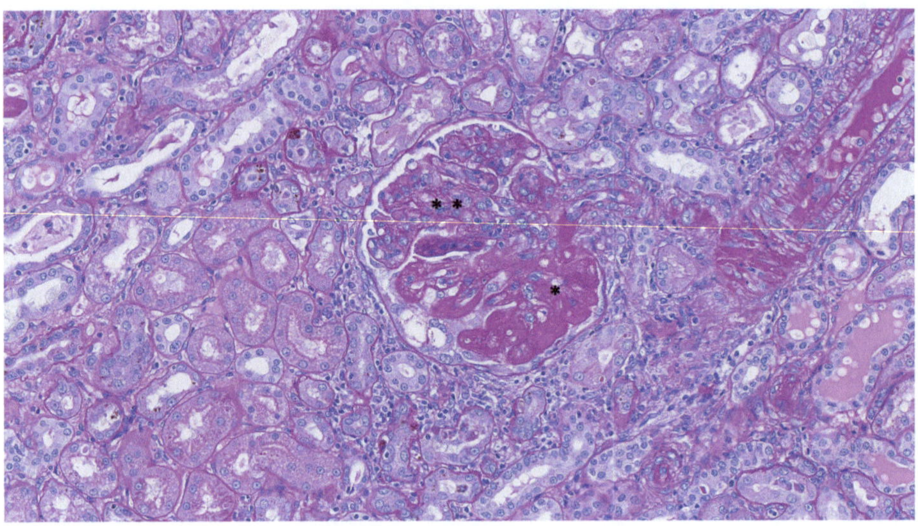

FIGURE 1.17B

Same specimen as shown in the previous image. Note the increased amount of PAS^{+} material in one major lobule (*) of the glomerular capillary tuft, with relative sparing the other major lobule in the glomerulus (**). Periodic acid-Schiff/hematoxylin (PASH) stain, 20 × objective magnification.

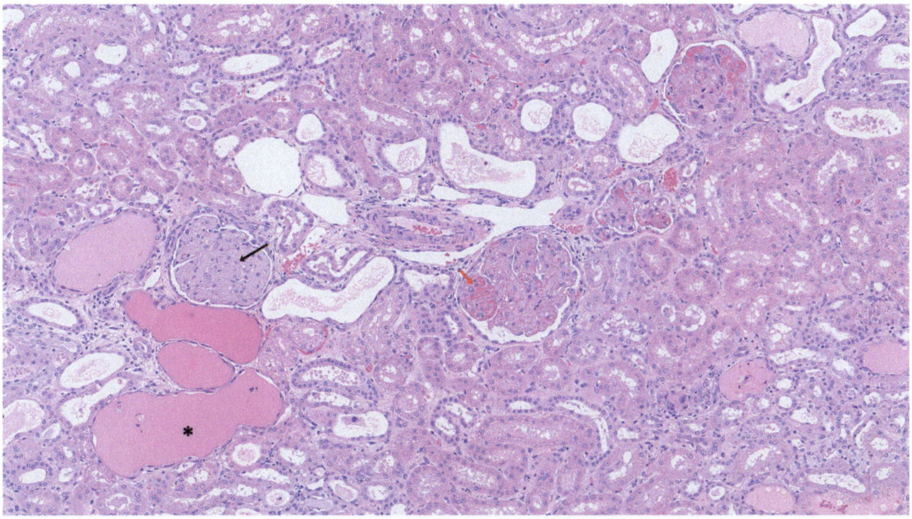

FIGURE 1.18

Renal cortex of a male rhesus macaque necropsied 105 days after irradiation at 11 Gy. Note the homogeneous eosinophilic cast in dilated tubules (*), increased mesangial matrix (black arrow) in a glomerulus, and thrombotic capillary (red arrow) in another glomerulus. Hematoxylin and eosin stain. 10 × objective magnification.

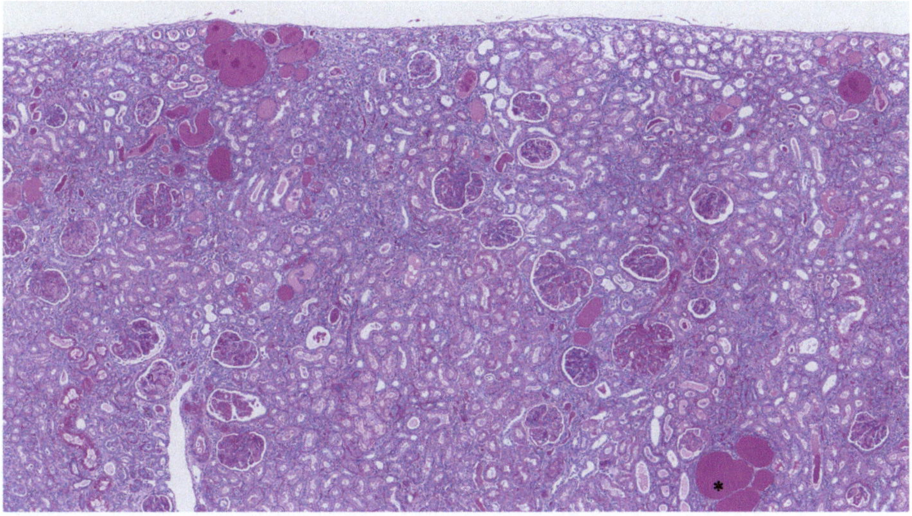

FIGURE 1.19

Periodic acid-Schiff/hematoxylin (PASH) stain on renal cortex of male rhesus macaque necropsied 138 days after irradiation at 10 Gy. Note the PAS$^+$ staining of casts in dilated tubules (*). The PAS staining indicates carbohydrate content of the cast material. 5 × objective magnification.

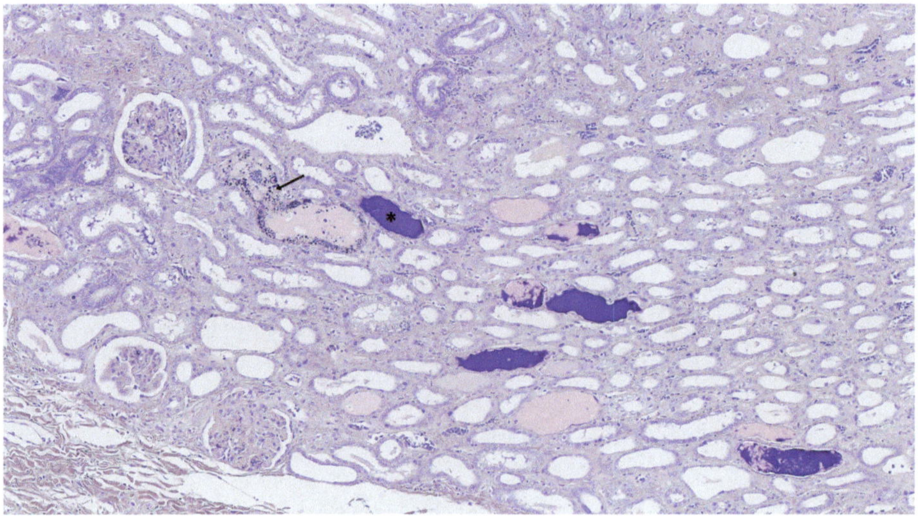

FIGURE 1.20

Phosphotungstic acid-hematoxylin (PTAH) stain on kidney of a male rhesus macaque necropsied 141 days after irradiation at 11 Gy. A dilated tubule (arrow) contains intact erythrocytes as well as $PTAH^+$ globules and punctate bodies that are consistent with erythrocyte fragments. An adjacent cast (*) that is uniformly $PTAH^+$ may represent an obstruction in outflow from the tubule. 10 × objective magnification.

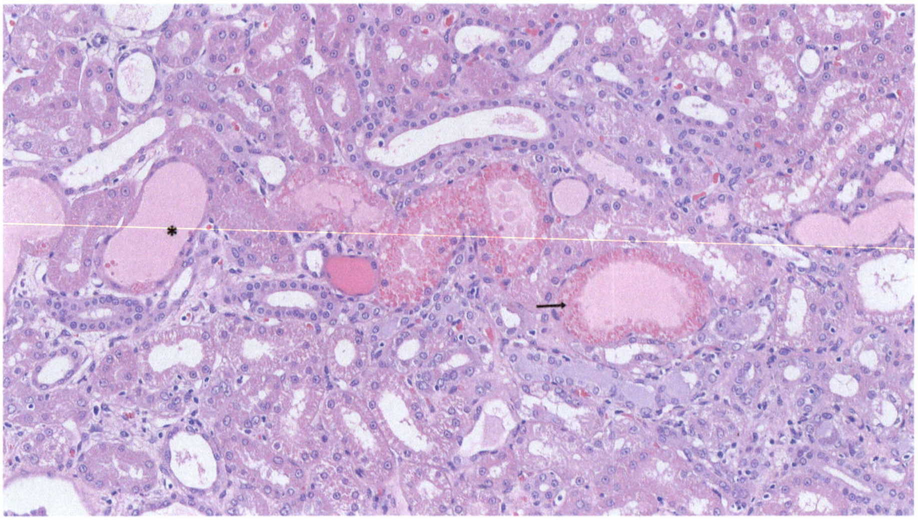

FIGURE 1.21

Renal cortex of a male rhesus macaque necropsied 182 days after irradiation at 10 Gy. Epithelial cells of a renal tubule (arrow) contain brightly eosinophilic hyaline droplets that indicate an accumulation of proteinic material subsequent to passage of proteinic material through glomeruli into tubular lumina (*). Hematoxylin and eosin stain, 20 × objective magnification.

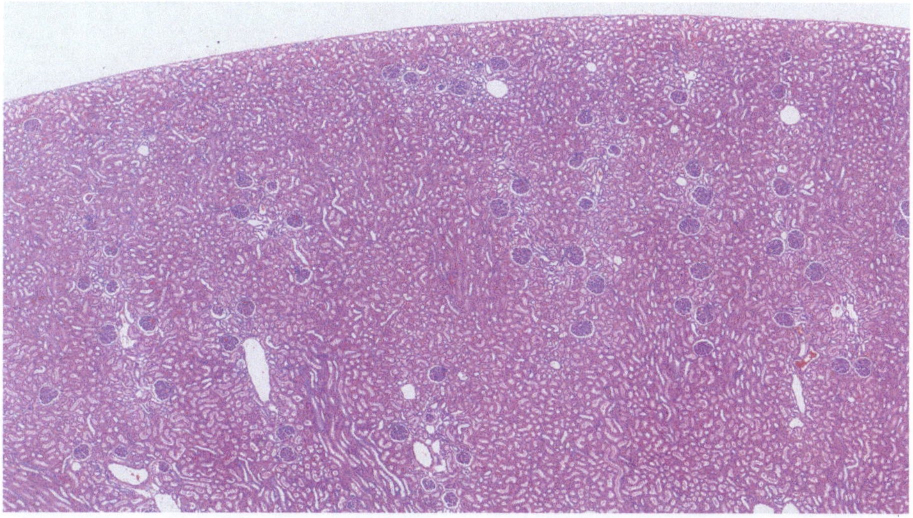

FIGURE 1.22A

Note the evenly distributed cortical tubules of uniform size in the renal cortex of a naive male rhesus macaque. Hematoxylin and eosin stain, 2.5 × objective magnification.

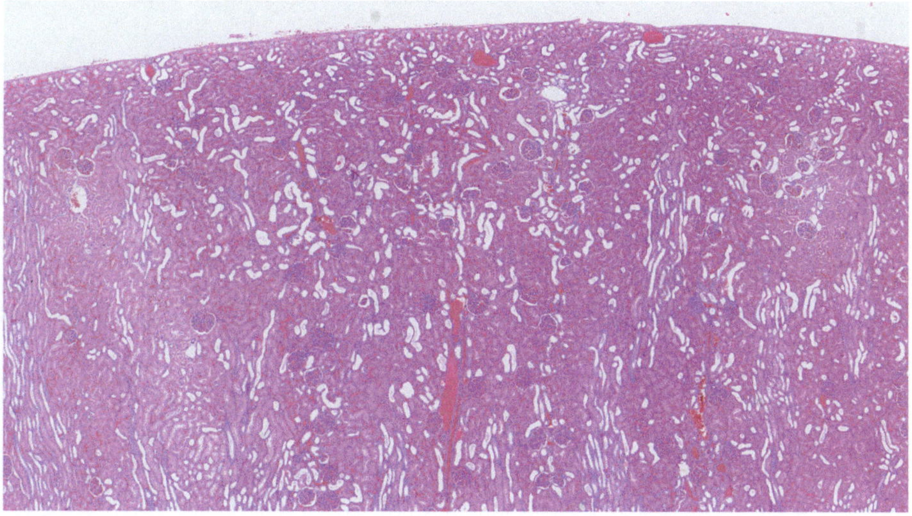

FIGURE 1.22B

Renal cortex of a male rhesus macaque necropsied 8 days after irradiation at 12 Gy. Note the pronounced dilatation (clear spaces) of numerous cortical tubules, suggesting an increased rate of fluid passage through glomeruli into the tubules. Hematoxylin and eosin stain, 2.5 × objective magnification.

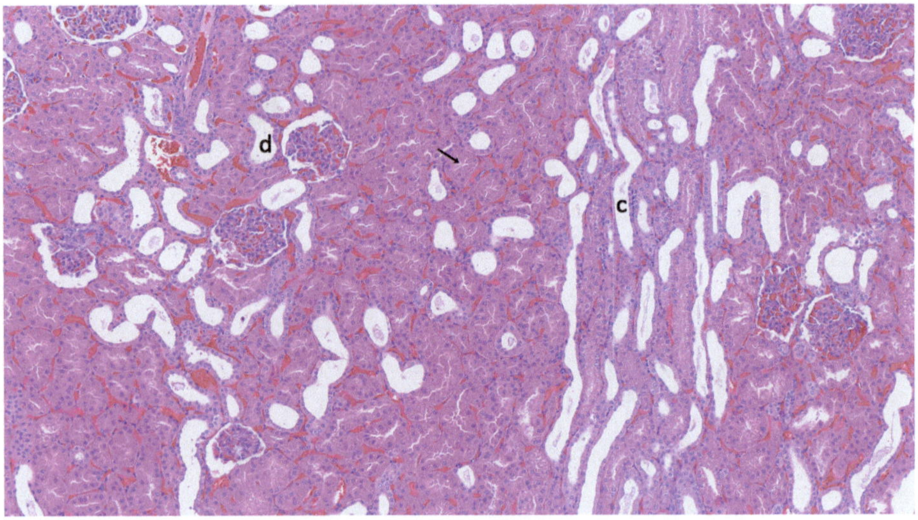

FIGURE 1.22C

Higher magnification of the previous specimen reveals the dilated tubules are primarily distal tubules (d) and collecting ducts (c) rather than proximal tubules (arrow). Hematoxylin and eosin stain, 5 × objective magnification.

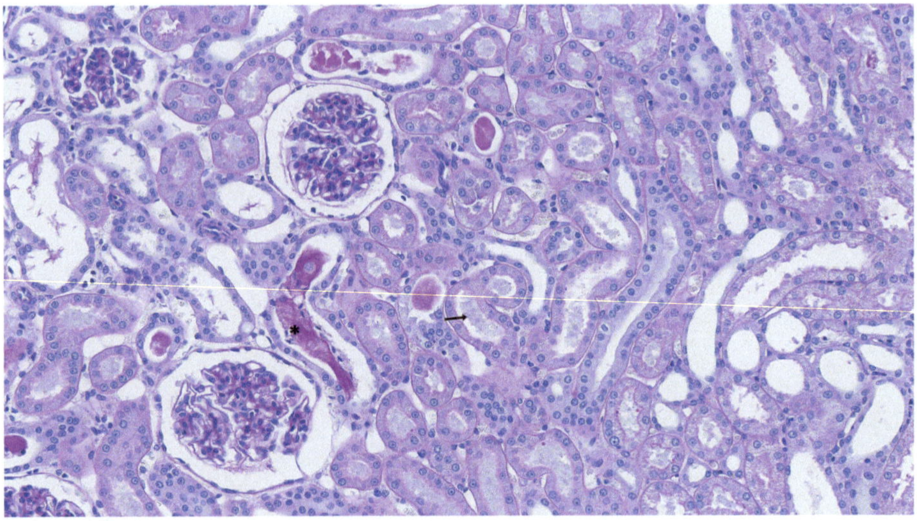

FIGURE 1.23

Periodic acid-Schiff/hematoxylin (PASH) stain on renal cortex of male rhesus macaque necropsied 9 days after irradiation at 12 Gy. Note the PAS^{+} contents of the distal tubules (*) compared to the negative PAS staining of the contents in proximal tubules (arrow). The PAS positivity results from introduction of carbohydrate-containing uromodulin (Tamm Horsfall protein) in the thick ascending loop of Henle. 20 × objective magnification.

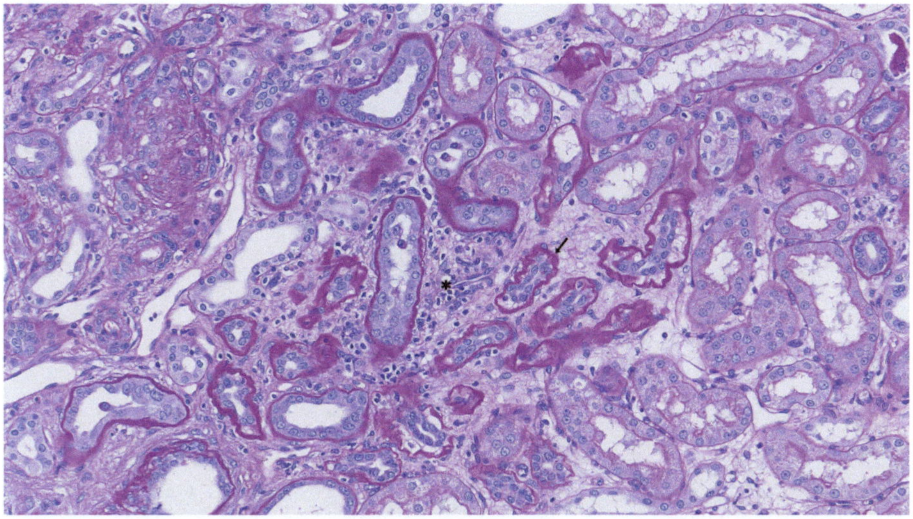

FIGURE 1.24

Periodic acid-Schiff/hematoxylin (PASH) stain on renal cortex of a male rhesus macaque necropsied 154 days after irradiation at 10 Gy has thickened PAS$^+$ tubular basement membranes (arrow) in an area of interstitial fibrosis. A mild infiltration of deeply stained infiltrating leukocytes (*) is present in the fibrotic area. This tubular change was present in only a small number of animals, and only in widely scattered clusters of tubules. 20 × objective magnification.

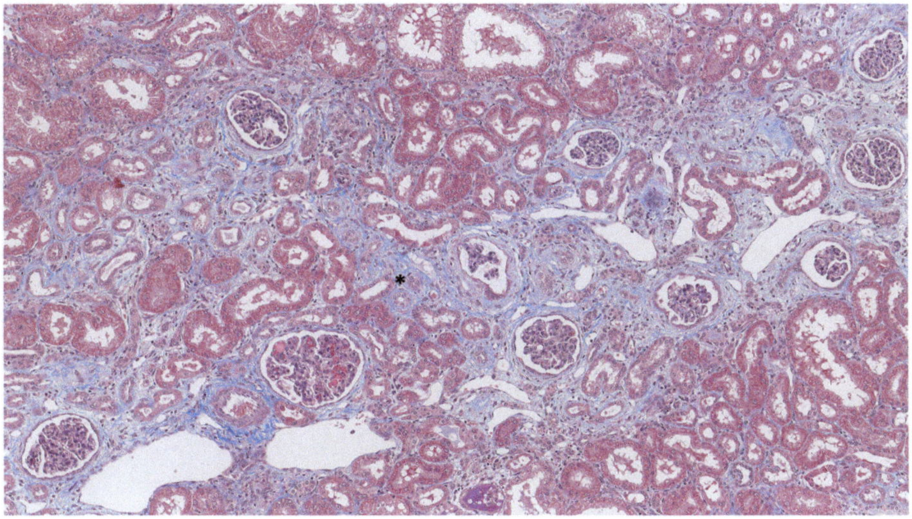

FIGURE 1.25

Interstitial fibrosis in the renal cortex of a male rhesus macaque necropsied 183 days following irradiation at 10 Gy. Note the roughly linear band of blue-stained fibrous connective tissue (*). Masson's trichrome stain, 10 × objective magnification.

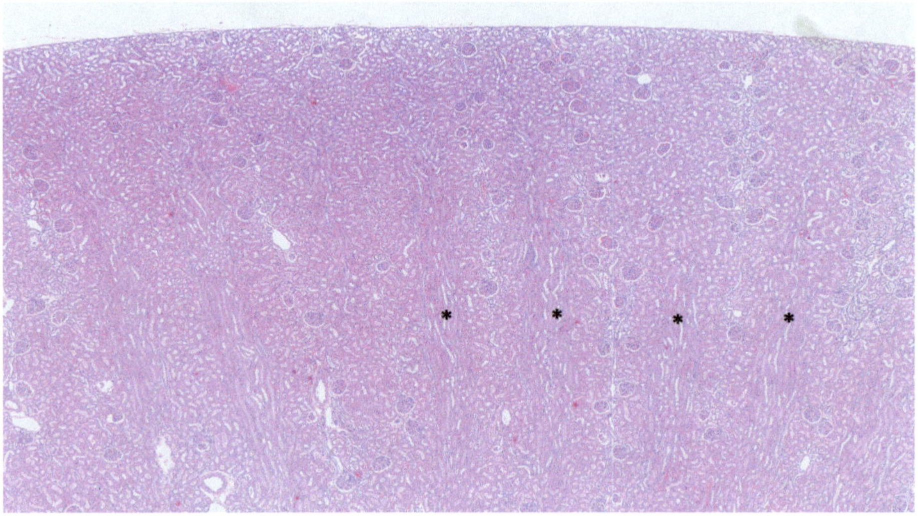

FIGURE 1.26A

Renal cortex of naive male rhesus macaque. Note the radially oriented arrays of medullary rays containing collecting ducts (*). Hematoxylin and eosin stain, 2.5 × objective magnification.

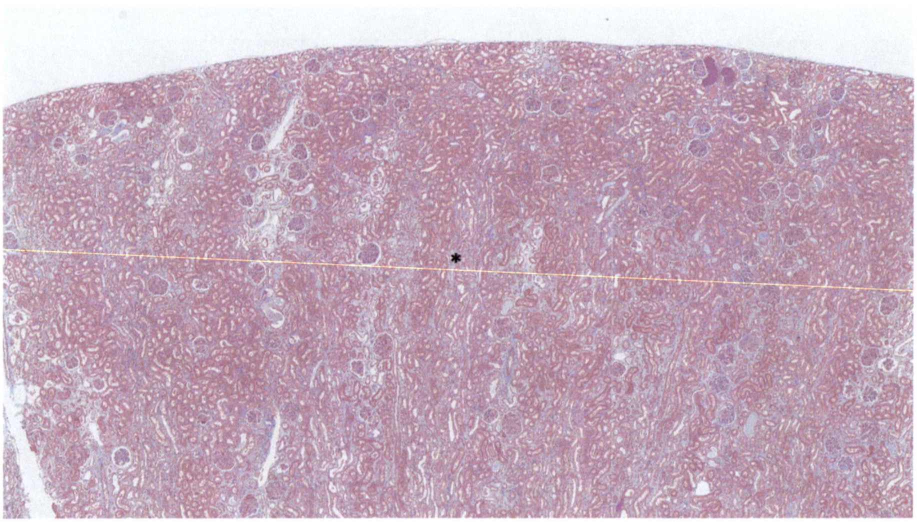

FIGURE 1.26B

Renal cortex of male rhesus macaque necropsied 141 days after irradiation at 11 Gy. Note the blue staining of increased fibrous connective tissue associated with the radially oriented medullary rays (*). Masson's trichrome stain, 2.5 × objective magnification.

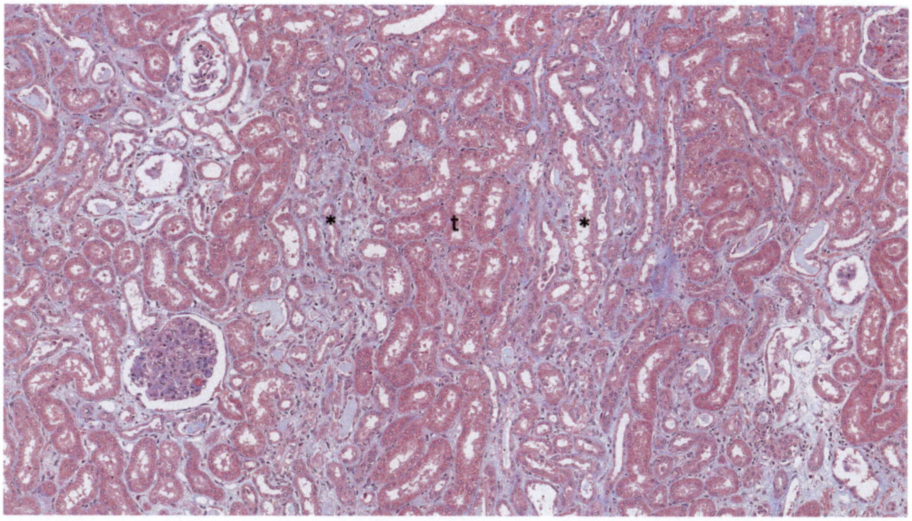

FIGURE 1.26C

Higher magnification of previous image showing concentration of blue-stained fibrous connective surrounding collecting ducts in medullary rays (*), with relative sparing of proximal tubular regions (t). Masson's trichrome stain, 10 × objective magnification.

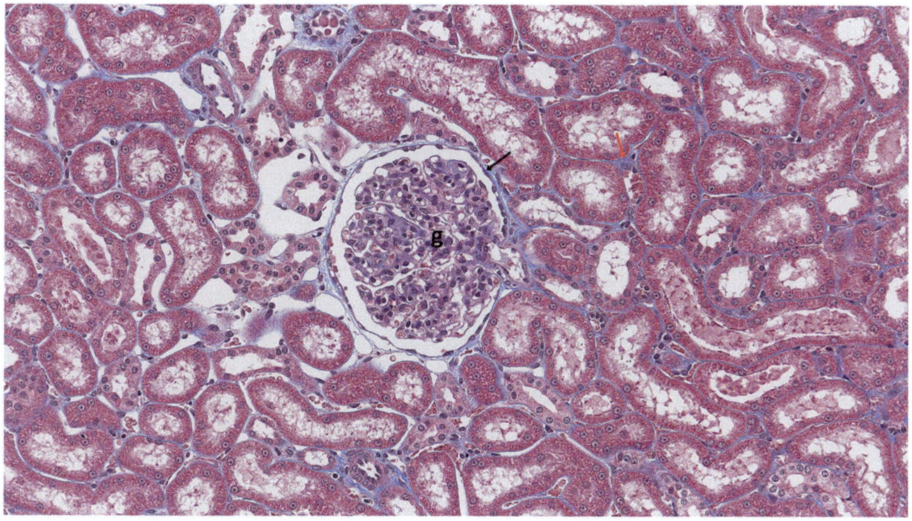

FIGURE 1.27A

Renal cortex with glomerulus (g) from naïve male rhesus macaque. Note the small amount of blue-stained fibrous connective tissue in the glomerular capsule (black arrow) and within the capillary tufts of the glomerulus. Cortical tubules are surrounded by thin blue lines (red arrow) representing interstitial fibrous connective tissue. Masson's trichrome stain, 20 × objective magnification.

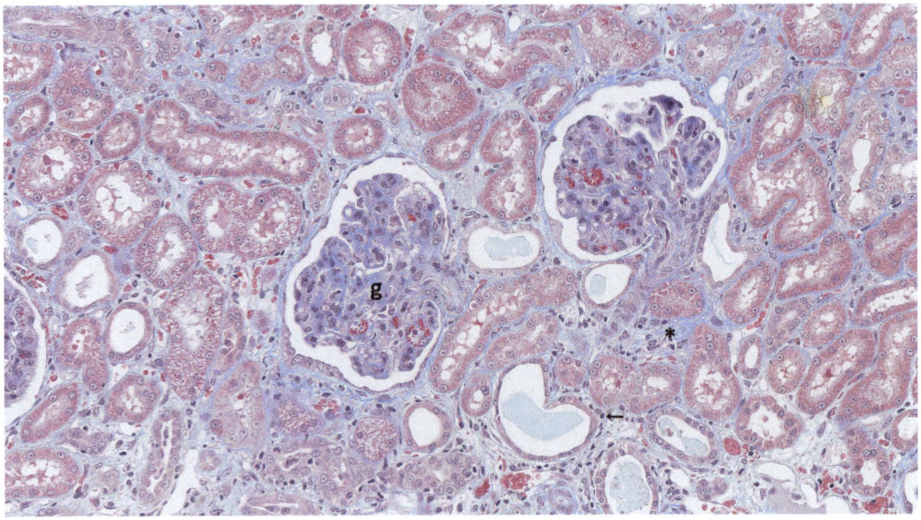

FIGURE 1.27B

Masson's trichrome stain on the kidney from a male rhesus macaque necropsied 171 days after irradiation at 10 Gy shows an increased amount of blue-stained fibrous connective tissue in the intertubular interstitium (*) and the glomerulus (g). The interstitium also has an increased population of small, densely stained leukocytes (arrow), indicating an inflammatory component of the pathological process at this stage. Masson's trichrome stain, 20 × objective magnification.

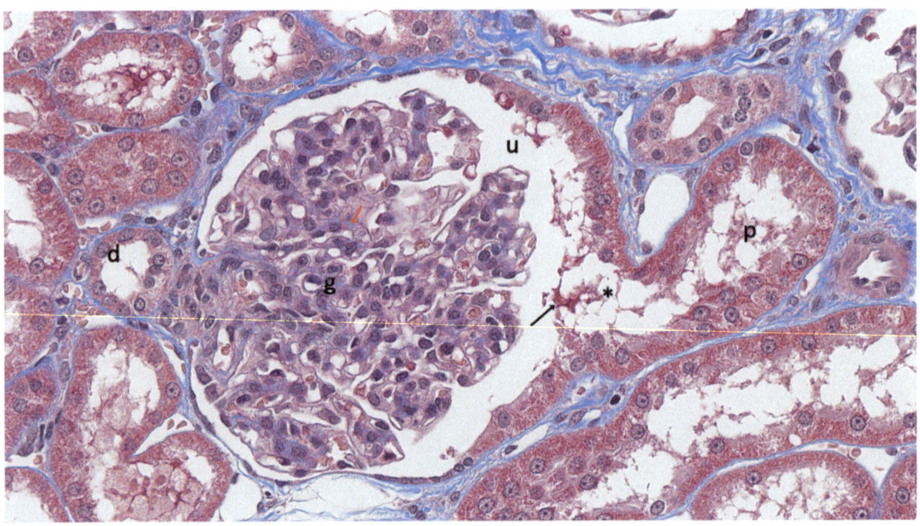

FIGURE 1.28A

The Masson's trichrome stain on kidney from a naïve control male rhesus macaque shows a small amount of blue-stained fibrous connective tissue in the capsule surrounding the glomerulus (g), proximal (p) tubules, and distal (d) tubules. A minimal amount of stained fibrous connective tissue is present in the glomerular matrix (red arrow). Note the open neck region (*) connecting the urinary space (u) of the glomerulus with the lumen of the proximal tubule. Also note the small amount of amorphous, presumed proteinic, material (arrow) extending from the urinary space through the tubular neck region into the lumen of the proximal tubule. Masson's trichrome stain, 40 × objective magnification.

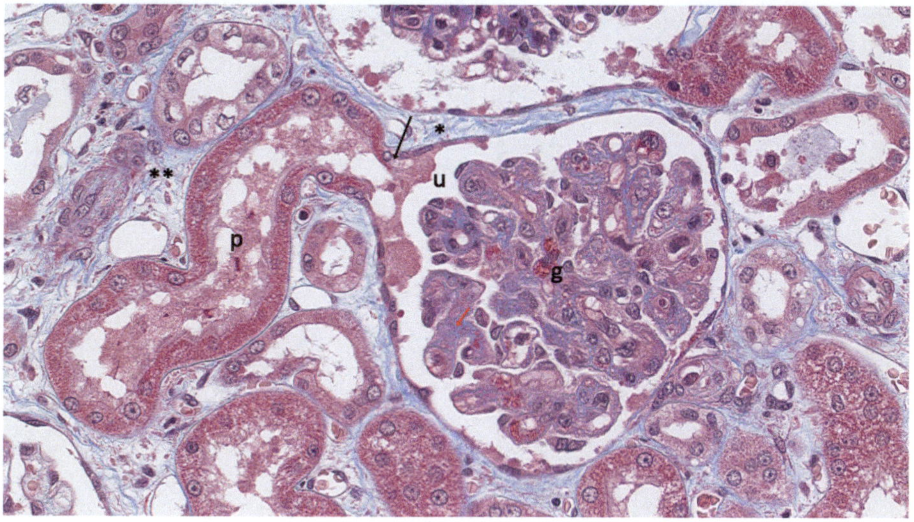

FIGURE 1.28B

The Masson's trichrome stain on kidney from a male rhesus macaque necropsied 141 days after irradiation at 11 Gy shows an increased amount of blue-stained fibrous connective tissue in the glomerular capsule (*), intertubular interstitium (**), and glomerulus (g). Note the constriction (black arrow) in the proximal tubule neck region. An increased amount of amorphous material within the urinary space (u) and proximal tubule (p) is consistent with passage of plasma proteins through the glomerular filtration apparatus. The glomerulus has an increased amount of blue-stained fibrous connective tissue (red arrow). Masson's trichrome stain, 30 × objective magnification.

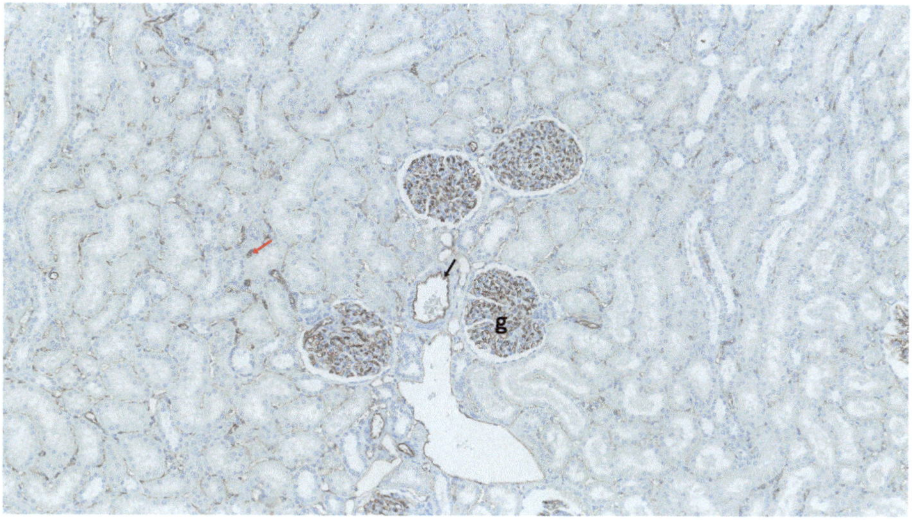

FIGURE 1.29A

Immunohistochemical stain for CD31 on renal cortex of a naive male rhesus macaque. Note the evenly dispersed CD31 staining of capillary endothelial cells throughout the glomeruli (g). Endothelial cells lining an artery (black arrow) are positively stained, as are arteriolar endothelial cells (red arrow) in the tubular interstitium. Stain performed with 3,3′-diaminobenzidine chromogen and hematoxylin counterstain. 10 × objective magnification.

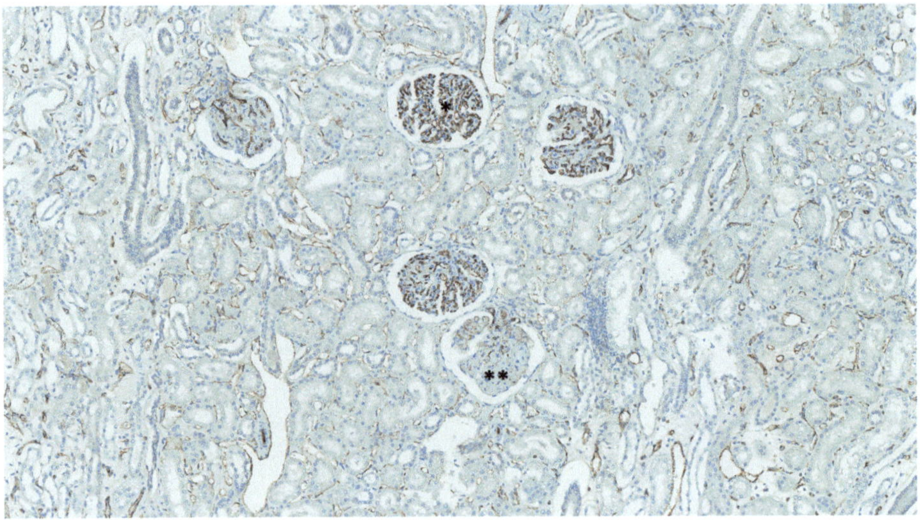

FIGURE 1.29B

CD31 immunohistochemical stain on renal cortex of a male rhesus macaque necropsied 180 days after irradiation at 10 Gy. Note the normal pattern of CD31 staining in one glomerulus (*), as compared to the markedly reduced CD31 staining in a radiation-damaged glomerulus (**). Other glomeruli in the section have varying degrees of CD31 staining. The reduced CD31 staining indicates a reduction in normal capillary endothelial cells as one facet of radiation-associated glomerular injury. Stain performed with 3,3′-diaminobenzidine chromogen and hematoxylin counterstain. 10 × objective magnification.

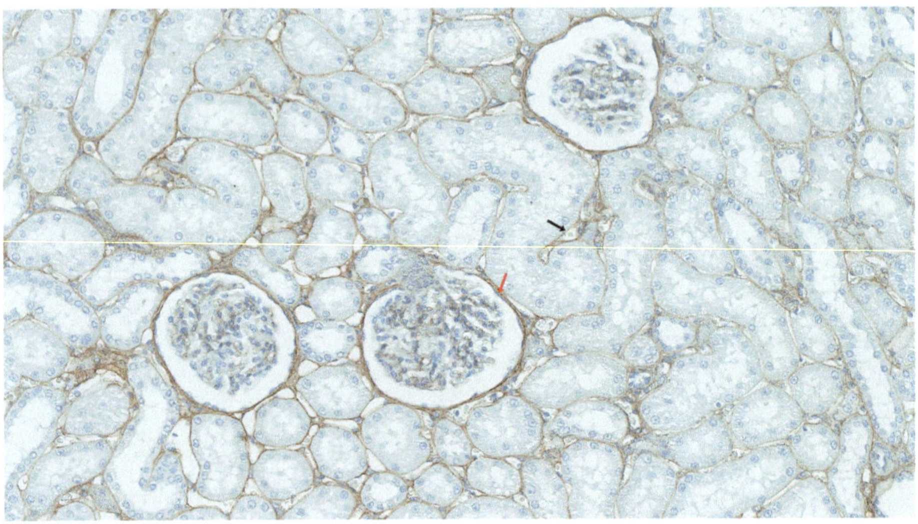

FIGURE 1.30A

Immunohistochemical stain for collagen-1 on renal cortex of a naive male rhesus macaque. Note the fine collagen-positive line surrounding cortical tubules (black arrow), with a slightly thicker band of collagen in the glomerular capsule (red arrow). Stain performed with 3,3′-diaminobenzidine chromogen and hematoxylin counterstain, 20 × objective magnification.

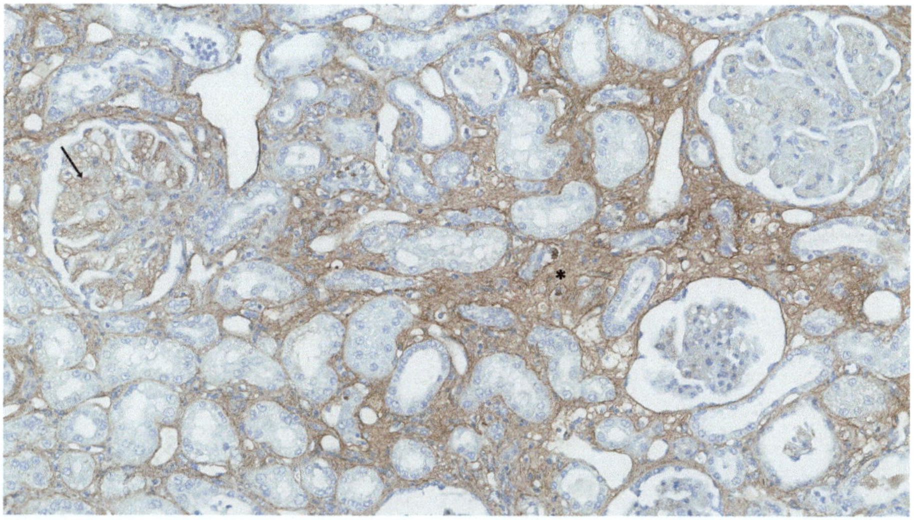

FIGURE 1.30B

Immunohistochemical stain for collagen-1 on renal cortex of a male rhesus macaque necropsied 174 days after irradiation at 11 Gy. Note the marked increase in collagen staining in the tubular interstitium (*) and glomeruli (arrow). Stain performed with 3,3′-diaminobenzidine chromogen and hematoxylin counterstain, 20 × objective magnification.

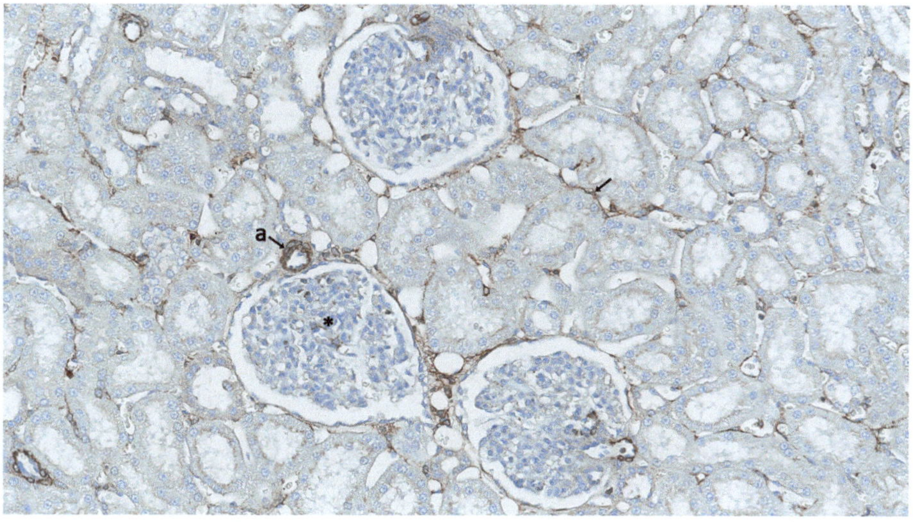

FIGURE 1.31A

Immunohistochemical stain for alpha-smooth muscle actin (αSMA) on renal cortex of naive male rhesus macaque. Note the minimal αSMA staining surrounding tubules (arrow) and within glomeruli (*). The smooth muscle component of arteriolar walls (a) is strongly stained. Stain performed with 3,3′-diaminobenzidine chromogen and hematoxylin counterstain, 20 × objective magnification.

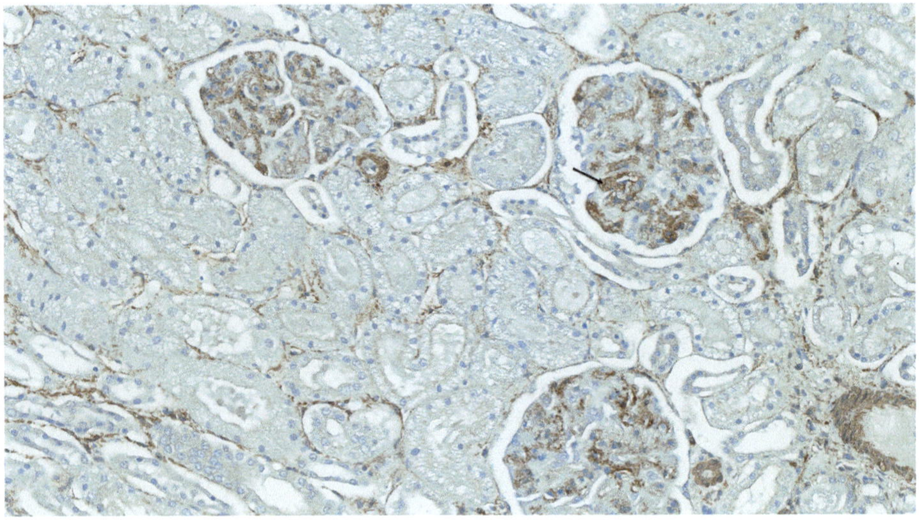

FIGURE 1.31B

Immunohistochemical stain for alpha-smooth muscle actin (αSMA) on renal cortex of a male rhesus macaque necropsied 105 days after irradiation at 10 Gy. Note the increased level of αSMA staining in glomeruli (arrow). Stain performed with 3,3′-diaminobenzidine chromogen and hematoxylin counterstain, 20× objective magnification.

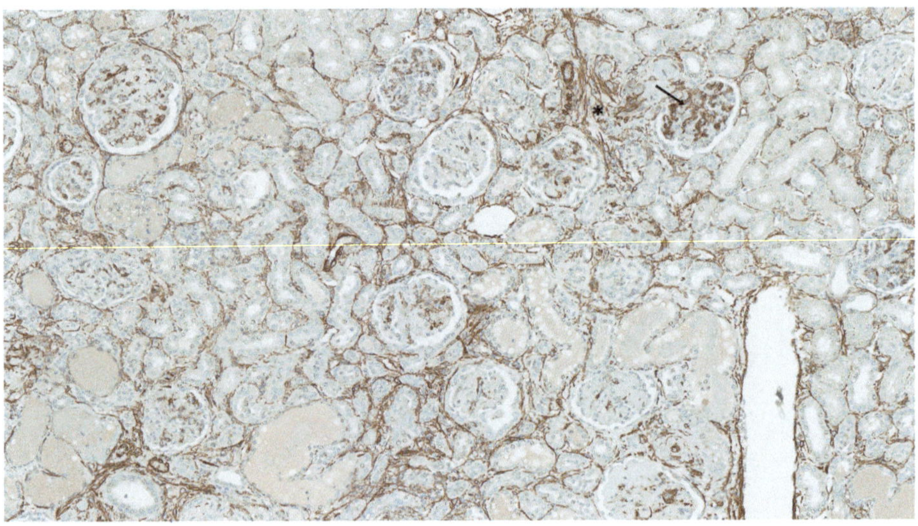

FIGURE 1.31C

Immunohistochemical stain for alpha-smooth muscle actin (αSMA) on renal cortex of a male rhesus macaque necropsied 95 days after irradiation at 10 Gy. Note the increased level of αSMA staining in the interstitium (*) and glomeruli (arrow). Stain performed with 3,3′-diaminobenzidine chromogen and hematoxylin counterstain, 10× objective magnification.

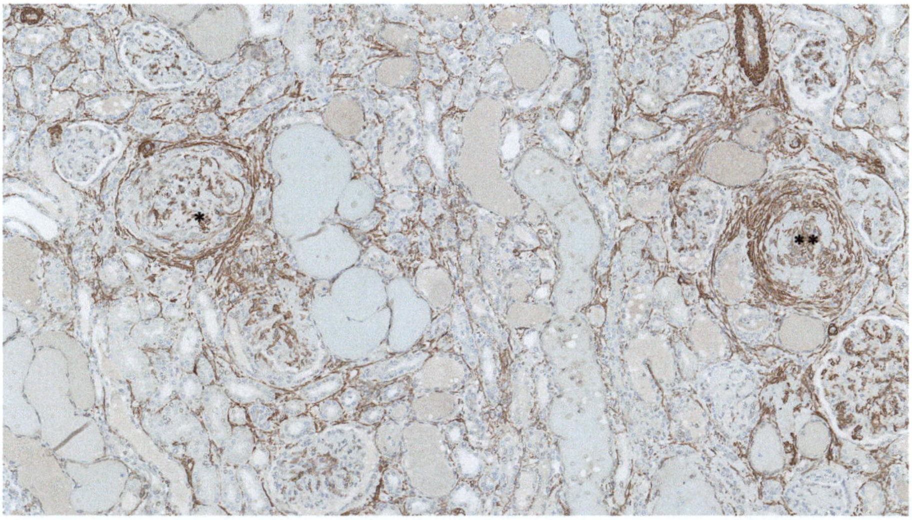

FIGURE 1.31D

Immunohistochemical stain for alpha-smooth muscle actin (αSMA) on renal cortex of a male rhesus macaque necropsied 95 days after irradiation at 10 Gy. Note the whorl of αSMA staining around glomeruli (*), with some glomeruli reduced to a nondescript clump of αSMA-positive tissue (**). Stain performed with 3,3′-diaminobenzidine chromogen and hematoxylin counterstain, 10 × objective magnification.

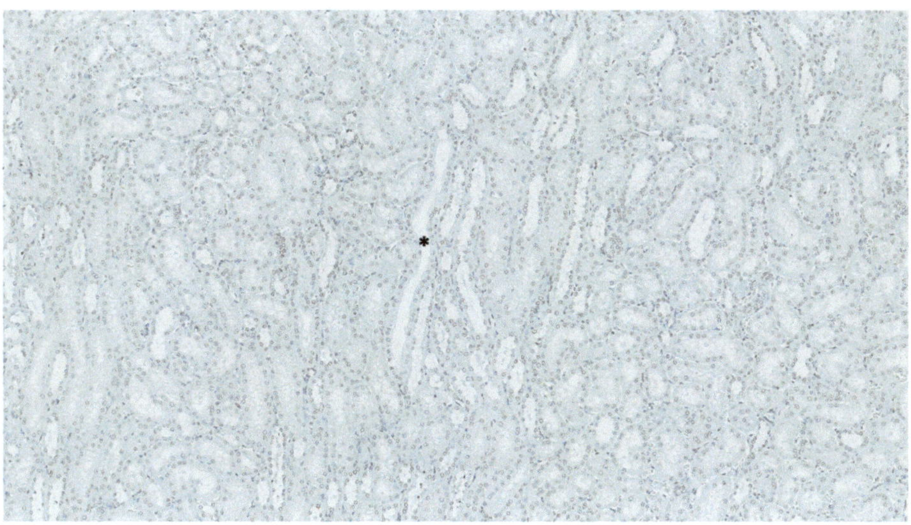

FIGURE 1.32A

Immunohistochemical stain for connective tissue growth factor (CTGF) on renal cortex of naive male rhesus macaque. Note the absence of CTGF staining in the radially oriented cluster of collecting ducts (*). Stain performed with 3,3′-diaminobenzidine chromogen and hematoxylin counterstain, 10 × objective magnification.

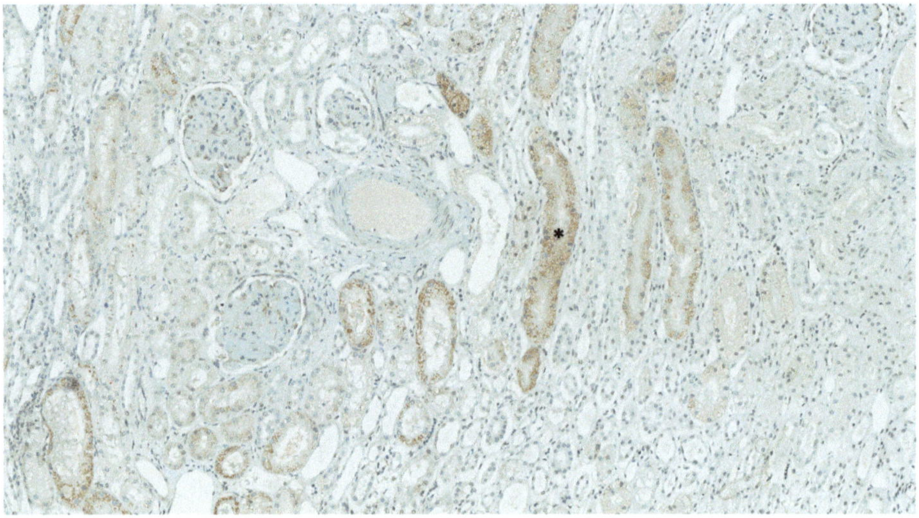

FIGURE 1.32B

Immunohistochemical staining for connective tissue growth factor (CTGF) on renal cortex of male rhesus macaque necropsied 105 days after irradiation at 10 Gy. Note positive CTGF staining of collecting ducts and tubules in a somewhat linear array (*). Stain performed with 3,3′-diaminobenzidine chromogen and hematoxylin counterstain. 10 × objective magnification.

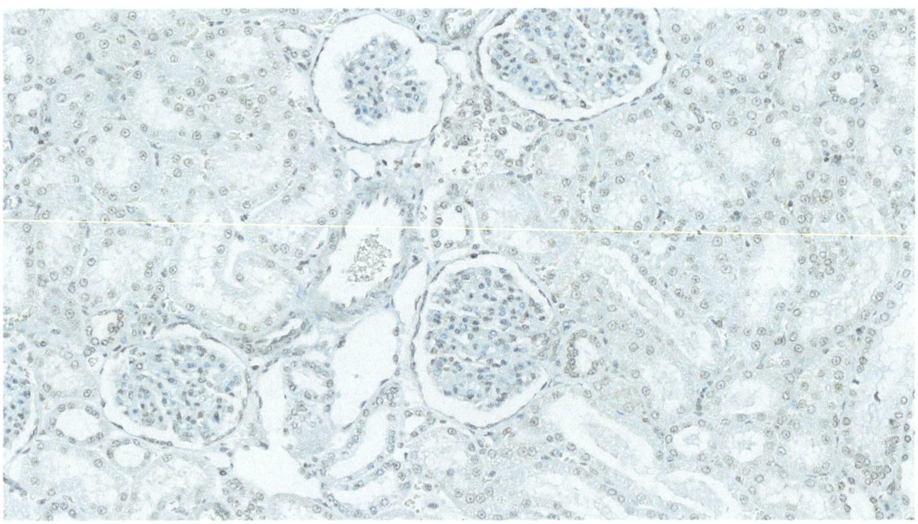

FIGURE 1.32C

Immunohistochemical stain for connective tissue growth factor (CTGF) on renal cortex of naive male rhesus macaque shows essentially no CTGF staining. Stain performed with 3,3′-diaminobenzidine chromogen and hematoxylin counterstain, 20 × objective magnification.

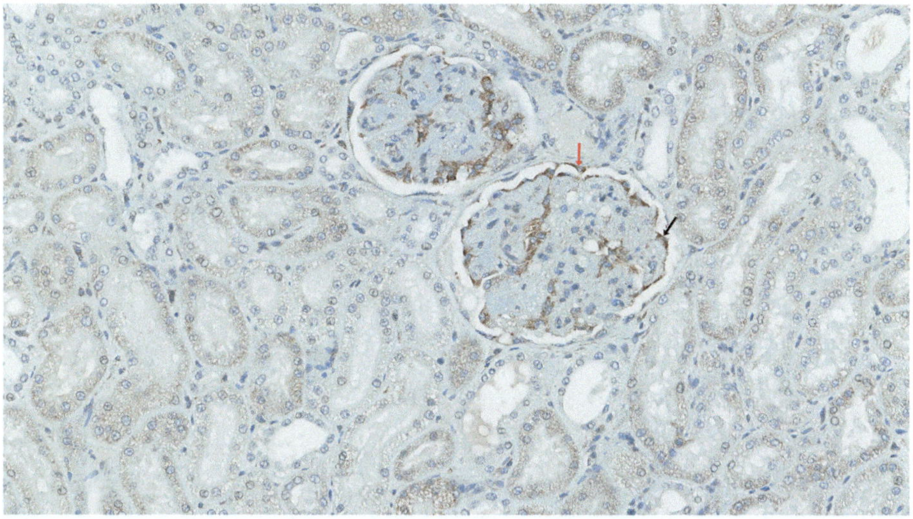

FIGURE 1.32D

Immunohistochemical staining for connective tissue growth factor (CTGF) in renal cortex of male rhesus macaque necropsied 105 days after irradiation at 10 Gy. Note positive staining of podocytes (black arrow) on surface of glomerular tufts, and slight staining (red arrow) of lining cell of glomerular capsule. Stain performed with 3,3′-diaminobenzidine chromogen and hematoxylin counterstain, 20 × objective magnification.

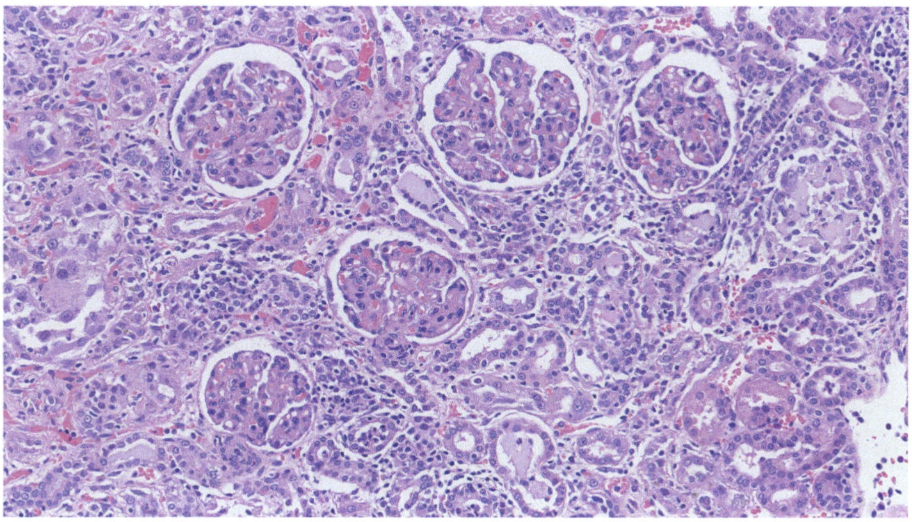

FIGURE 1.33

Renal cortex of male rhesus macaque necropsied 183 days after irradiation at 11 Gy. Note the diffuse interstitial infiltration of deeply stained inflammatory cells, which subsequent IHC stains revealed to be largely lymphocytes and macrophages. The inflammatory cell infiltration in this animal was unusually intense, as the majority of irradiated animals had minimal inflammatory cell infiltration. Hematoxylin and eosin stain, 20 × objective magnification.

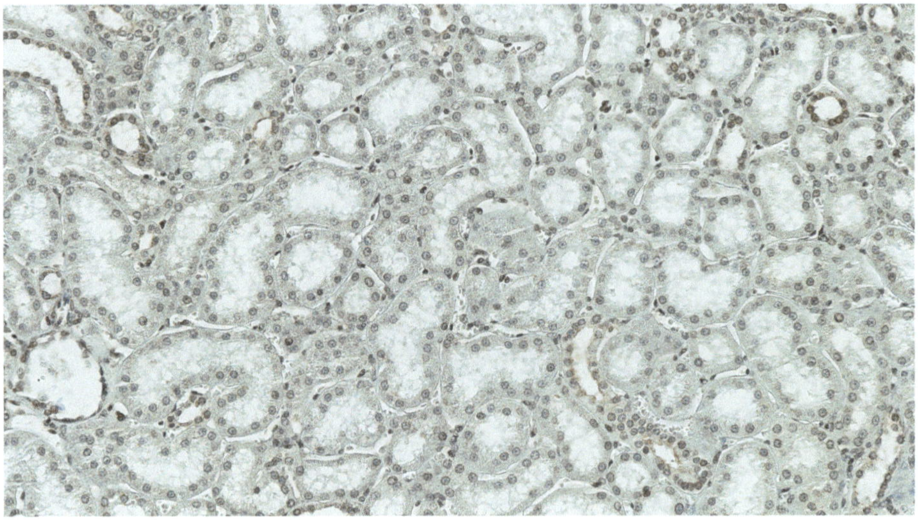

FIGURE 1.34A

Immunohistochemical stain for CD38 on renal cortex of naïve male rhesus macaque. CD38 is found on the surface of many leukocytes, including lymphocytes and natural killer (NK) cells. Note the virtual absence of $CD38^+$ cells in this image. Stain performed with 3,3′-diaminobenzidine chromogen and hematoxylin counterstain, 20 × objective magnification.

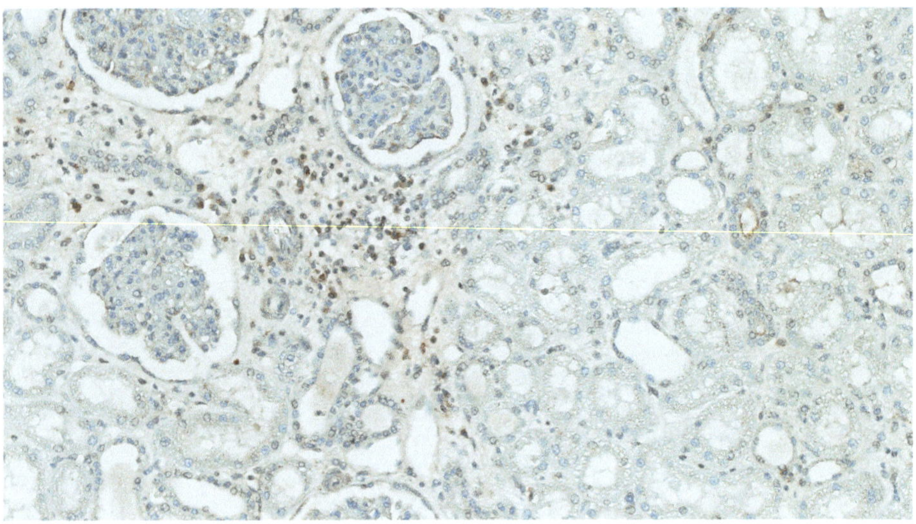

FIGURE 1.34B

Immunohistochemical stain for CD38 on renal cortex of male rhesus macaque necropsied 177 days after irradiation at 11 Gy. Note the population of brown stained $CD38^+$ cells in the interstitium. Stain performed with 3,3′diaminobenzidine chromogen and hematoxylin counterstain, 20 × objective magnification.

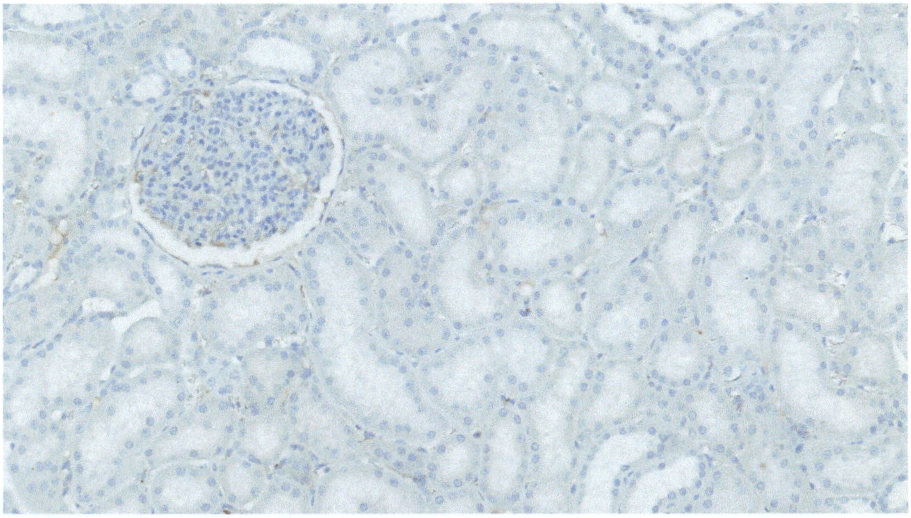

FIGURE 1.35A

Immunohistochemical stain for CD68 on renal cortex of naïve male rhesus macaque. CD68 is a protein expressed by cells of the monocyte lineage, including macrophages. Note the virtual absence of $CD68^+$ cells in this image. Stain performed with 3,3′-diaminobenzidine chromogen and hematoxylin counterstain, 20 × objective magnification.

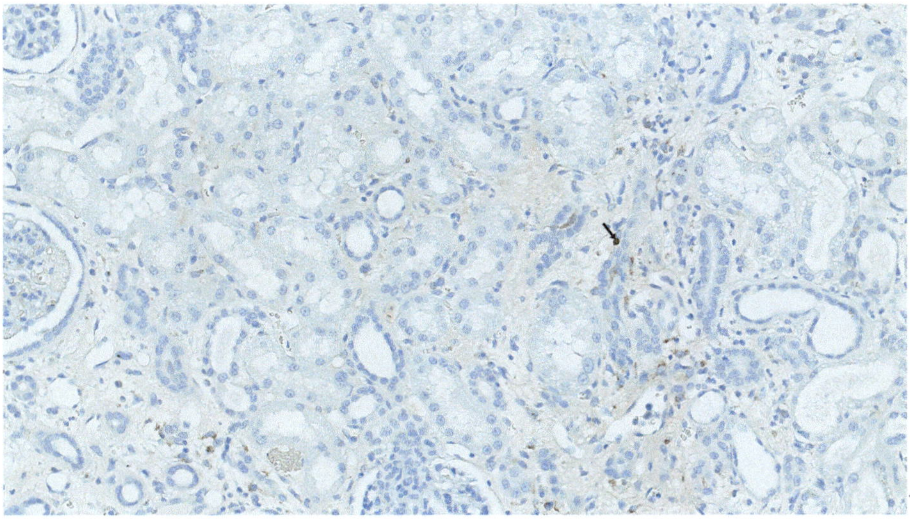

FIGURE 1.35B

Immunohistochemical stain for CD68 on renal cortex of a male rhesus macaque necropsied 177 days after irradiation at 11 Gy. Note the sparse population (arrow) of $CD68^+$ cells in this image. This image is from the same animal shown for CD38 staining. Stain performed with 3,3′-diaminobenzidine chromogen and hematoxylin counterstain, 20 × objective magnification.

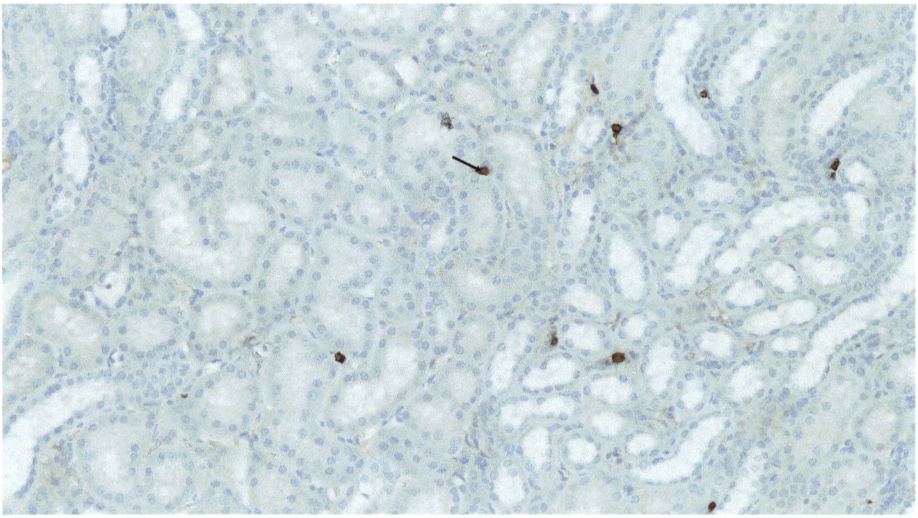

FIGURE 1.36A

Immunohistochemical stain for CD20, a B lymphocyte marker, on renal cortex of naïve male rhesus macaque. Note the sparse population of $CD20^+$ cells (arrow) in this image. Stain performed with 3,3′-diaminobenzidine chromogen and hematoxylin counterstain, 20 × objective magnification.

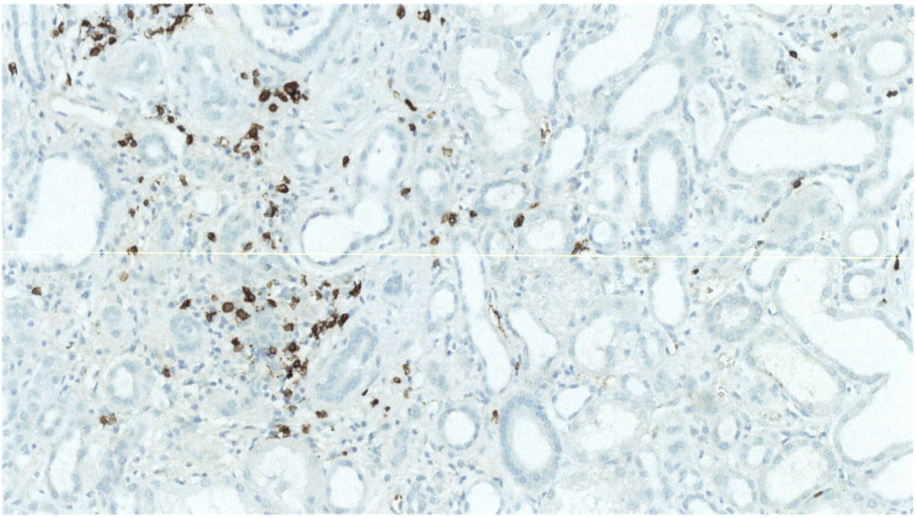

FIGURE 1.36B

Immunohistochemical stain for CD20 on renal cortex of male rhesus macaque necropsied 177 days after irradiation at 11 Gy. Note the increased population of brown stained $CD20^+$ B lymphocytes. This image is from the same animal shown for CD38 staining. Stain performed with 3,3′-diaminobenzidine chromogen and hematoxylin counterstain, 20 × objective magnification.

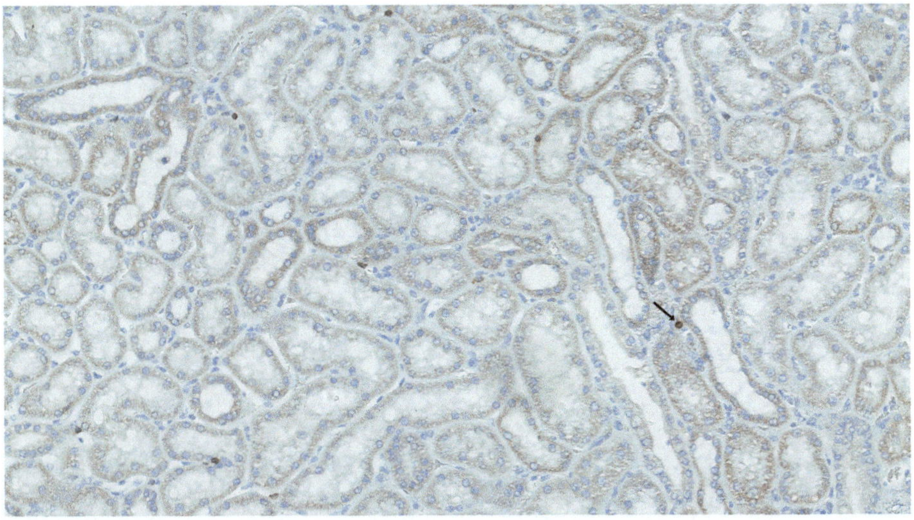

FIGURE 1.37A

Immunohistochemical stain for CD3, a T lymphocyte marker, on renal cortex of naïve male rhesus macaque. Note the sparse population of $CD3^+$ cells (arrow) in this image. Stain performed with 3,3′-diaminobenzidine chromogen and hematoxylin counterstain, 20 × objective magnification.

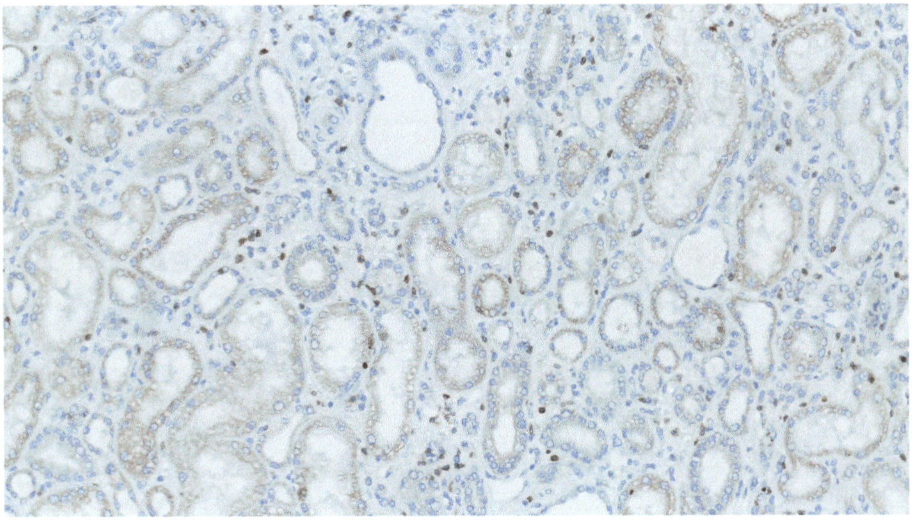

FIGURE 1.37B

Immunohistochemical stain for CD3 on renal cortex of male rhesus macaque necropsied 177 days after irradiation at 11 Gy. Note the increased population of brown stained $CD3^+$ T lymphocytes. This image is from the same animal shown for CD38 staining. Stain performed with 3,3′-diaminobenzidine chromogen and hematoxylin counterstain, 20 × objective magnification.

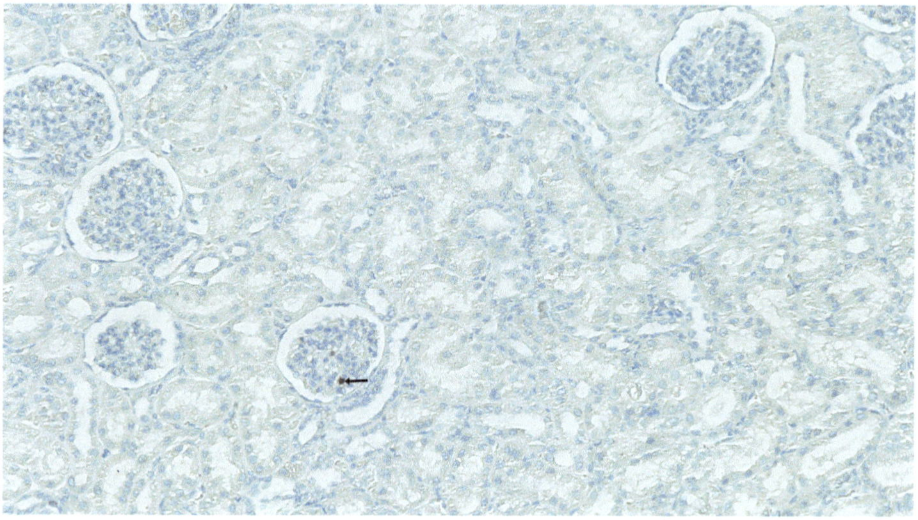

FIGURE 1.38A

Immunohistochemical stain for myeloperoxidase (MPO) on renal cortex of naïve male rhesus macaque. Myeloperoxidase is expressed by neutrophils and macrophages of the monocytic line. Note the sparse population of MPO^+ cells (arrow) in this image. Stain performed with 3,3′-diaminobenzidine chromogen and hematoxylin counterstain, 20 × objective magnification.

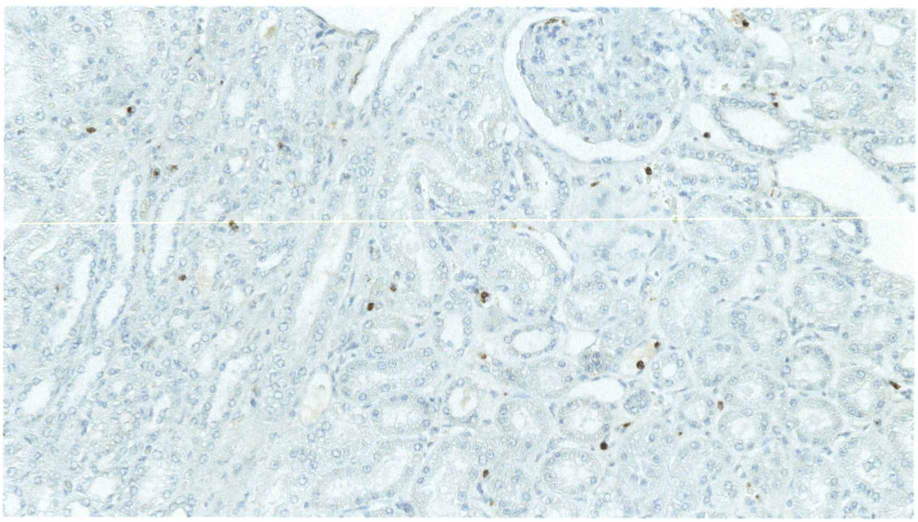

FIGURE 1.38B

Immunohistochemical stain for myeloperoxidase (MPO) on renal cortex of male rhesus macaque necropsied 115 days after irradiation at 11 Gy. Note the increased population of brown stained MPO^+ cells. The size and microscopic features of the stained cells were consistent with macrophages. Stain performed with 3,3′ diaminobenzidine chromogen and hematoxylin counterstain, 20 × objective magnification.

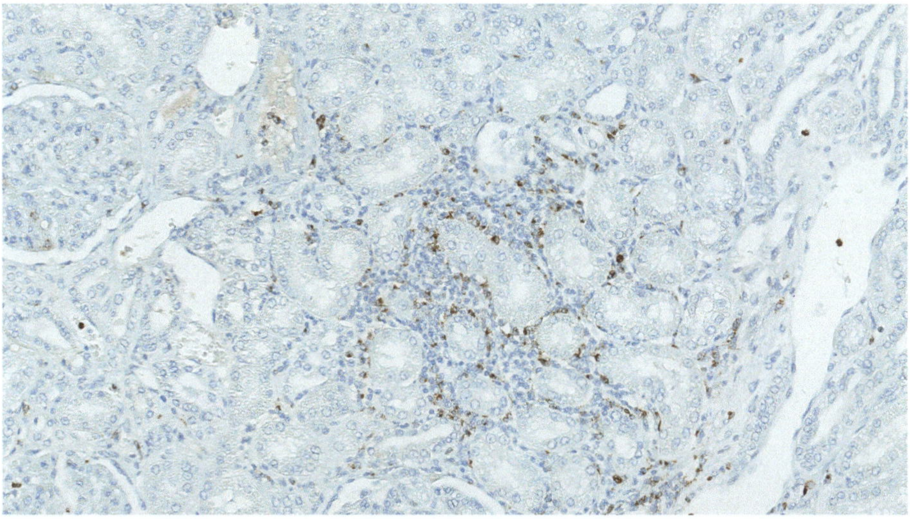

FIGURE 1.38C

Immunohistochemical stain for myeloperoxidase (MPO) on renal cortex of male rhesus macaque necropsied 115 days after irradiation at 11 Gy. Note the increased population of brown stained MPO$^+$ cells in an area of inflammatory cell infiltration. The size and microscopic features indicated the MPO$^+$ cells were primarily macrophages, with a minimal number of neutrophils. Stain performed with 3,3′-diaminobenzidine chromogen and hematoxylin counterstain, 20 × objective magnification.

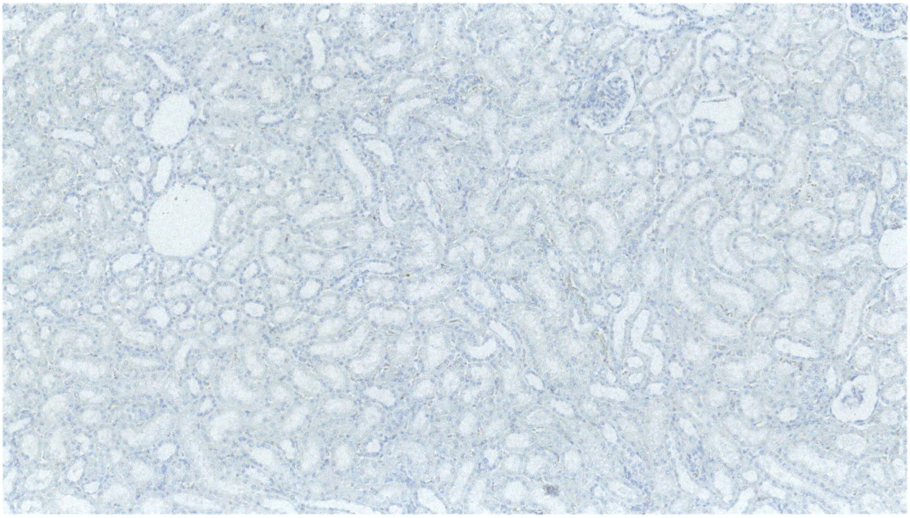

FIGURE 1.39A

Immunohistochemical stain for CD163, a marker for cells of the monocyte/macrophage lineage, on renal cortex of naïve male rhesus macaque. Note the virtual absence of brown stained CD163$^+$ cells. Stain performed with 3,3′-diaminobenzidine chromogen and hematoxylin counterstain, 10 × objective magnification.

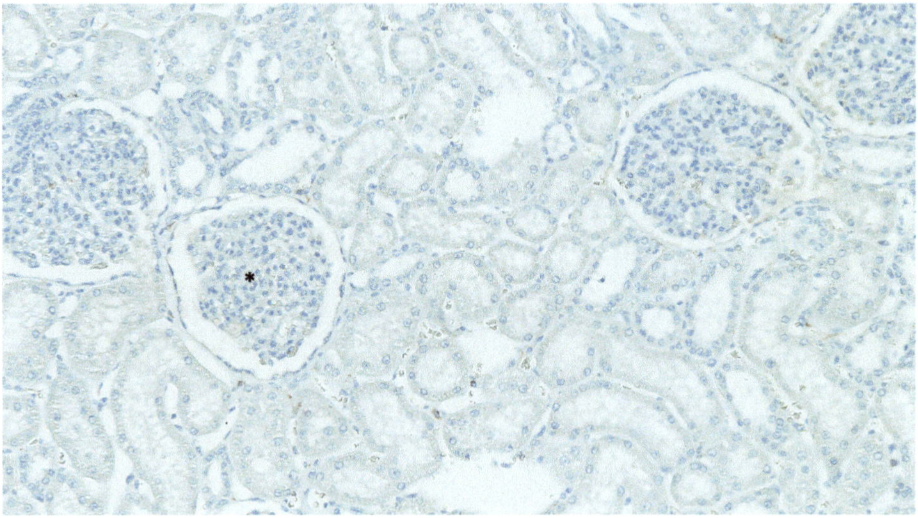

FIGURE 1.39B

Higher magnification of immunohistochemical stain for CD163 on renal cortex of naïve male rhesus macaque. Note the virtual absence of brown stained CD163$^+$ cells in glomeruli (*) and intertubular interstitium. Stain performed with 3,3′-diaminobenzidine chromogen and hematoxylin counterstain, 20 × objective magnification.

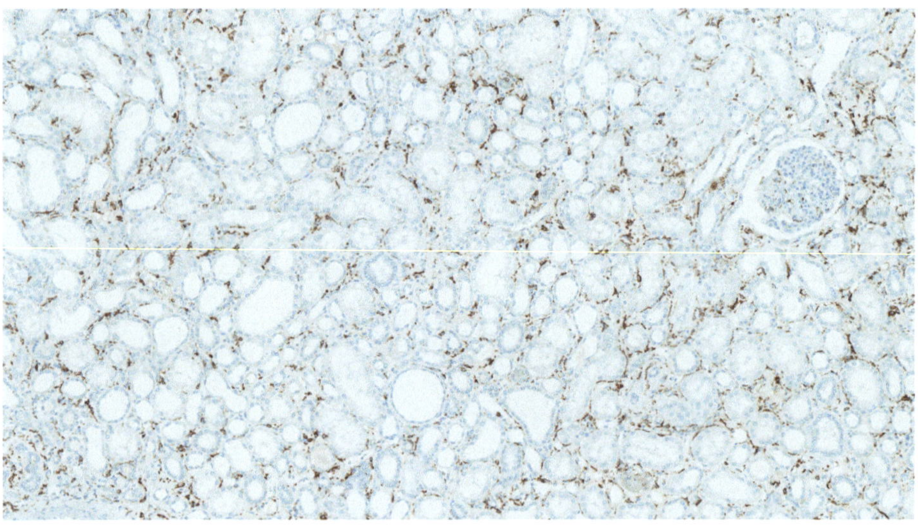

FIGURE 1.39C

Immunohistochemical stain for CD163 on renal cortex of a male rhesus macaque necropsied 177 days after irradiation at 11 Gy. Note the marked increase in the population of brown stained CD163$^+$ cells. This image is from the same animal shown for CD38, CD68, CD20, and CD3 staining. Stain performed with 3,3′-diaminobenzidine chromogen and hematoxylin counterstain. 10 × objective magnification.

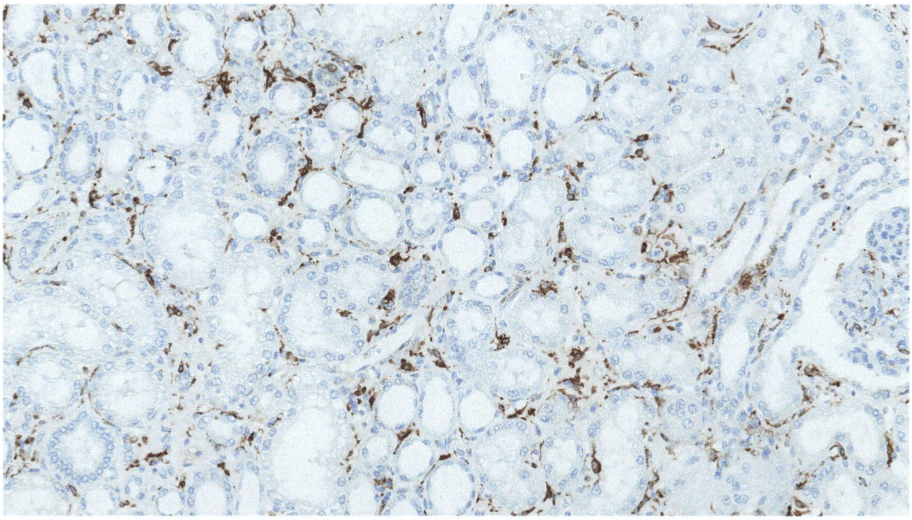

FIGURE 1.39D

Higher magnification of the previous specimen. The increased population of brown stained CD163[+] cells is located primarily in the intertubular interstitium. This image is from the same animal shown for CD38, CD68, CD20, and CD3 staining. Stain performed with 3,3′-diaminobenzidine chromogen and hematoxylin counterstain, 20× objective magnification.

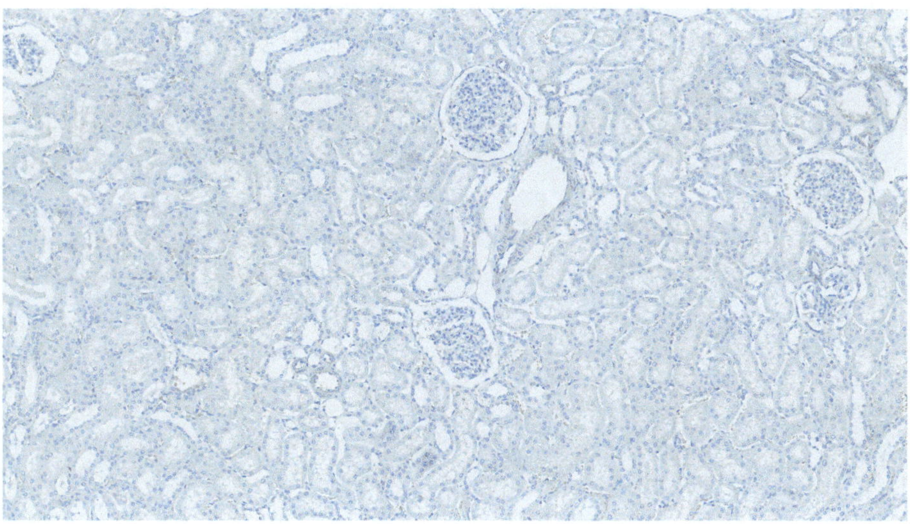

FIGURE 1.40A

Immunohistochemical stain for tryptase on renal cortex of naïve male rhesus macaque. Tryptase is a marker for mast cell granules. Note the virtual absence of brown-stained tryptase[+] cells. Stain performed with 3,3′-diaminobenzidine chromogen and hematoxylin counterstain, 10× objective magnification.

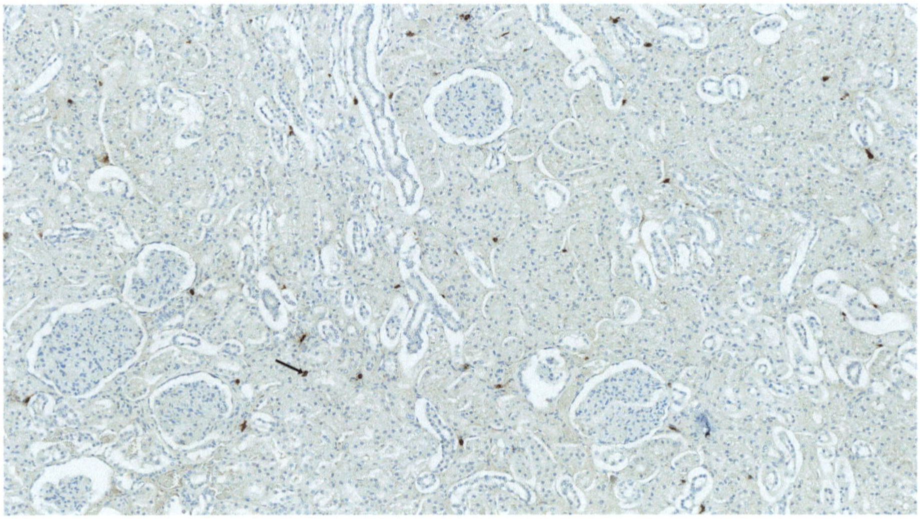

FIGURE 1.40B

Immunohistochemical stain for tryptase on renal cortex of a male rhesus macaque necropsied 181 days after irradiation at 11 Gy. Note the increased population of brown-stained tryptase^{+} cells (arrow). Stain performed with 3,3′-diaminobenzidine chromogen and hematoxylin counterstain, 10 × objective magnification.

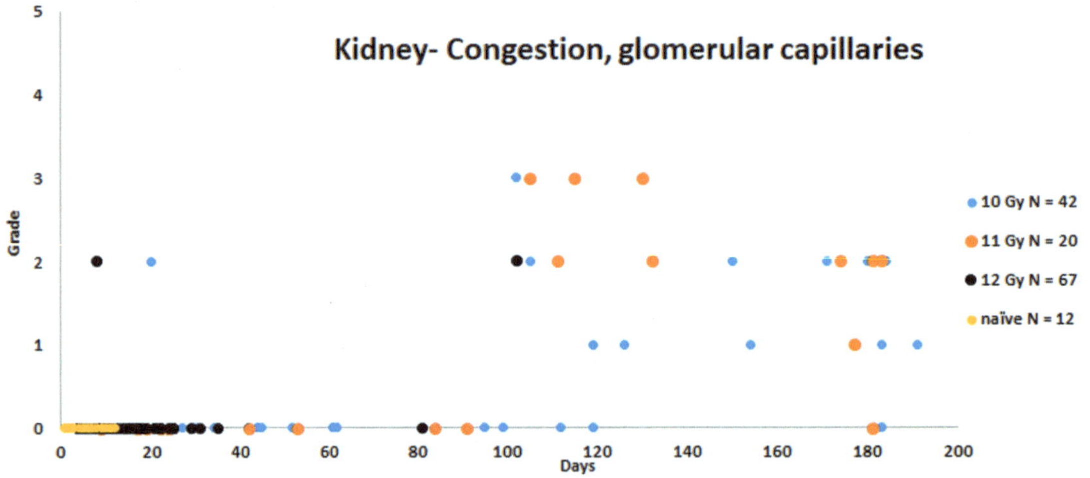

FIGURE 1.41

Scatter plot: kidney—congestion, glomerular capillaries. Observations based on light microscopic examination of paraffin-embedded specimens stained with hematoxylin and eosin.

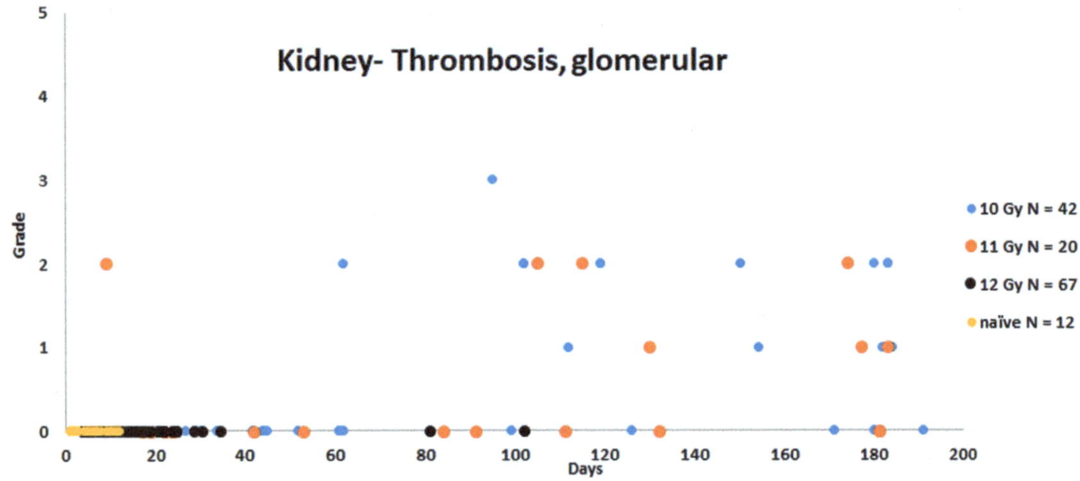

FIGURE 1.42

Scatter plot: kidney—thrombosis, glomerular. Observations based on light microscopic examination of paraffin-embedded specimens stained with hematoxylin and eosin.

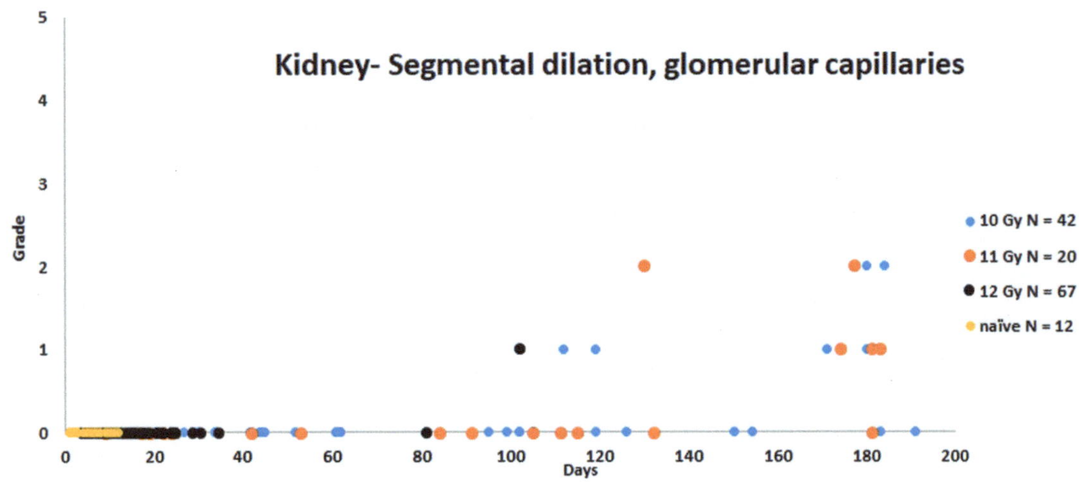

FIGURE 1.43

Scatter plot: kidney—segmental dilation, glomerular capillaries. Observations based on light microscopic examination of paraffin-embedded specimens stained with hematoxylin and eosin.

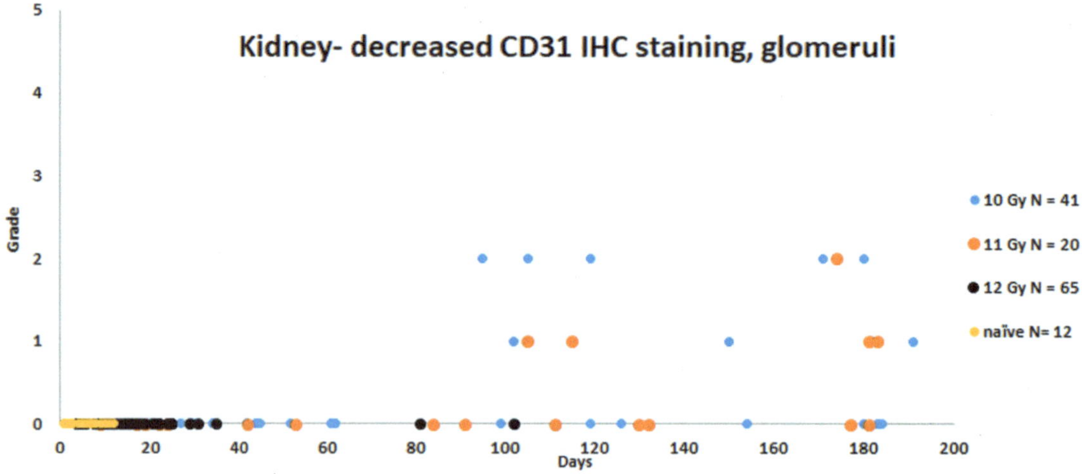

FIGURE 1.44

Scatter plot: kidney—decreased CD31 immunohistochemical staining, glomeruli. Observations based on light microscopic examination of paraffin-embedded specimens stained by immunohistochemical procedures for CD31 with 3,3′-diaminobenzidine chromogen and hematoxylin counterstain.

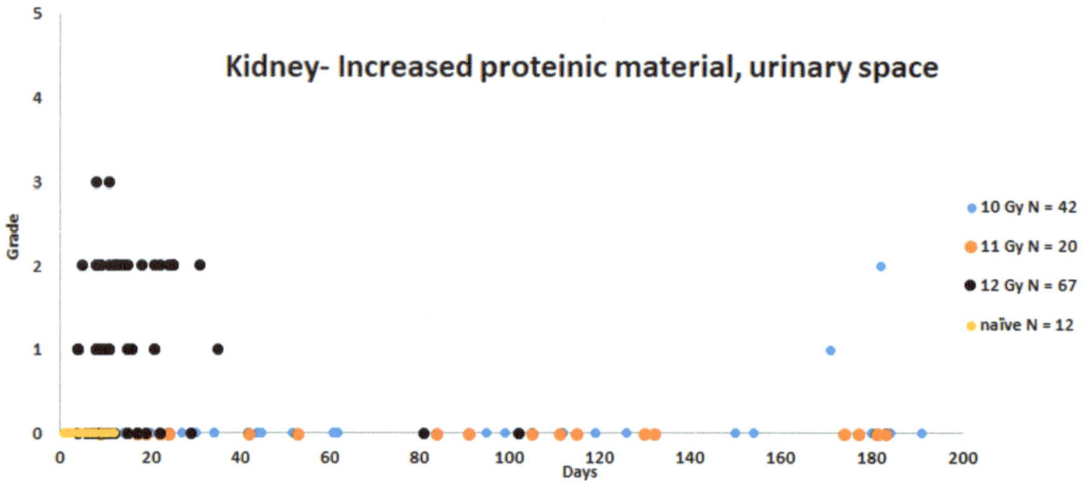

FIGURE 1.45

Scatter plot: kidney—increased proteinic material, urinary space. Observations based on light microscopic examination of paraffin-embedded specimens stained with hematoxylin and eosin.

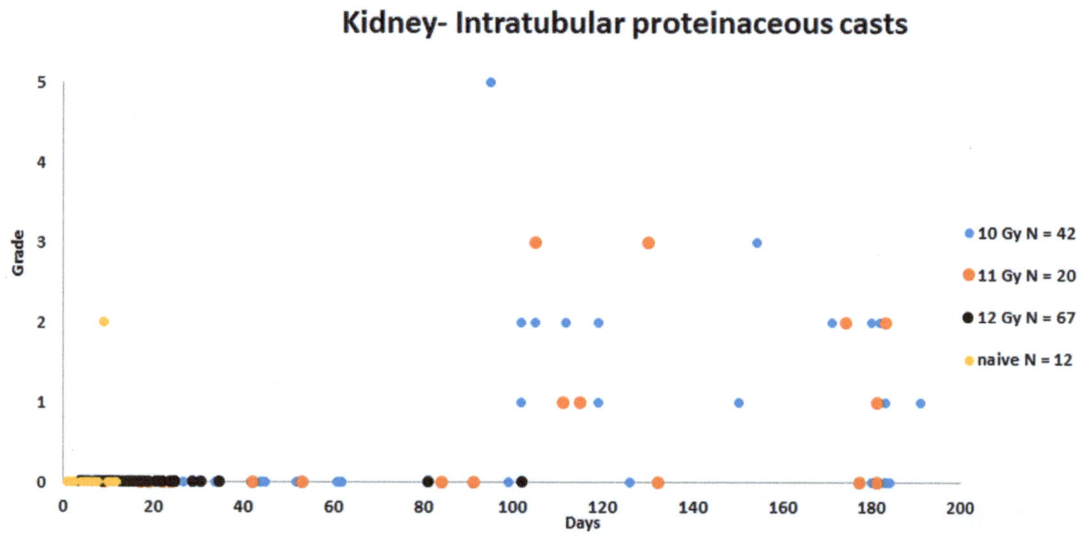

FIGURE 1.46

Scatter plot: kidney—intratubular proteinaceous casts. Observations based on light microscopic examination of paraffin-embedded specimens stained with hematoxylin and eosin.

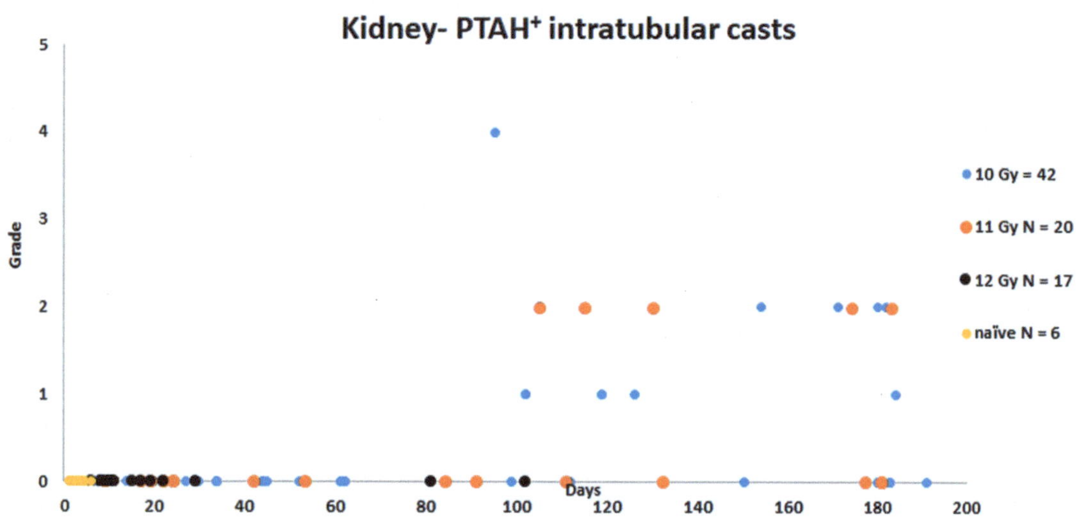

FIGURE 1.47

Scatter plot: kidney—PTAH$^+$ intratubular casts. Observations based on light microscopic examination of paraffin-embedded specimens stained with phosphotungstic acid-hematoxylin.

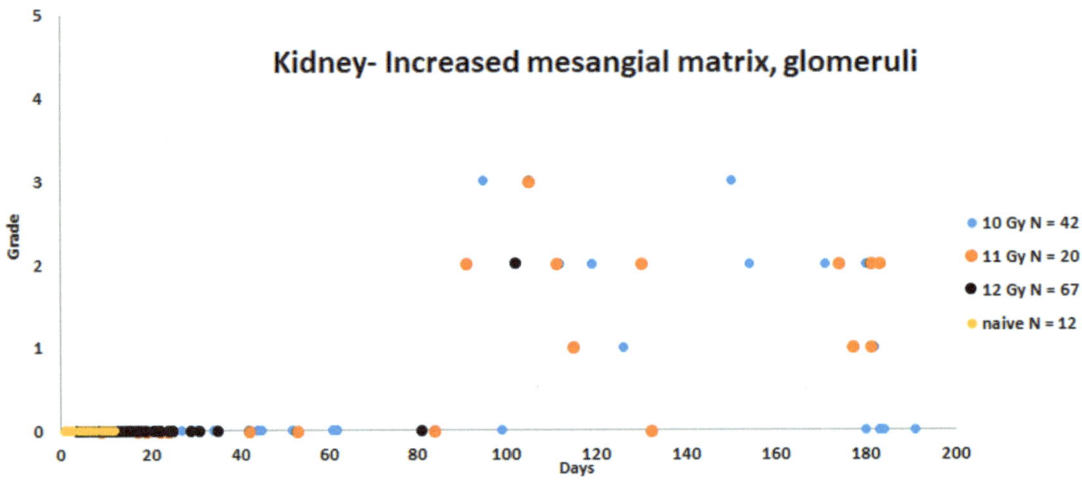

FIGURE 1.48

Scatter plot: kidney—increased mesangial matrix, glomeruli. Observations based on light microscopic examination of paraffin-embedded specimens stained with hematoxylin and eosin.

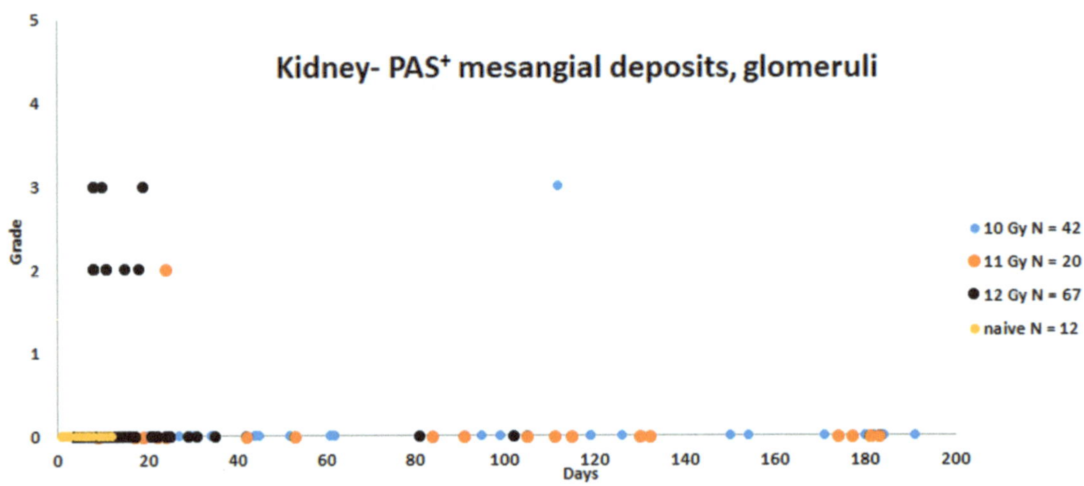

FIGURE 1.49

Scatter plot: kidney—PAS$^+$ mesangial deposits, glomeruli. Observations based on light microscopic examination of paraffin-embedded specimens stained with periodic acid/Schiff (PAS) stain with hematoxylin counterstain.

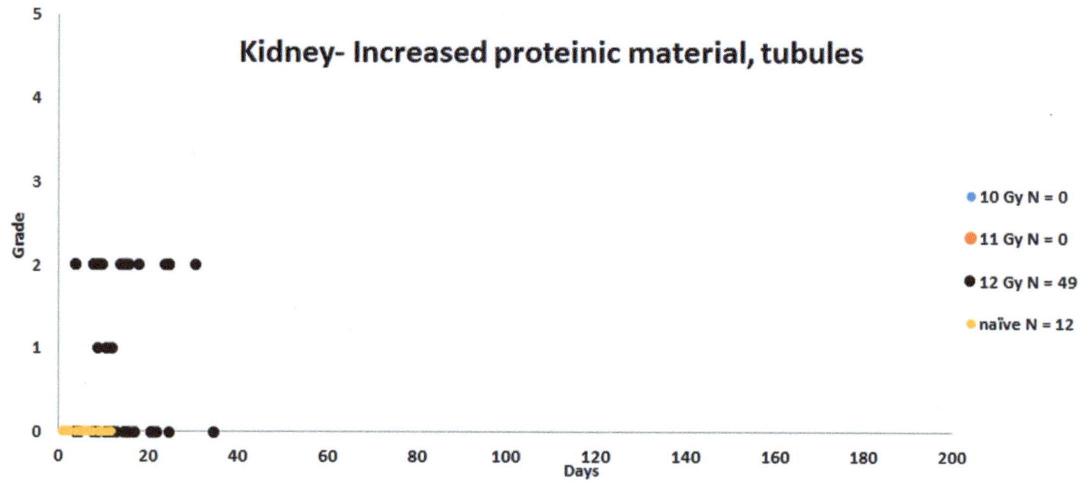

FIGURE 1.50

Scatter plot: kidney—increased proteinic material, tubules. Observations based on light microscopic examination of paraffin-embedded specimens stained with hematoxylin and eosin.

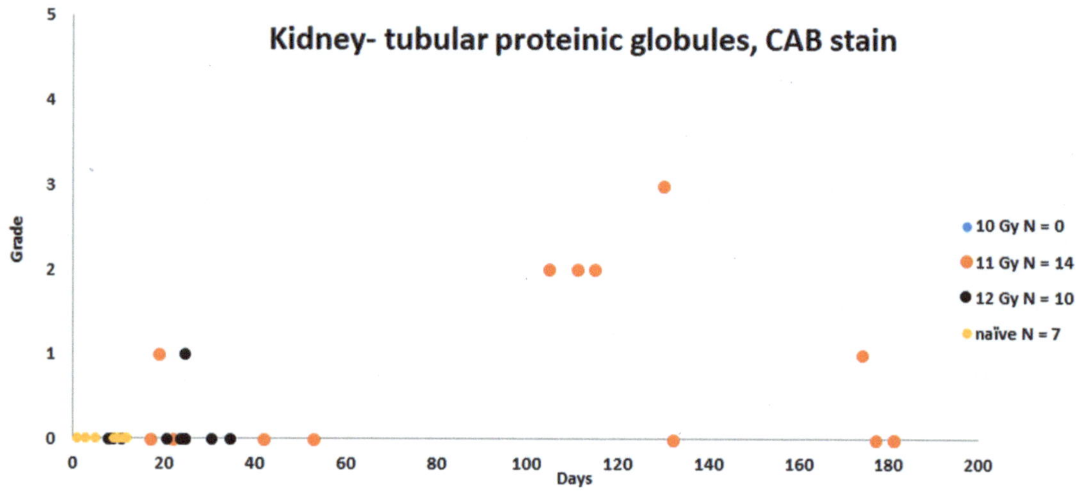

FIGURE 1.51

Scatter plot: kidney—tubular proteinic globules, chromotrope/aniline blue (CAB) stain. Observations based on light microscopic examination of paraffin-embedded specimens stained with CAB stain.

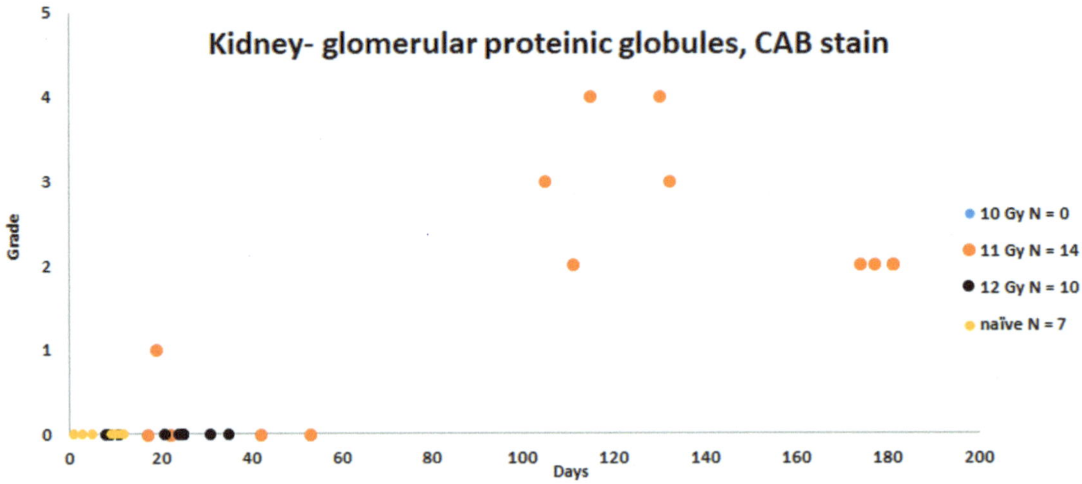

FIGURE 1.52

Scatter plot: kidney—glomerular proteinic globules, chromotrope/aniline blue (CAB) stain. Observations based on light microscopic examination of paraffin-embedded specimens stained with CAB stain.

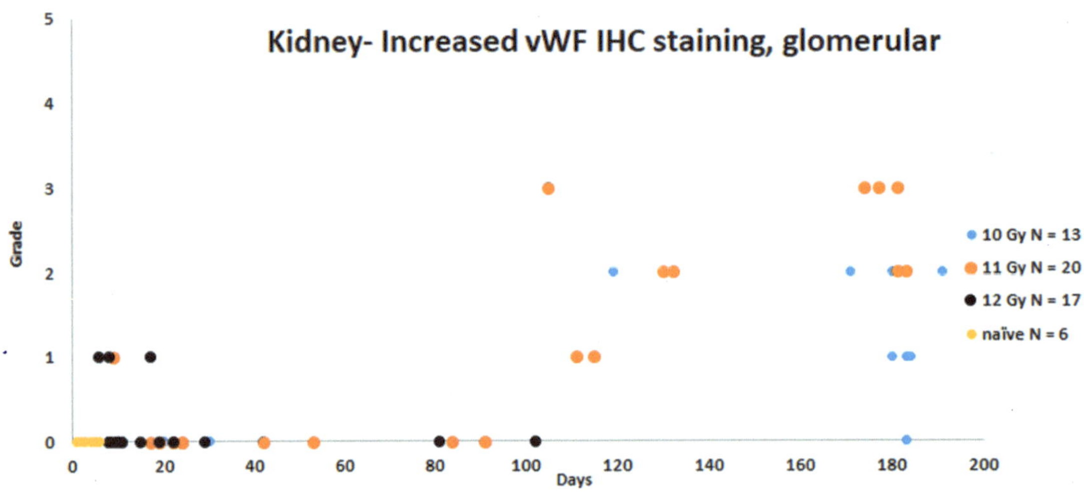

FIGURE 1.53

Scatter plot: kidney—increased von Willebrand factor (vWF) staining, glomerular. Observations based on light microscopic examination of paraffin-embedded specimens stained by immunohistochemical procedures for vWF with 3,3′-diaminobenzidine chromogen and hematoxylin counterstain.

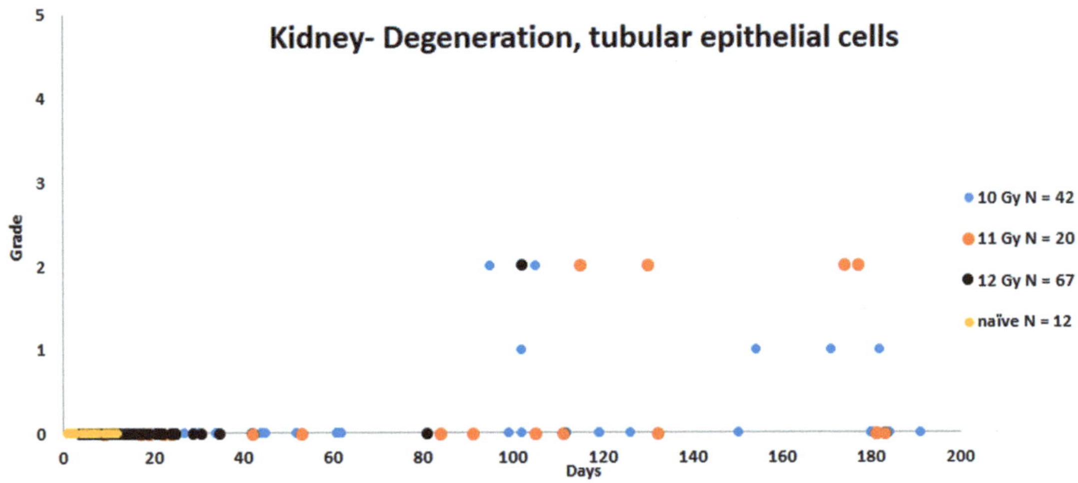

FIGURE 1.54

Scatter plot: kidney—degeneration, tubular epithelial cells. Observations based on light microscopic examination of paraffin-embedded specimens stained with hematoxylin and eosin.

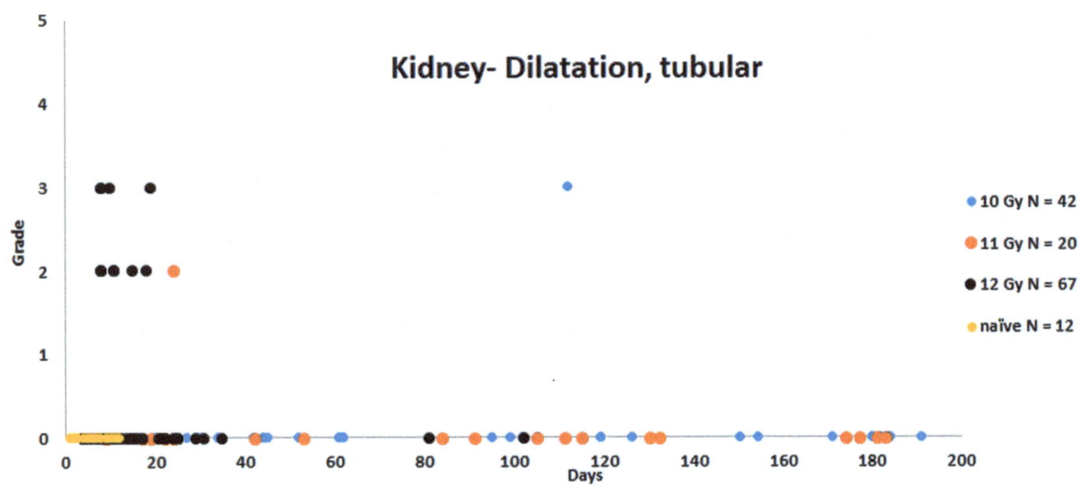

FIGURE 1.55

Scatter plot: kidney—dilatation, tubular. Observations based on light microscopic examination of paraffin-embedded specimens stained with hematoxylin and eosin.

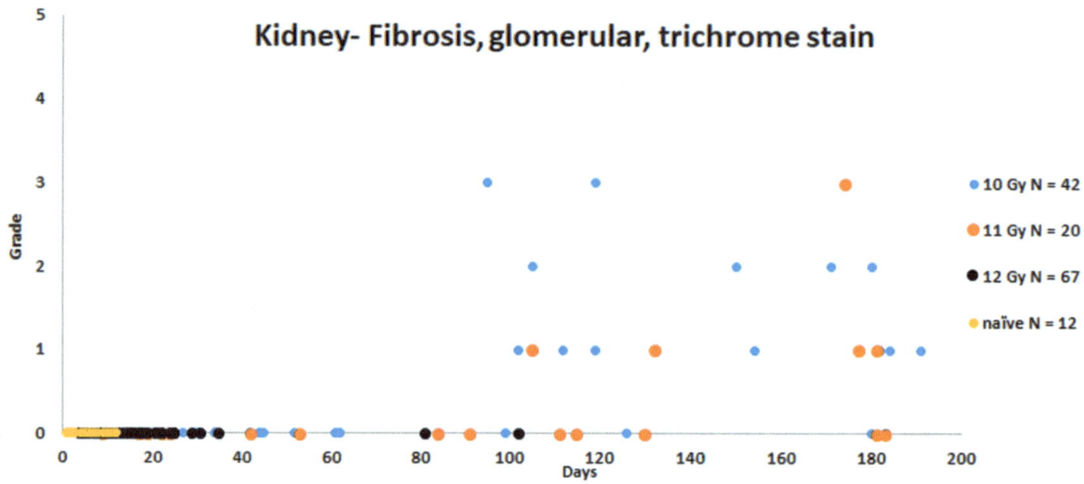

FIGURE 1.56

Scatter plot: kidney—fibrosis, glomerular, trichrome stain. Observations based on light microscopic examination of paraffin-embedded specimens stained with Masson's trichrome stain.

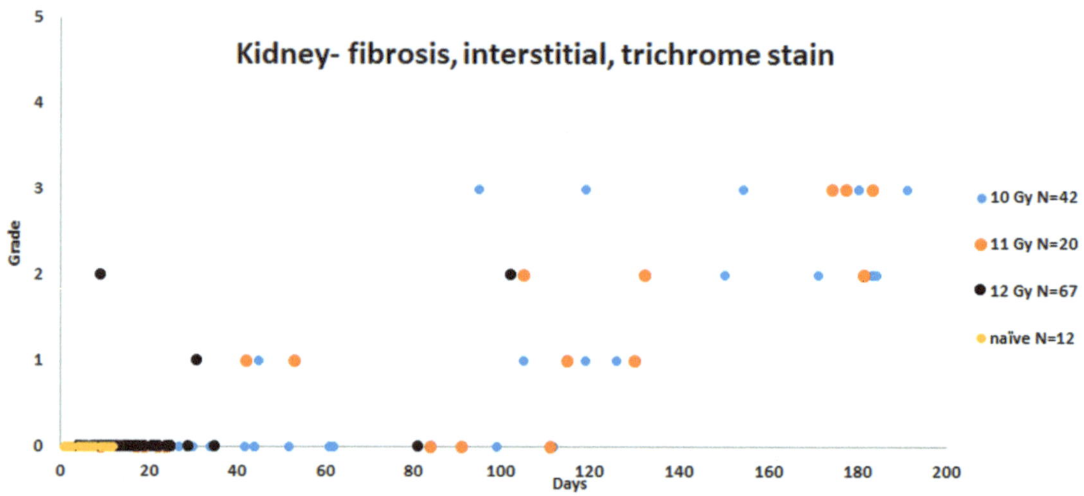

FIGURE 1.57

Scatter plot: kidney—fibrosis, interstitial, trichrome stain. Observations based on light microscopic examination of paraffin-embedded specimens stained with Masson's trichrome stain.

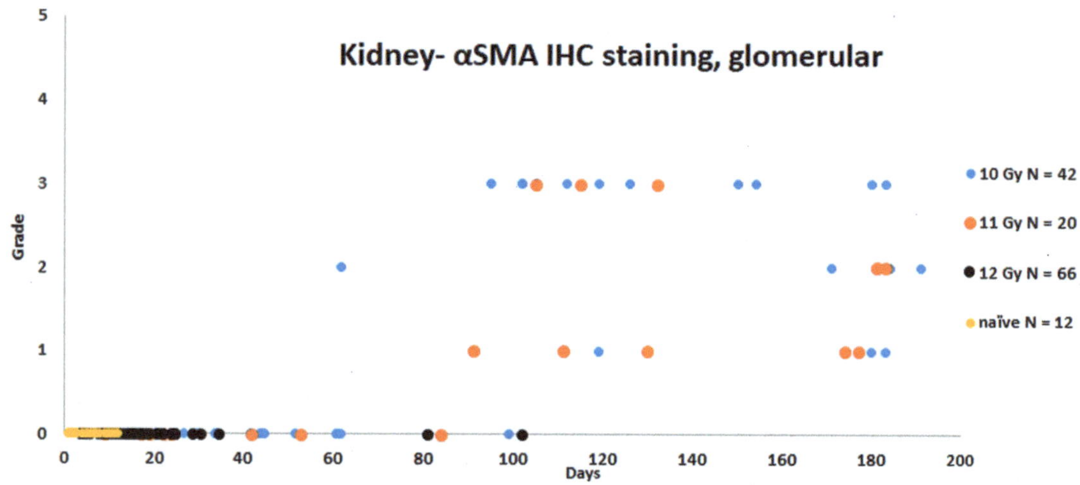

FIGURE 1.58

Scatter plot: kidney—alpha smooth muscle actin staining, glomerular. Observations based on light microscopic examination of paraffin-embedded specimens stained by immunohistochemical procedures for αSMA with 3,3′-diaminobenzidine chromogen and hematoxylin counterstain.

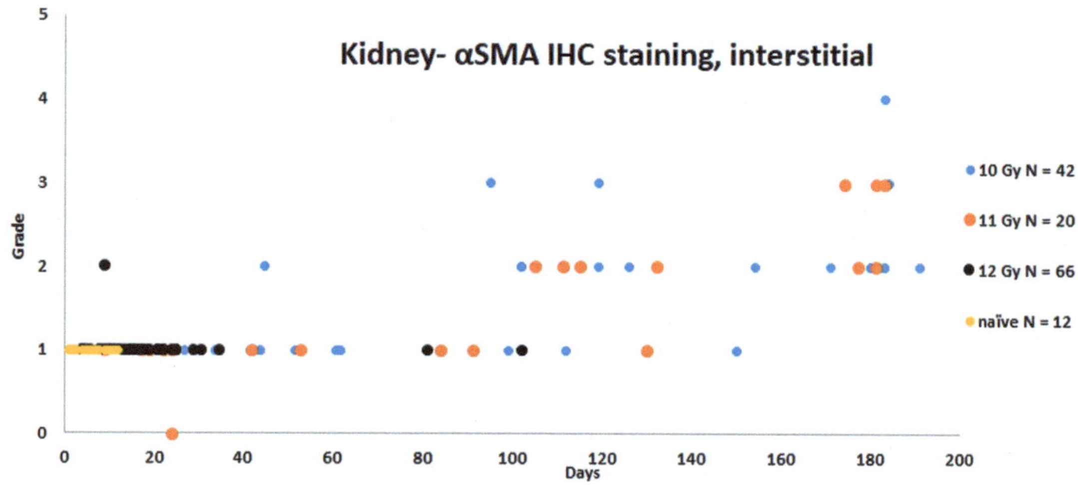

FIGURE 1.59

Scatter plot: kidney—alpha smooth muscle actin (αSMA) staining, interstitial. Observations based on light microscopic examination of paraffin-embedded specimens stained by immunohistochemical procedures for αSMA with 3,3′-diaminobenzidine chromogen and hematoxylin counterstain.

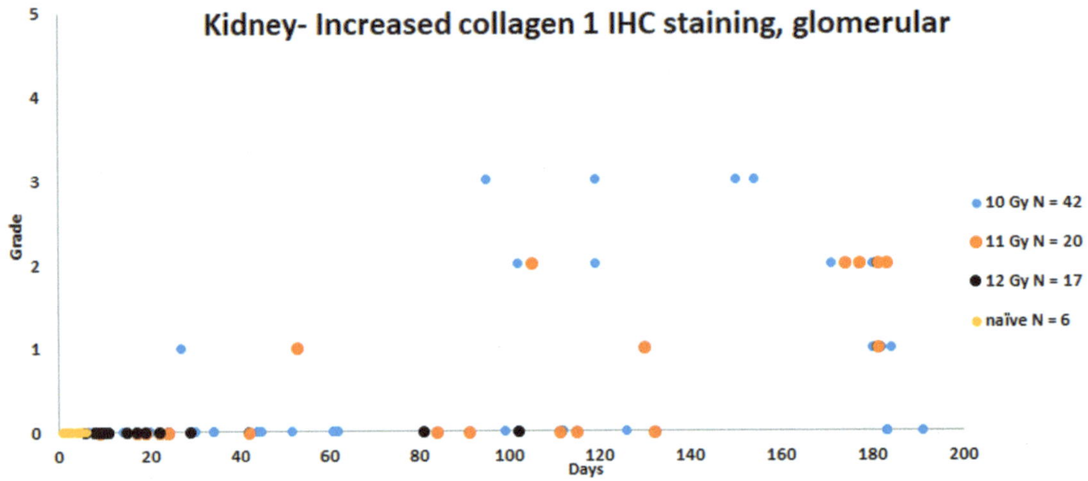

FIGURE 1.60

Scatter plot: kidney—increased collage 1 staining, glomerular. Observations based on light microscopic examination of paraffin-embedded specimens stained by immunohistochemical procedures for collagen 1 with 3,3′-diaminobenzidine chromogen and hematoxylin counterstain.

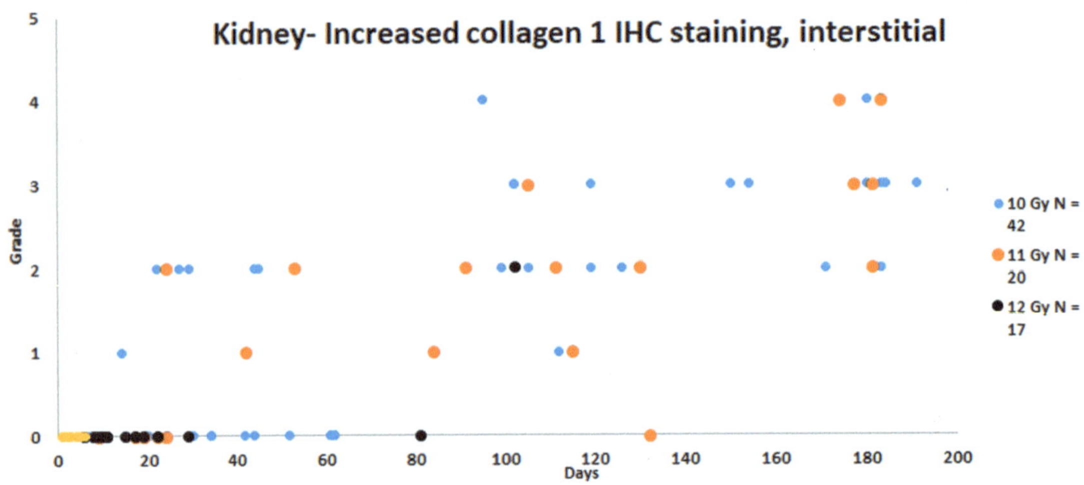

FIGURE 1.61

Scatter plot: kidney—increased collage 1 staining, interstitial. Observations based on light microscopic examination of paraffin-embedded specimens stained by immunohistochemical procedures for collagen 1 with 3,3′-diaminobenzidine chromogen and hematoxylin counterstain.

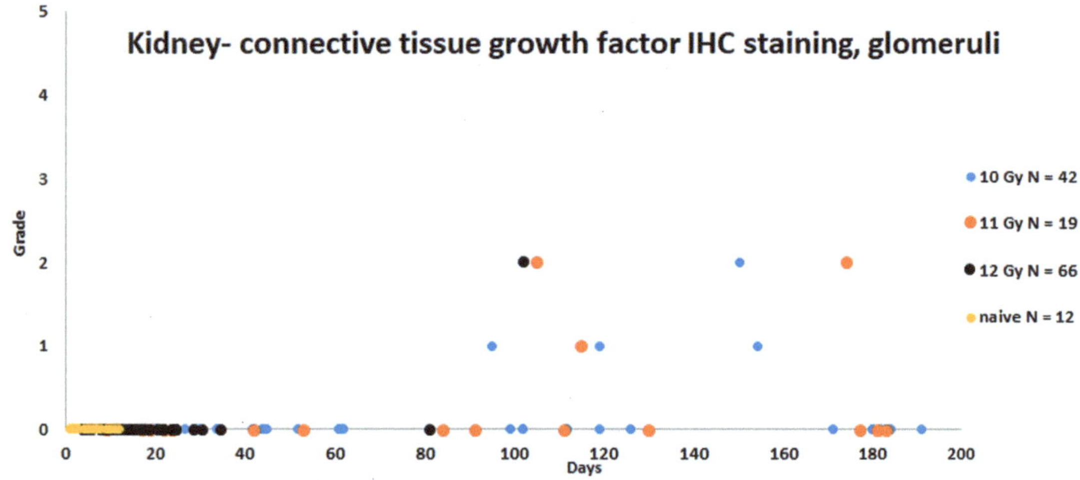

FIGURE 1.62

Scatter plot: Kidney—connective tissue growth factor (CTGF) staining, glomeruli. Observations based on light microscopic examination of paraffin-embedded specimens stained by immunohistochemical procedures for CTGF with 3,3′-diaminobenzidine chromogen and hematoxylin counterstain.

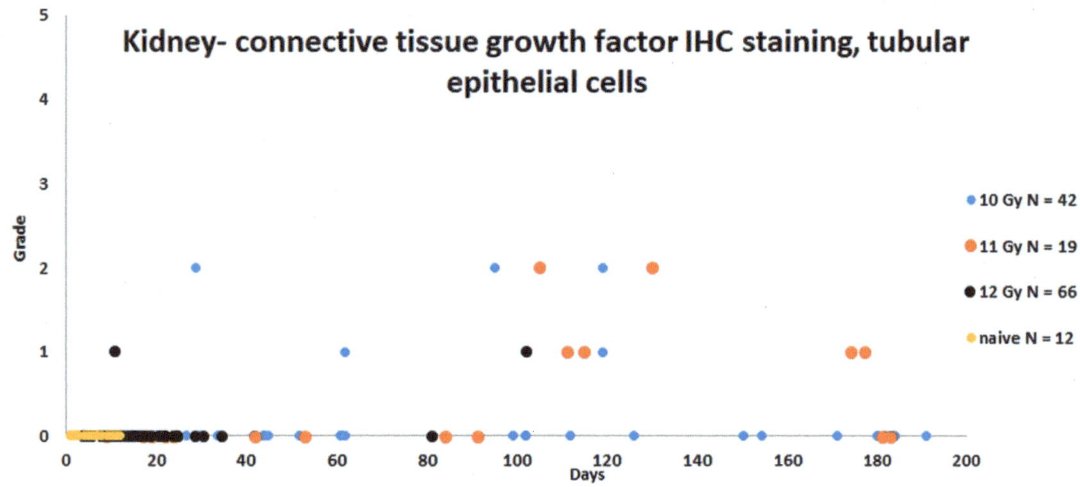

FIGURE 1.63

Scatter plot: Kidney—connective tissue growth factor (CTGF) staining, tubular epithelial cells. Observations based on light microscopic examination of paraffin-embedded specimens stained by immunohistochemical procedures for CTGF with 3,3′-diaminobenzidine chromogen and hematoxylin counterstain.

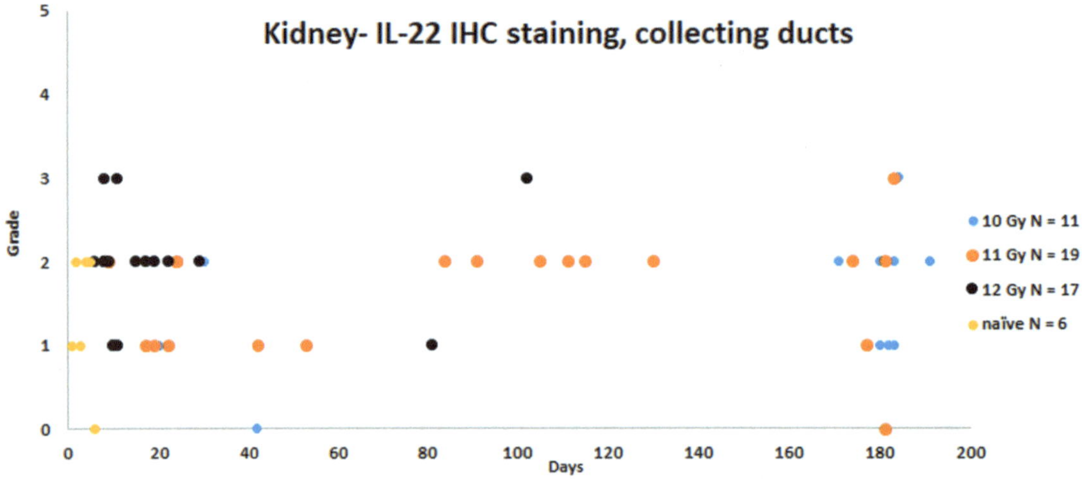

FIGURE 1.64

Scatter plot: Kidney—IL-22 staining, collecting ducts. Observations based on light microscopic examination of paraffin-embedded specimens stained by immunohistochemical procedures for IL-22 with 3,3′-diaminobenzidine chromogen and hematoxylin counterstain.

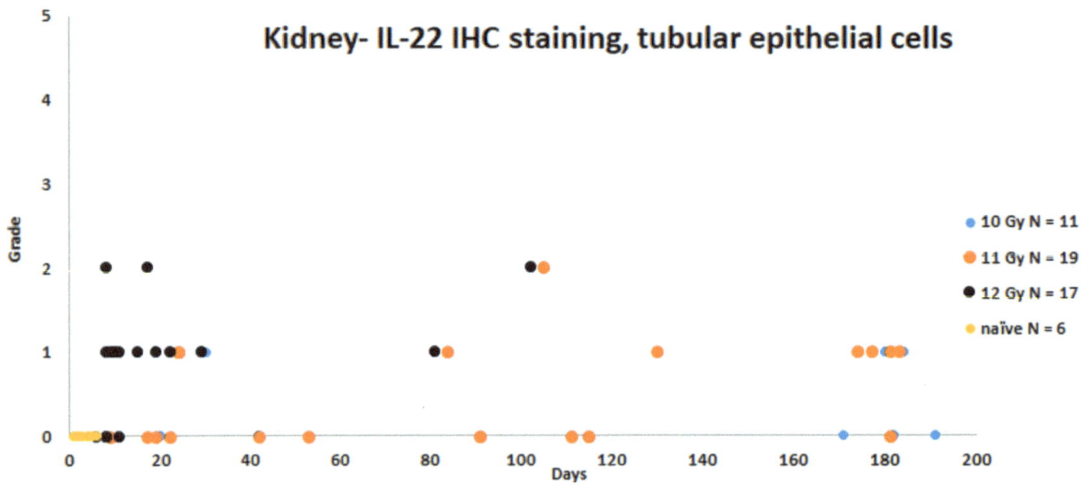

FIGURE 1.65

Scatter plot: Kidney—IL-22 staining, tubular epithelial cells. Observations based on light microscopic examination of paraffin-embedded specimens stained by immunohistochemical procedures for IL-22 with 3,3′-diaminobenzidine chromogen and hematoxylin counterstain.

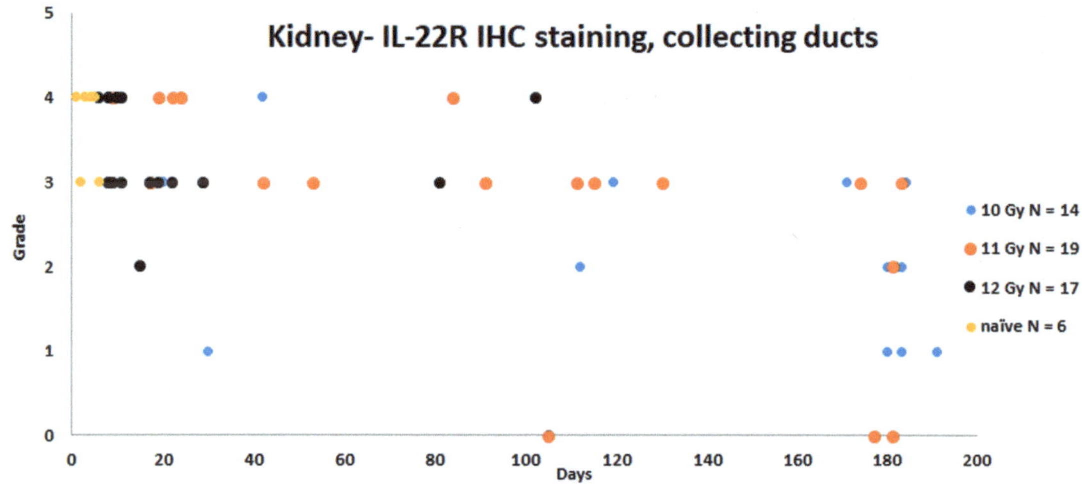

FIGURE 1.66

Scatter plot: Kidney—IL-22R staining, collecting ducts. Observations based on light microscopic examination of paraffin-embedded specimens stained by immunohistochemical procedures for IL-22R with 3,3′-diaminobenzidine chromogen and hematoxylin counterstain.

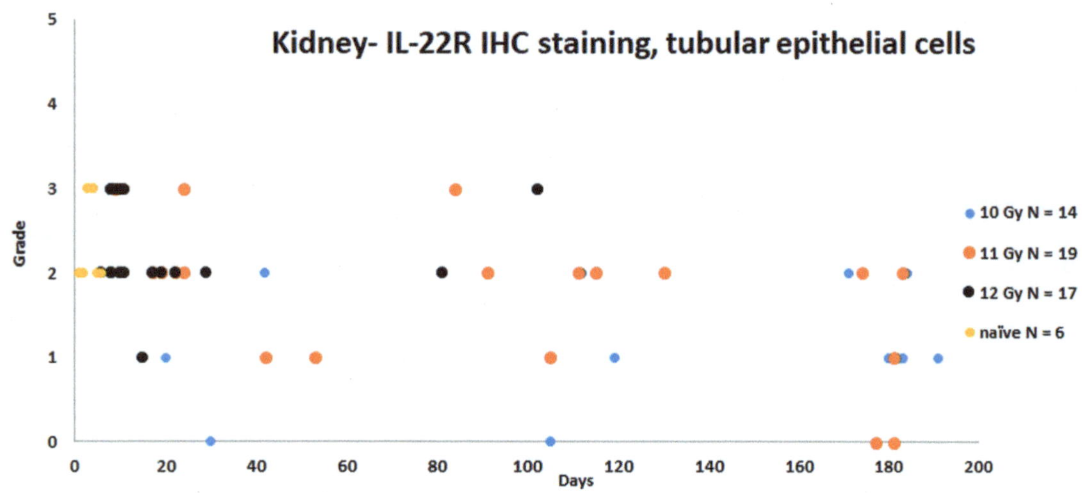

FIGURE 1.67

Scatter plot: Kidney—IL-22R staining, tubular epithelial cells. Observations based on light microscopic examination of paraffin-embedded specimens stained by immunohistochemical procedures for IL-22R with 3,3′-diaminobenzidine chromogen and hematoxylin counterstain.

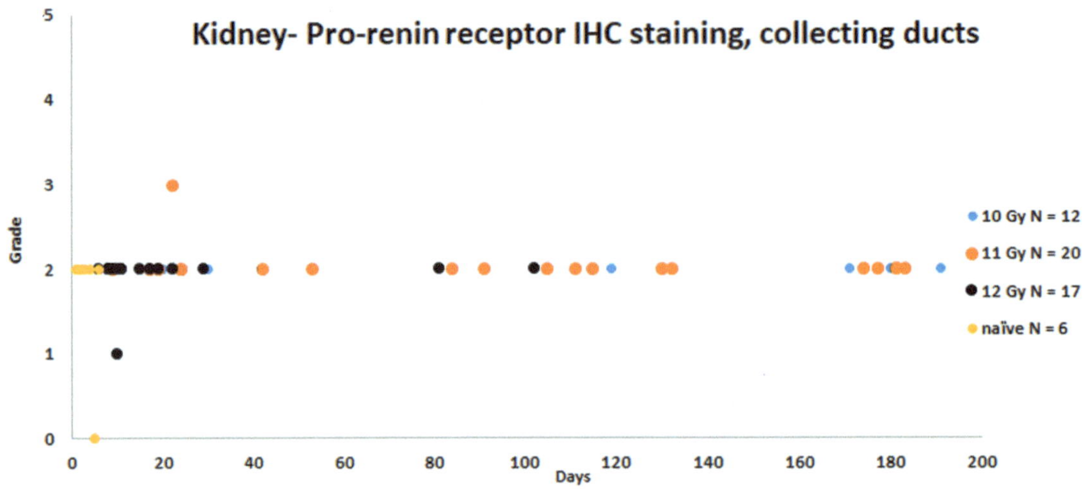

FIGURE 1.68

Scatter plot: Kidney—pro-renin receptor staining, collecting ducts. Observations based on light microscopic examination of paraffin-embedded specimens stained by immunohistochemical procedures for pro-renin receptor with 3,3′-diaminobenzidine chromogen and hematoxylin counterstain.

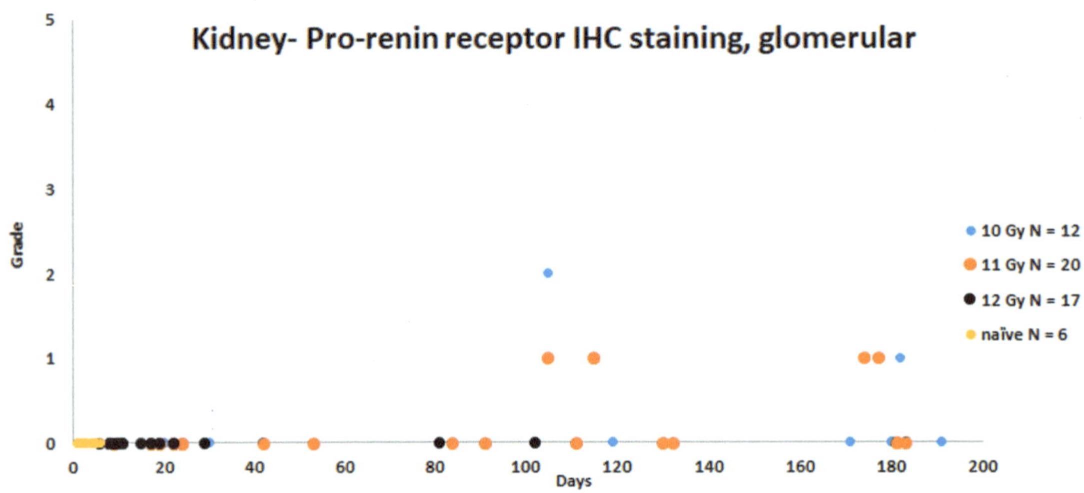

FIGURE 1.69

Scatter plot: Kidney—pro-renin receptor staining, glomerular. Observations based on light microscopic examination of paraffin-embedded specimens stained by immunohistochemical procedures for pro-renin receptor with 3,3′-diaminobenzidine chromogen and hematoxylin counterstain.

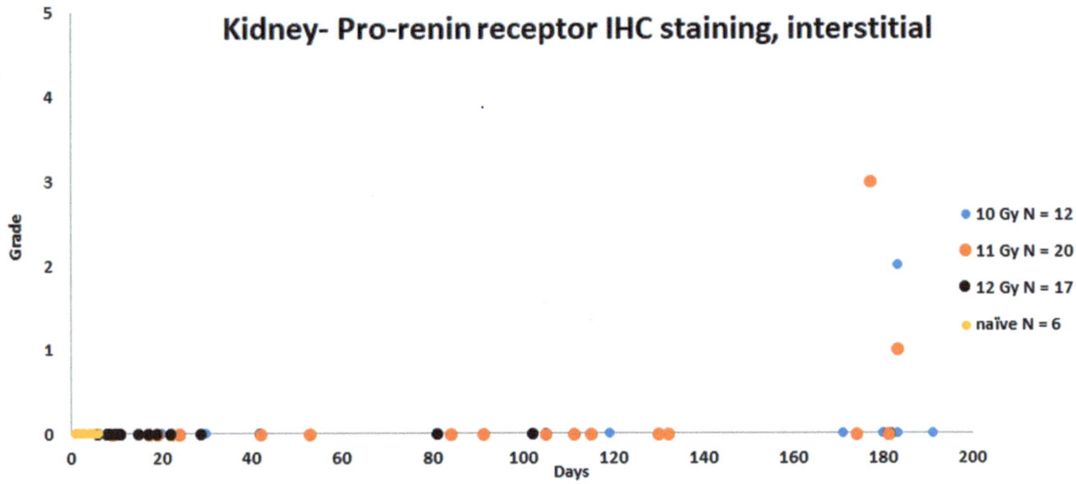

FIGURE 1.70

Scatter plot: Kidney—Pro-renin receptor staining, interstitial. Observations based on light microscopic examination of paraffin-embedded specimens stained by immunohistochemical procedures for pro-renin receptor with 3,3′-diaminobenzidine chromogen and hematoxylin counterstain.

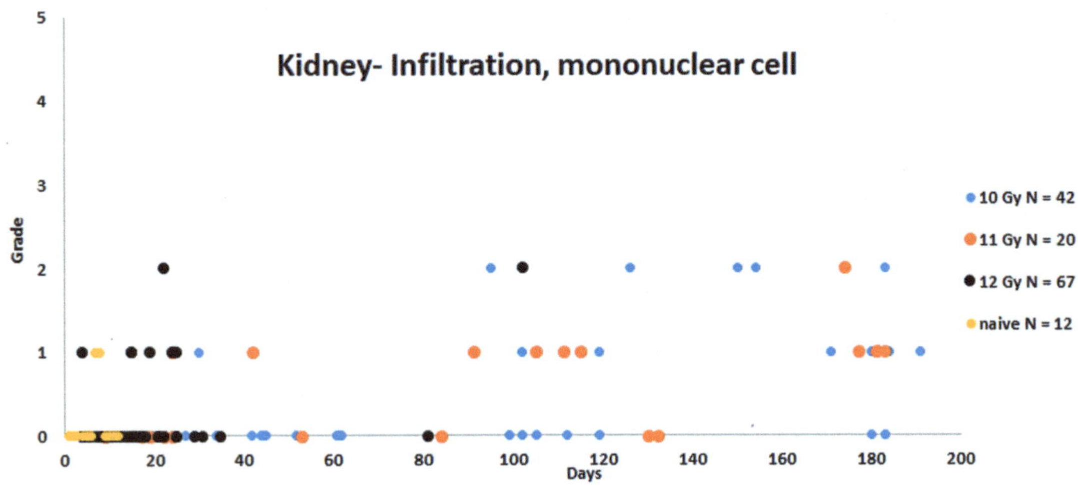

FIGURE 1.71

Scatter plot: Kidney—infiltration, mononuclear cell. Observations based on light microscopic examination of paraffin-embedded specimens stained with hematoxylin and eosin.

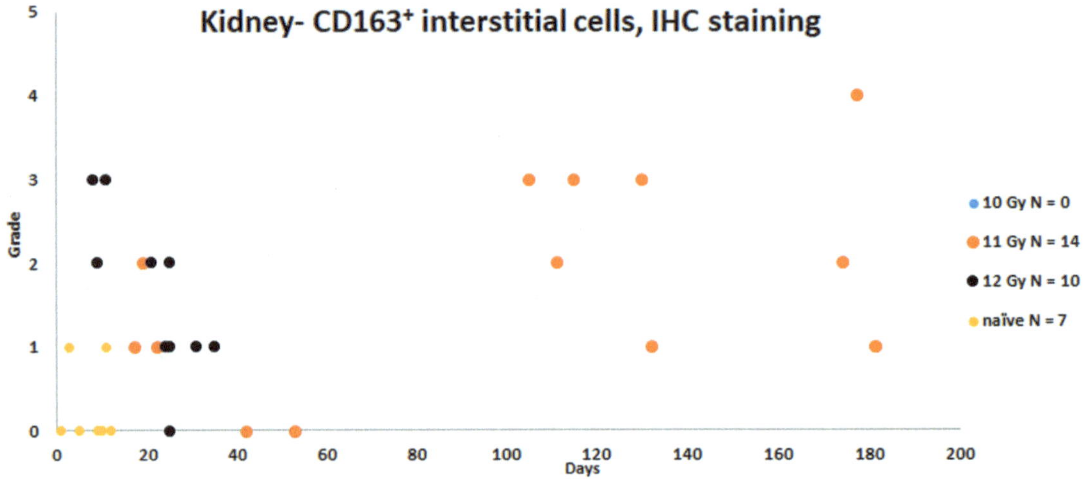

FIGURE 1.72

Scatter plot: Kidney—CD163^{+} interstitial cells. Observations based on light microscopic examination of paraffin-embedded specimens stained by immunohistochemical procedures for CD163 with 3,3′-diaminobenzidine chromogen and hematoxylin counterstain.

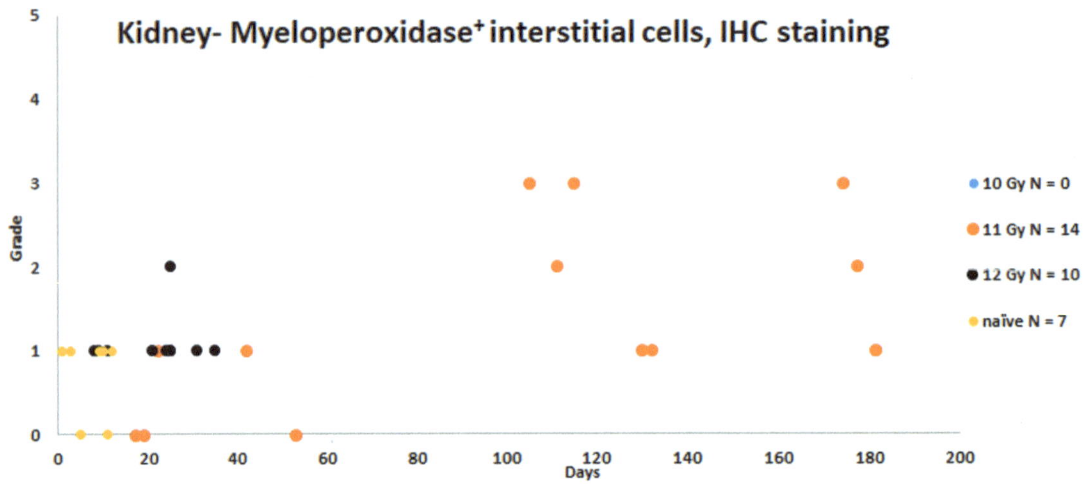

FIGURE 1.73

Scatter plot: Kidney—myeloperoxidase^{+} interstitial cells. Observations based on light microscopic examination of paraffin-embedded specimens stained by immunohistochemical procedures for myeloperoxidase with 3,3′-diaminobenzidine chromogen and hematoxylin counterstain.

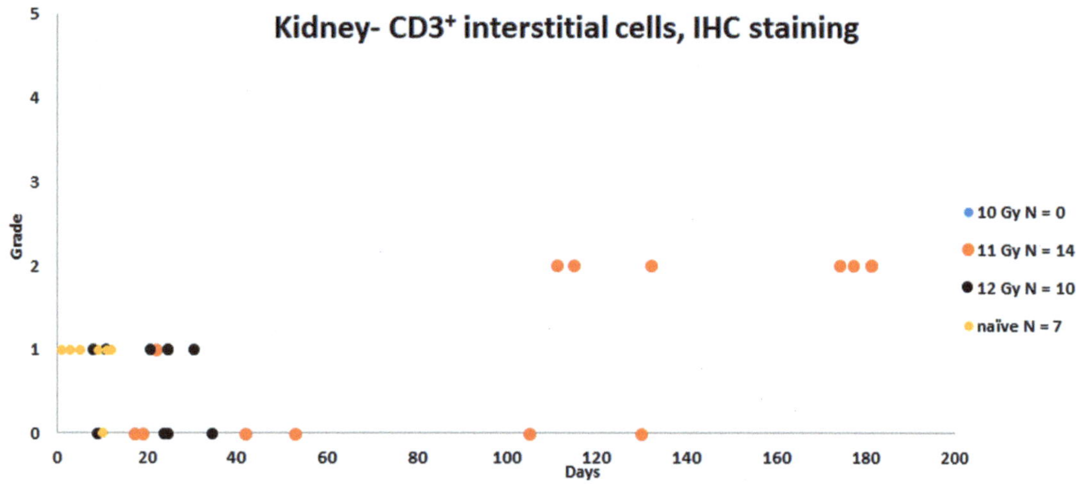

FIGURE 1.74

Scatter plot: Kidney—CD3$^+$ interstitial cells. Observations based on light microscopic examination of paraffin-embedded specimens stained by immunohistochemical procedures for CD3 with 3,3′-diaminobenzidine chromogen and hematoxylin counterstain.

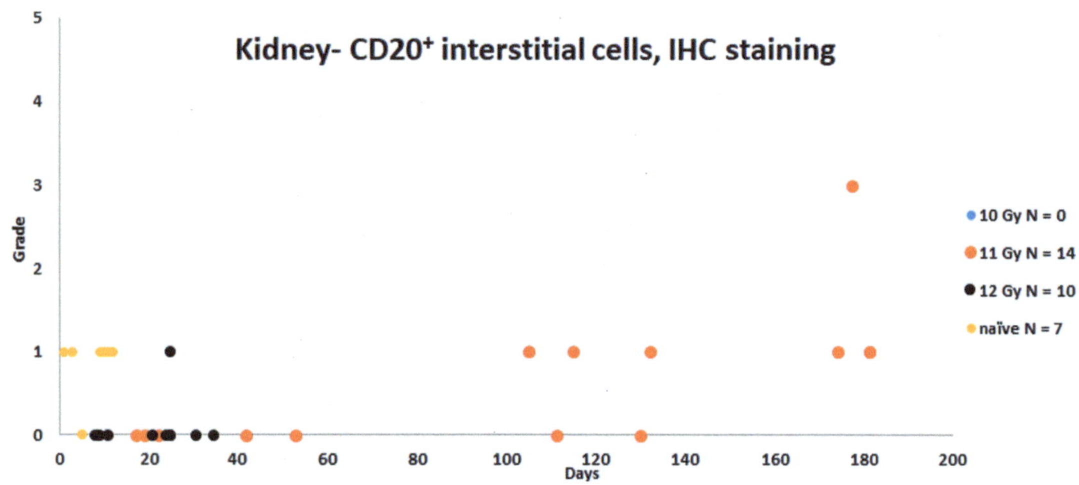

FIGURE 1.75

Scatter plot: Kidney—CD20$^+$ interstitial cells. Observations based on light microscopic examination of paraffin-embedded specimens stained by immunohistochemical procedures for CD20 with 3,3′-diaminobenzidine chromogen and hematoxylin counterstain.

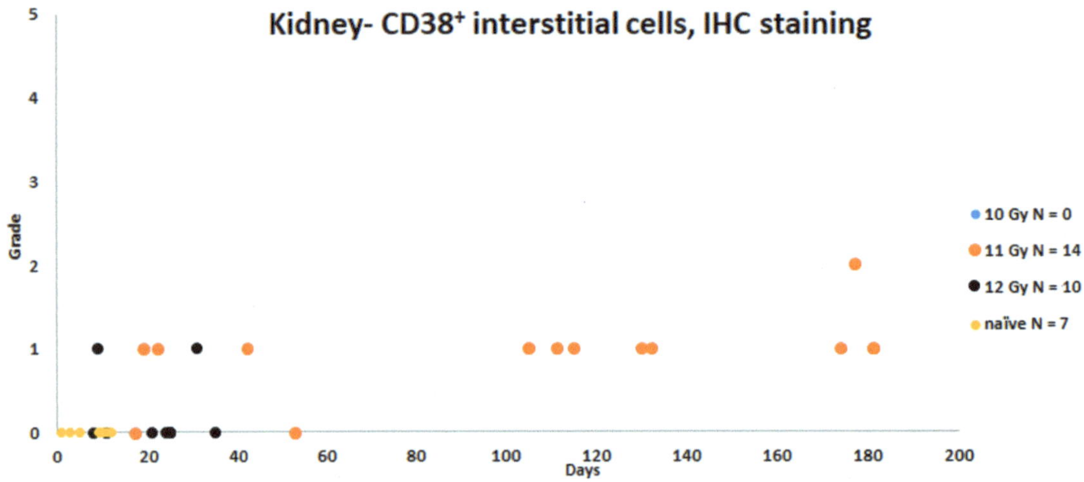

FIGURE 1.76

Scatter plot: Kidney—CD38$^+$ interstitial cells. Observations based on light microscopic examination of paraffin-embedded specimens stained by immunohistochemical procedures for CD38 with 3,3′-diaminobenzidine chromogen and hematoxylin counterstain.

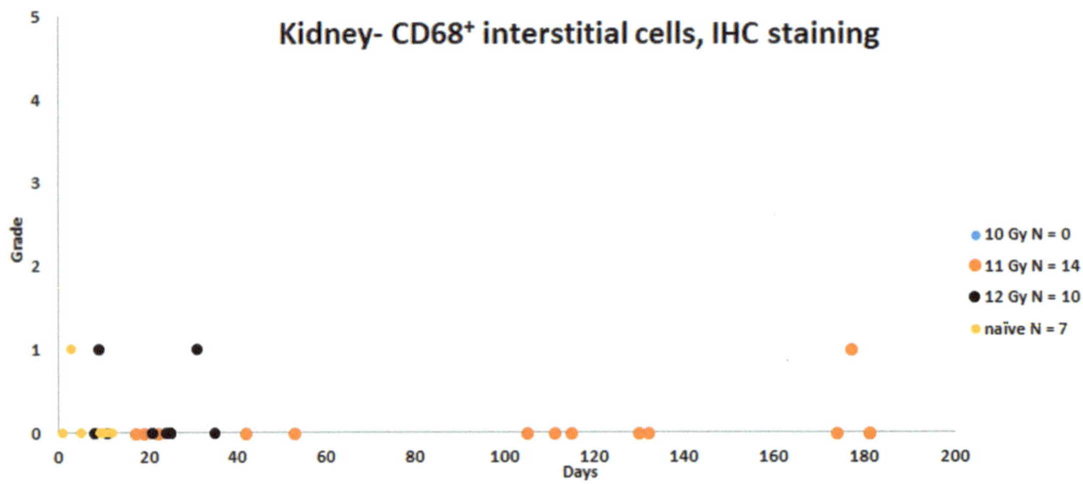

FIGURE 1.77

Scatter plot: Kidney—CD68$^+$ interstitial cells. Observations based on light microscopic examination of paraffin-embedded specimens stained by immunohistochemical procedures for CD68 with 3,3′-diaminobenzidine chromogen and hematoxylin counterstain.

CHAPTER

Radiation-associated changes in the lung of rhesus macaques

2

Overview

Histopathological alterations in the lung were centered on fibrosis, inflammation, and reactive/proliferative changes in pneumocytes in animals necropsied from approximately 85 days postirradiation through the 180-day observation period. The radiation-associated pulmonary changes were multifocally distributed in the lung with some lobes being completely spared. Interstitial fibrosis was commonly observed, and pleural fibrosis was commonly associated with underlying areas of interstitial pulmonary fibrosis. Interstitial and pleural fibrosis were apparent with Masson's trichrome and collagen 1 immunohistochemical (IHC) staining and were commonly associated with increased α-smooth muscle actin (α-SMA) staining. The more pronounced areas of pulmonary fibrosis commonly had reduced expression of CD31, an endothelial cell marker, suggesting a vicious cycle in which fibrosis resulted in reduced microvascular perfusion, which could lead to additional pulmonary fibrosis.

In the early phases of the pulmonary reaction, IHC staining for myeloperoxidase (MPO) revealed an equivocal increase in pulmonary macrophage populations in irradiated animals. IHC staining for CD38, a marker for pro-inflammatory macrophages, revealed no change in the population of $CD38^+$ macrophages in irradiated animals. By contrast, IHC staining for CD163 revealed a substantial increase in the population of $CD163^+$ pulmonary macrophages. In the less severely involved areas of the lung, the population of $CD163^+$ macrophages was largely contained within alveolar walls, and positively stained cells were evenly distributed throughout the lung, suggesting the population of $CD163^+$ cells represented the "mesenchymal" macrophage population rather than the bone marrow-origin macrophage population. In more advanced lesions, the accumulations of alveolar macrophages were also $CD163^+$. The combination of observations suggested a phenotype switch to a pro-fibrotic macrophage phenotype in the absence of a significant increase in the overall macrophage population in the early stages of the radiation-associated pulmonary lesion complex.

Alveolar edema was commonly present in irradiated animals throughout the observation period, suggesting radiation-associated endothelial injury. Alveolar walls in edematous areas sometimes contained an atypical cell population, termed "myxoid cells," which suggested an endothelial-, epithelial-, or pericyte-mesenchymal transition.

In the later phases of the pulmonary reaction, the most pronounced cellular component of pulmonary inflammation was the accumulation of alveolar macrophages, with less intense infiltrations of lymphocytes and neutrophils. It was not clear whether the macrophage accumulation was a cause or a result of the interstitial fibrosis. Alveolar macrophages exhibited positive IHC staining for CD163, CD206, transforming growth factor beta (TGF-β), and MPO, suggesting activated alveolar macrophages may be involved in the generation and/or perpetuation of the inflammatory and fibrotic process in the lung. Proliferation of type II pneumocytes was noted in areas of lung injury, and the proliferating pneumocytes expressed connective tissue growth factor (CTGF) that could contribute to the fibrogenic response.

Histopathology Atlas of Acute Radiation Syndrome and Delayed Effects in Rhesus Macaques. DOI: https://doi.org/10.1016/B978-0-323-91393-5.00002-5

Focally extensive vacuolation of pleural mesothelial cells was considered to be a direct effect of the radiation beam. The visceral pleura of a few animals had coagulation-type changes, also consistent with direct radiation injury. In a few animals, these focally pronounced pleural lesions were associated with subpleural areas wherein the alveolar walls were composed primarily of myxoid-type cells. These areas had a plaque-like presentation beneath the pleural surface and were notably devoid of inflammatory cell infiltration. The absence of inflammatory cell infiltration in these lesions indicated direct radiation injury and morphological alteration of the pulmonary alveolar structure, with no indication of associated inflammatory processes. The exact nature of this myxoid cell population was unclear, but it was suspected to be a predecessor of interstitial fibrosis.

Distinctive focal or multifocal alveolar–bronchiolar hyperplasia was noted in a subset of animals necropsied from approximately day 105 postirradiation through the remaining observation period. The hyperplastic lesions had many features of preneoplastic change, but overt neoplastic proliferations were not present.

Toluidine blue and tryptase IHC staining revealed no alteration of mast cell populations relative to irradiation level or postirradiation interval. Tryptase IHC staining revealed a greater number of mast cells than toluidine blue staining. Mast cell populations were more prominent in areas of interstitial fibrosis, but it was unclear whether the mast cells were a cause or effect of the fibrosis. The toluidine blue staining revealed distinct metachromasia in the myxoid cell population.

Staining and figures list

Routine histology stain-

Hematoxylin and eosin (H&E)—Figs. 2.1A, 2.1B, 2.2A, 2.2B, 2.3A, 2.3B, 2.4, 2.5, 2.6, 2.7A, 2.8A, 2.8B, 2.10A, 2.11, 2.12A, 2.20A, 2.21, 2.22A, 2.22B, 2.23, 2.24, 2.25, 2.26A, 2.26B

Histochemical stains-

Masson's trichrome—Figs. 2.9A, 2.9B, 2.10B, 2.12B,
Toluidine blue—Figs. 2.7B, 2.22C,

Immunohistochemical stains-

Alpha smooth muscle actin—Figs. 2.7C, 2.15A, 2.15B, 2.15C, 2.15D, 2.15E, 2.15F
CD31—Figs. 2.8C, 2.18A, 2.18B, 2.18C
CD163—Figs. 2.28A, 2.28B, 2.28C, 2.28D
CD206—Figs. 2.29A, 2.29B
Caspase 3—Fig. 2.31
Collagen 1—Figs. 2.13A, 2.13B, 2.13C, 2.14
Connective tissue growth factor—Figs. 2.17A, 2.17B
Galectin-3—Figs. 2.19A, 2.19B, 2.20B,
Myeloperoxidase—Figs. 2.27A, 2.27B, 2.27C
Transforming growth factor-beta—Figs. 2.16A, 2.16B, 2.16C, 2.16D
Tryptase—Figs. 2.30A, 2.30B, 2.30C

Graphic presentations (scatter plots) of individual histologic observations in the kidney are presented in the following figures:

Fig. 2.32. Lung—edema, alveolar
Fig. 2.33. Lung—accumulation, alveolar macrophages
Fig. 2.34. Lung—hyperplasia, alveolar/bronchiolar
Fig. 2.35. Lung—hyperplasia, type 2 pneumocytes
Fig. 2.36. Lung—fibrosis, interstitium
Fig. 2.37. Lung—fibrosis, pleura
Fig. 2.38. Lung—alpha smooth muscle actin (αSMA) staining, interstitium
Fig. 2.39. Lung—alpha smooth muscle actin (αSMA) staining, pleura
Fig. 2.40. Lung—collagen 1 staining, interstitium
Fig. 2.41. Lung—collagen 1 staining, pleura
Fig. 2.42. Lung—infiltration, lymphocytes
Fig. 2.43. Lung—myeloperoxidase staining, macrophages
Fig. 2.44. Lung—CD163 staining, macrophages
Fig. 2.45. Lung—CD206 staining, macrophages
Fig. 2.46. Lung—CD38 staining, macrophages
Fig. 2.47. Lung—caspase 3, macrophages
Fig. 2.48. Lung—decreased CD31 staining
Fig. 2.49. Lung—connective tissue growth factor (CTGF) staining, alveolar/duct
Fig. 2.50. Lung—myxoid cells

Detailed microscopic observations

Alveolar macrophage accumulation

Macrophages normally traffic through the lung from the interstitium into the alveolar lumina and exit the lung via the mucociliary "elevator." A minor population of alveolar macrophages is normally visible microscopically. Pathological accumulation of alveolar macrophages was sporadically present in animals necropsied up to 112 days postirradiation, after which it was present in all irradiated animals. Accumulations of alveolar macrophages were most commonly associated with areas of interstitial fibrosis or foci of alveolar-bronchiolar hyperplasia.

Alveolar edema

Alveolar edema was visualized in histologic sections by the presence of lightly eosinophilic material that represents edema fluid protein. In some cases, the edema fluid contained large, extracellular vacuoles that represent air admixed with the edema fluid.

Type 2 pneumocyte hyperplasia

Proliferation of type 2 pneumocytes typically occurs following loss of mature type I pneumocytes from the surface of pulmonary alveoli. The plump proliferating cells have a "tombstone" contour that resembles the epithelial lining of bronchioles. Type 2 pneumocyte hyperplasia was present in the great majority of animals necropsied after day 105 postirradiation but was rarely present in animals necropsied before that time.

Hyperplasia, alveolar–bronchiolar

Focal or multifocal hyperplasia of alveolar and/or bronchiolar epithelial cells was present in a substantial subset of animals necropsied after 111 days postirradiation. The proliferative lesions were most commonly located near major airways. The proliferative lesions tended to radiate outward into the surrounding alveolar tissue with no evidence of tissue compression and often contained numerous entrapped alveolar macrophages.

Lymphocyte infiltration

Infiltrations of inflammatory cells into the lung consisted primarily of mononuclear cells that were microscopically consistent with mature lymphocytes, though it should be noted that some of these cells may be immature macrophages that have yet to attain the microscopic features that permit their identification as macrophages. Lymphocytic infiltration was commonly observed in animals necropsied after 105 days postirradiation but was uncommonly present in animals necropsied prior to that time. The lymphocytic infiltrations were most commonly noted in areas that had other histological changes such as interstitial fibrosis, type II pneumocyte hyperplasia, or alveolar–bronchiolar hyperplasia.

Neutrophil infiltration

Infiltration by neutrophils was noted less commonly than lymphocytic infiltration. As was seen with lymphocytic infiltrations, the neutrophilic infiltrations were most commonly noted in areas that had other histological changes such as interstitial fibrosis, type II pneumocyte hyperplasia, or alveolar–bronchiolar hyperplasia.

Emphysema

Alveolar emphysema consisted of focal or multifocal overexpansion of alveolar spaces with an associated breakdown of alveolar walls. The alveolar emphysema was suspected to be due to extreme respiratory excursions associated with other radiation-associated pathological alterations in the lung.

Myxoid cells

Alveolar walls of a subset of animals had unidentified polygonal cells, termed "myxoid cells," that were characterized by amphophilic cytoplasm and other features similar to the cells seen in embryonal tissues or umbilical cord. This unusual cell population was noted sporadically in animals necropsied throughout the observation period, with an apparent relationship to postirradiation interval or radiation dose. The basic investigative techniques employed in this study did not permit determination of the exact nature of the myxoid cells, but they are suspected to represent a cellular transition such as epithelial–mesenchymal, endothelial–mesenchymal, or pericyte–mesenchymal transition.

Interstitial and pleural fibrosis

Interstitial and pleural fibrosis were discernible in multiple stains, including the routine H&E stains. The trichrome staining provided a more precise indication of the extent and severity of the fibrosis. Interstitial fibrosis tended to be multifocally distributed within the lung, with no distribution pattern relative to airways or vasculature. The most pronounced interstitial fibrosis was typically located near pleural surfaces.

Pleural fibrosis was focally or multifocally distributed and was invariably associated with underlying interstitial fibrosis.

Toluidine blue stain

Low-level mast cell populations were present in the lungs of all naïve control animals as well as the majority of the irradiated animals, but microscopically discernible mast cell populations were essentially absent from the lungs of a substantial number of animals that received 12 Gy radiation exposure. It should be noted that the toluidine blue staining, as well as the tryptase IHC staining presented below, is based on the contents of mast cell granules. As a result, neither stain would reveal degranulated mast cells.

Alpha smooth muscle actin immunohistochemical stain

Staining for α-SMA generally paralleled the histologic evidence of interstitial or pleural fibrosis, as recorded under trichrome stains.

Tryptase immunohistochemical stain

Tryptase staining revealed a more pronounced population of pulmonary mast cells than was revealed by toluidine blue staining. This reflects the known population of mucosal mast cells in the lung, which are known to be refractory to toluidine blue staining unless the tissue specimens are preserved in special fixatives. Tryptase$^+$ mast cell populations were less pronounced in the lungs of animals that received 12 Gy irradiation. The tryptase-positive mast cell population was more prominent in areas of interstitial fibrosis. The nonmucosal population of mast cells, also known as connective tissue mast cells, is known to preferentially localize in areas containing collagenous tissue; therefore it was not possible to determine whether the mast cell population was a cause or a result of the pulmonary fibrosis.

CD31 immunohistochemical stain

A reduction in microvascular density, as revealed by a reduced number of CD31-positive endothelial cells, was present in animals with pronounced interstitial fibrosis in the lungs. The reduced CD-31 staining was noted in areas of the lung that had interstitial fibrosis, but not in areas of the same histologic specimens that did not have interstitial fibrosis. It was not clear whether the reduced microvascular density in fibrotic areas was an initial cause or a result of the interstitial fibrosis, but it is suspected the reduced microvascular density would contribute to local hypoxia and could promote additional interstitial fibrosis.

Myeloperoxidase immunohistochemical stain

Positive MPO immunostaining in the lung was largely a reflection of the population of alveolar macrophages in the lung, as well as any infiltrating neutrophils that accompanied inflammatory reactions. Positive MPO staining was noted in virtually all animals in the study, but there was an apparent reduction in the population of MPO^+ macrophages in animals irradiated at 12 Gy (all but two of which were necropsied within 30 days of irradiation). There was no appreciable difference in the population of MPO-positive pulmonary cells in animals irradiated at 11 Gy versus 10 Gy.

CD163 immunohistochemical stain

Given the normal level of cycling of macrophages through the lung, a low-level alveolar macrophage population is expected in histologic sections. The population of $CD163^+$ alveolar macrophages was increased above this baseline level in a moderate number of irradiated animals. The most pronounced populations of $CD163^+$ alveolar macrophages were associated with areas of interstitial fibrosis.

Compared to the $CD163^+$ alveolar macrophage population, a smaller number of $CD163^+$ interstitial macrophages were noted. The population of $CD163^+$ interstitial macrophages was increased above the baseline level in a moderate number of irradiated animals.

CD206 immunohistochemical stain

The pulmonary population of $CD206^+$ alveolar macrophages was commonly elevated above baseline levels in irradiated animals and occasionally reduced below the baseline level. Similar to the situation with $CD163^+$ macrophages, more $CD206^+$ macrophages were present in areas of interstitial fibrosis.

The population of $CD206^+$ macrophages in the pulmonary interstitium was sparse relative to the population of $CD206^+$ alveolar macrophages. The population of $CD206^+$ interstitial macrophages was commonly elevated above baseline levels in irradiated animals, but not to the degree seen with the expanded population of $CD163^+$ interstitial macrophages.

CD38 immunohistochemical stain

CD38 staining of cellular populations was rarely present in the lung. One animal irradiated at 12 Gy had a substantial area of necrosis and macrophage accumulation in one of the lung sections. A substantial population of $CD38^+$ macrophages was associated with that lesion, confirming the involvement of $CD38^+$ macrophages in overt inflammatory lesions.

Collagen 1 immunohistochemical stain

An increased amount of interstitial collagen was noted in animals from all irradiated groups, with severity related primarily to the postirradiation interval. Increased collagen deposition tended to be focal or multifocal within the lung sections and tended to be most pronounced near the pleural surface.

Increased collagen deposition on the pleural surface of the lung was also noted in animals from all irradiated treatment groups but increased pleural collagen deposition was noted in a lower incidence than interstitial collagen deposition. The increased pleural collagen deposition was focal or multifocal within the lung sections and was invariably associated with a subjacent area of increased interstitial collagen deposition.

The collagen IHC staining results closely paralleled results obtained from the Masson's trichrome stains, though the collagen IHC stains tended to reveal a greater degree of collagen deposition than was discernible in the trichrome stains. The difference was most apparent with subtle increases in collagen deposition, which was variably apparent in the trichrome-stained sections.

Transforming growth factor β immunohistochemical stain

Very little TGFβ staining was noted in the lungs. The TGFβ pleural staining tended to be located on the superficial surface of the immature fibrous connective tissue associated with pleural fibrosis, suggesting that TGF-β signaling may be involved in perpetuation of the pleural fibrosis process. Positive TGF-β staining of macrophages was most commonly noted in areas of interstitial fibrosis.

Connective tissue growth factor immunohistochemical stain

Alveolus/duct CTGF IHC staining involved alveolar epithelial cells, primarily proliferating type II pneumocytes. Bronchial gland staining for CTGF was present whenever bronchial glands were present in the section and were considered to represent normal biology or consistent off-target staining of no relevance to the present study.

Additional details of lung observations have been previously presented (MacVittie et al., 2019; Parker et al., 2019).

References

MacVittie TJ, Farese AM, Parker GA, Jackson W 3rd: The Time Course of Radiation-induced Lung Injury in a Nonhuman Primate Model of Partial-body Irradiation With Minimal Bone Marrow Sparing: Clinical and Radiographic Evidence and the Effect of Neupogen Administration, *Health Phys* 116(3):366–382, 2019.

Parker GA, Li N, Takayama K, Farese AM, MacVittie TJ: Lung and Heart Injury in a Nonhuman Primate Model of Partial-body Irradiation with Minimal Bone Marrow Sparing: Histopathological Evidence of Lung and Heart Injury, *Health Phys* 116(3):383–400, 2019.

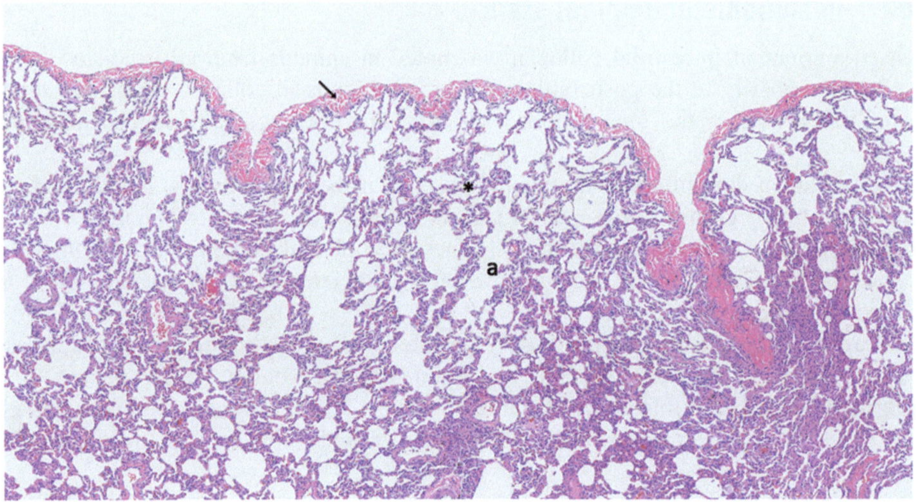

FIGURE 2.1A

The lung specimen from a naive male rhesus macaque has a thick layer of fibrous connective tissue (arrow) in the visceral pleura. Branching airways (a) and alveoli with delicate walls (*) constitute the bulk of the lung structure. Hematoxylin and eosin stain, 5 × objective magnification.

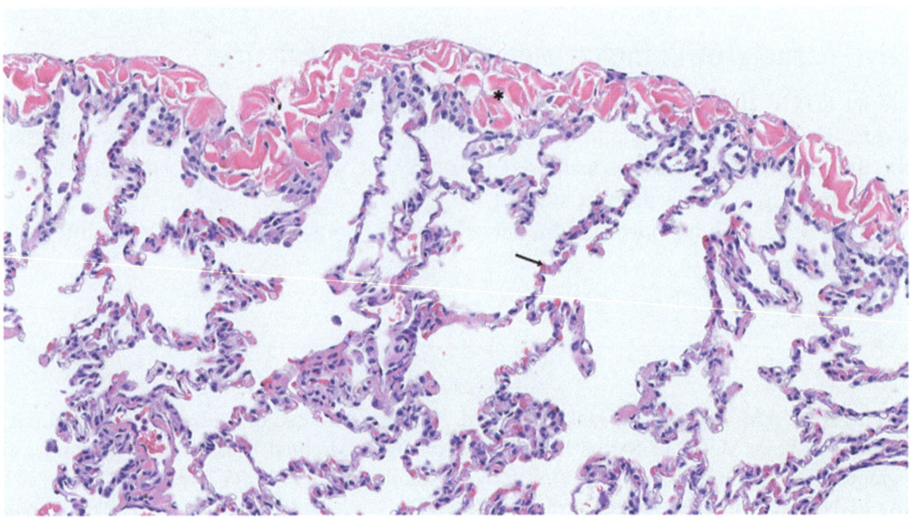

FIGURE 2.1B

Higher magnification of the previous specimen shows the dense fibrous connective tissue layer of the visceral pleura (*) and delicate alveolar walls (arrow) with capillaries containing brightly eosinophilic (red) erythrocytes. Hematoxylin and eosin stain, 20 × objective magnification.

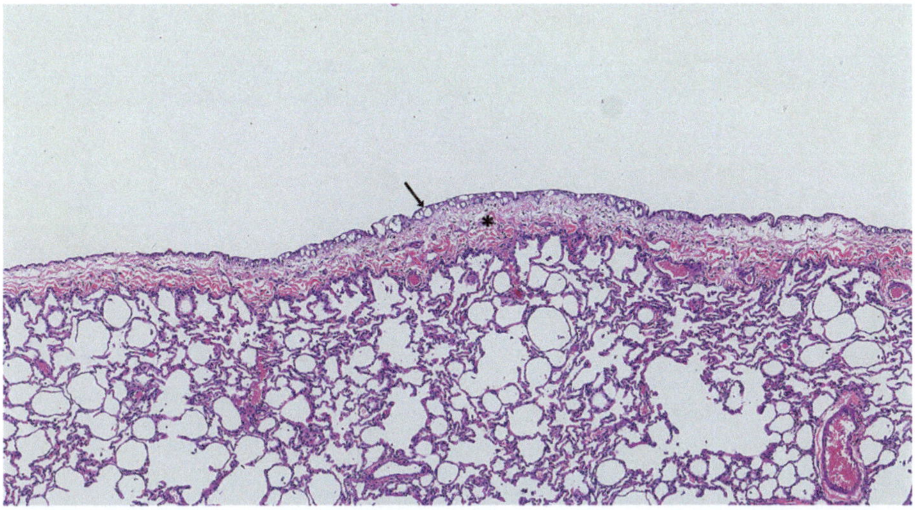

FIGURE 2.2A

Lung specimen from a male rhesus macaque necropsied five days after irradiation at 12 Gy. The pleura (*) on the surface of the lung is edematous and the superficial mesothelial cells are vacuolated (arrow). Hematoxylin and eosin stain, 5 × objective magnification.

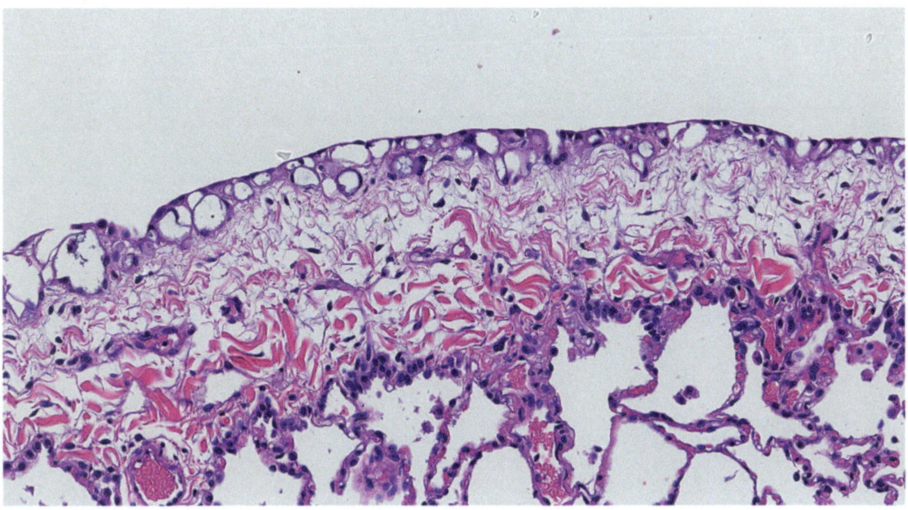

FIGURE 2.2B

Higher magnification of the previous specimen shows details of the pleural changes. Hematoxylin and eosin stain, 20 × objective magnification.

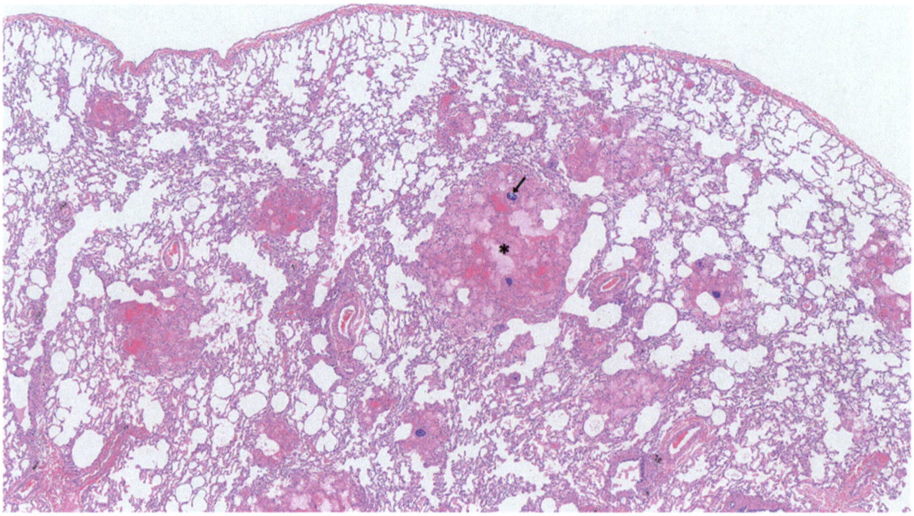

FIGURE 2.3A

Lung specimen from a male rhesus macaque necropsied 17 days after irradiation at 12 Gy. Note the multifocal areas of hemorrhage (*), some of which contain deeply stained bacterial colonies (arrow). Hematoxylin and eosin stain, 2.5 × objective magnification.

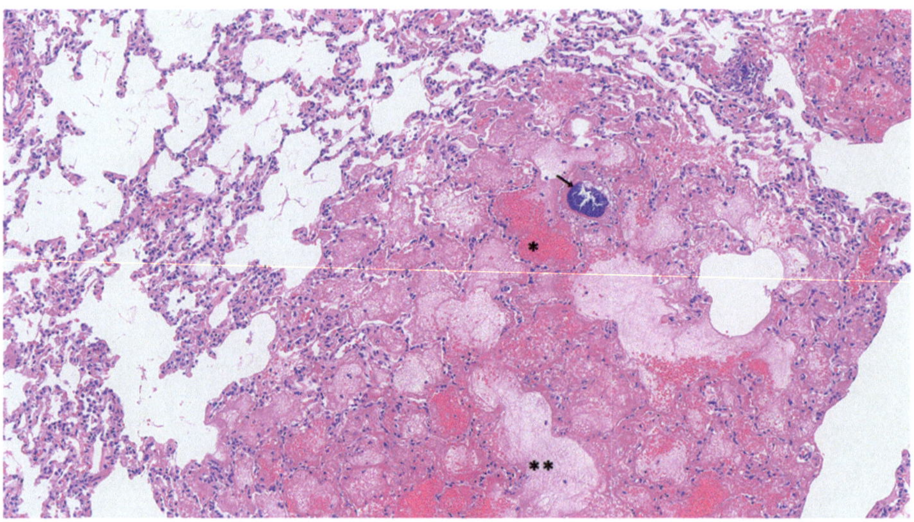

FIGURE 2.3B

Higher magnification of the previous specimen shows intermixed areas of intact erythrocytes (*) and slightly fibrillar clotted plasma (**). Absence of inflammatory cell infiltration associated with the bacterial colony (arrow) indicates a pronounced radiation-associated reduction in bone marrow production of leukocytes. Hematoxylin and eosin stain, 10 × objective magnification.

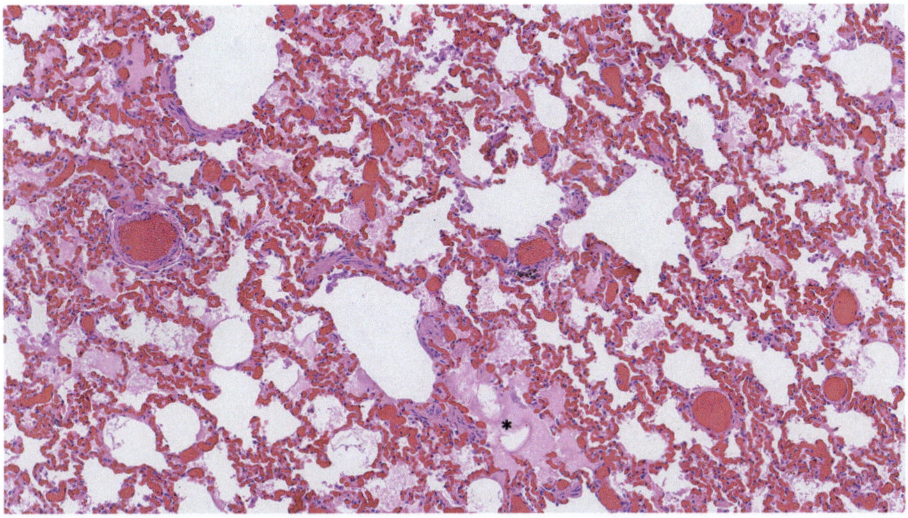

FIGURE 2.4

Lung specimen of a male rhesus macaque necropsied six days after irradiation at 12 Gy. Note the engorgement of alveolar capillaries with red-stained erythrocytes and accumulations of lightly stained edema fluid protein (*) in alveolar lumina. Histologic changes of this type are consistent with terminal cardiovascular dysfunction. Hematoxylin and eosin stain. 10 × objective magnification.

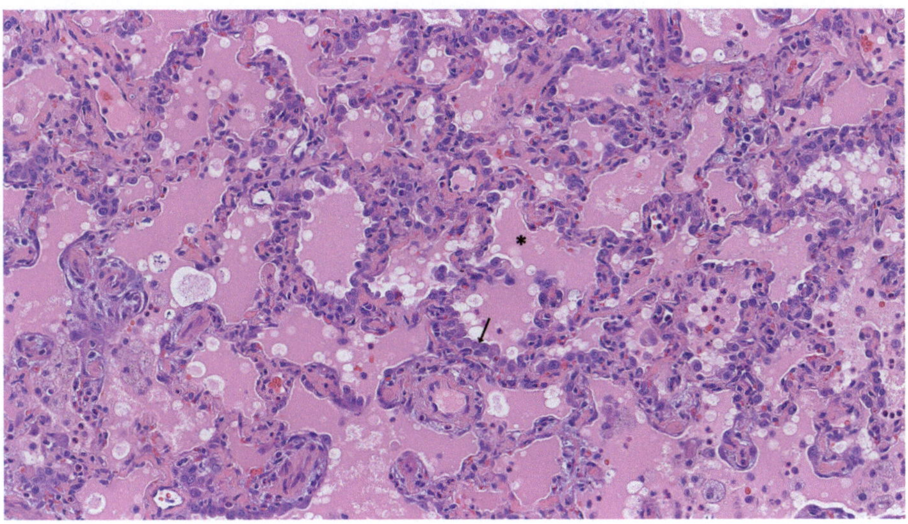

FIGURE 2.5

Lung specimen from a male rhesus macaque necropsied 182 days after irradiation at 10 Gy has light pink edema fluid protein (*) in alveolar lumina, with clear vacuoles indicating antemortem admixture of air. Note the plump, proliferative type II pneumocytes (arrow). Hematoxylin and eosin stain, 20 × objective magnification.

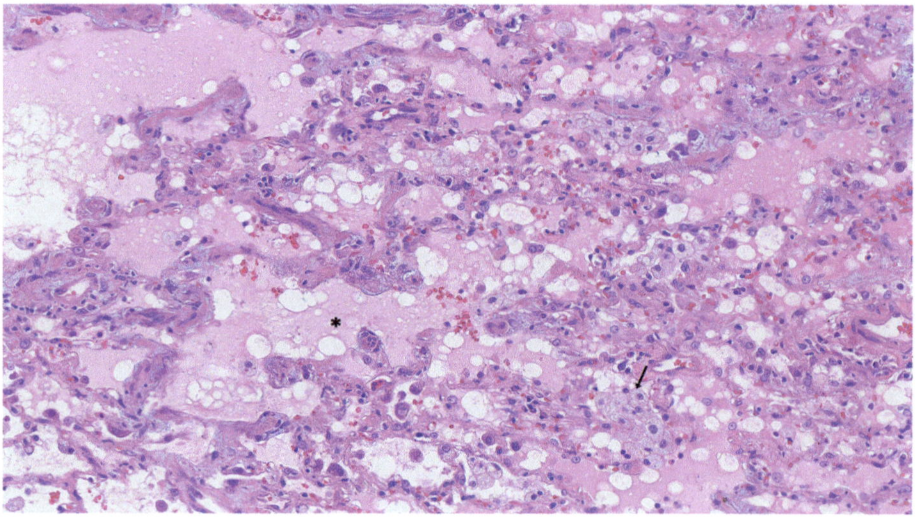

FIGURE 2.6

Lung of a male rhesus macaque necropsied 77 days after irradiation at 11 Gy. Note the accumulations of lightly stained edema fluid protein (*) and aggregations of alveolar macrophages (arrow). Clear spaces within the alveolar edema fluid represent admixture of inhaled air. Hematoxylin and eosin stain, 20 × objective magnification.

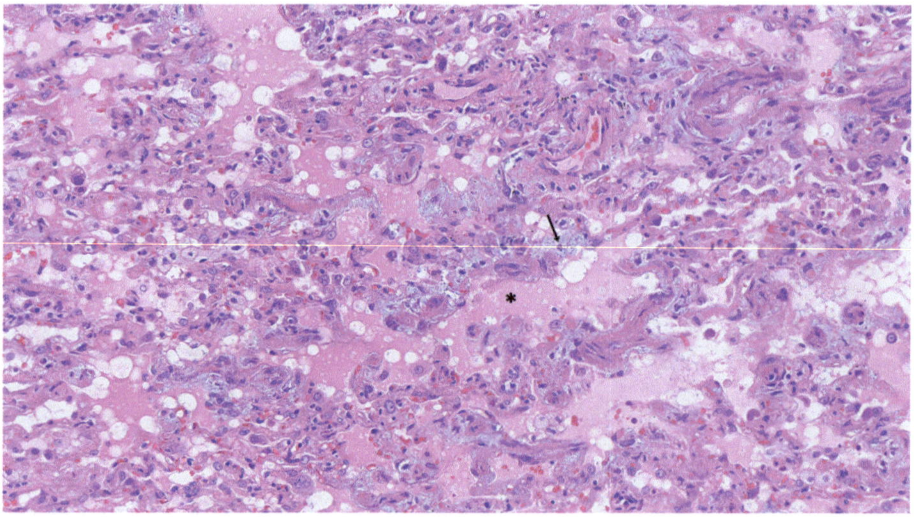

FIGURE 2.7A

Lung of a male rhesus macaque necropsied 77 days after irradiation at 11 Gy. Alveolar walls contain numerous myxoid cells (arrow) characterized by amphophilic, indistinctly vacuolated cytoplasm. Areas of the lung containing myxoid cells commonly had alveolar edema (*). Hematoxylin and eosin stain, 20 × objective magnification.

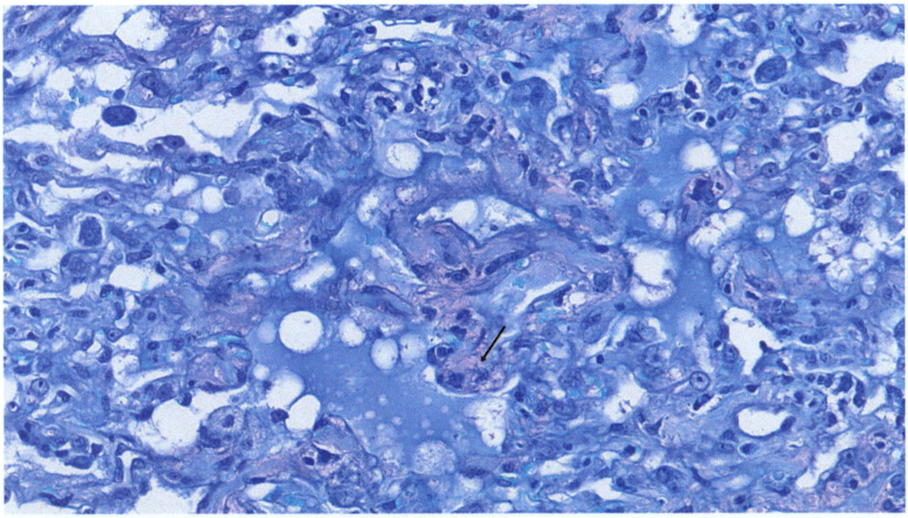

FIGURE 2.7B

Toluidine blue staining of the previous lung specimen shows metachromasia (purple color) in myxoid cells (arrow), indicating carbohydrate content of the cellular cytoplasm. This cytological change suggests a shift in the cellular constituents of the alveolar walls, possibly an epithelial-, endothelial—or pericyte-to-mesenchymal transition. 40 × objective magnification.

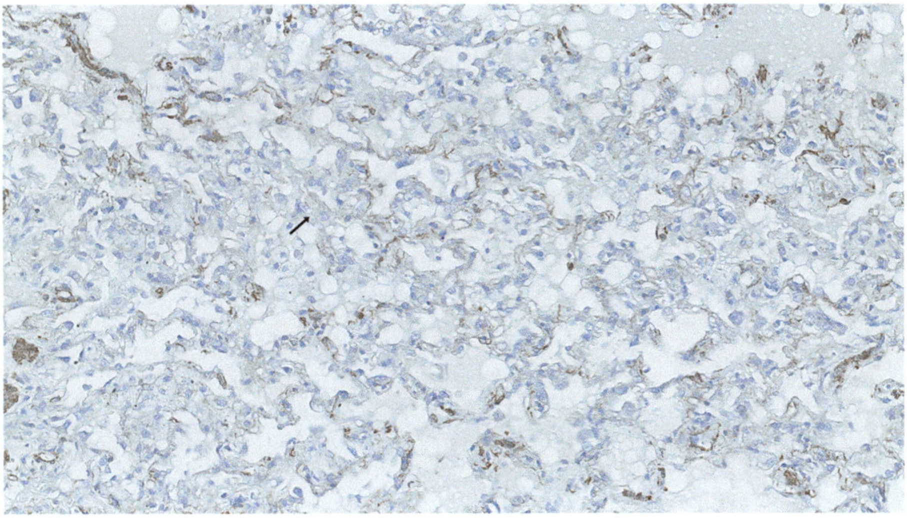

FIGURE 2.7C

Immunohistochemical stain for alpha smooth muscle actin (αSMA) performed on the lung from a male rhesus macaque necropsied 77 days after irradiation at 10 Gy. Note the patchy absence of αSMA staining in segments of alveolar walls that have myxoid cell populations (arrow), suggesting a phenotypic change in undefined cell populations of the alveolar walls. Stain performed with 3,3′-diaminobenzidine chromogen and hematoxylin counterstain, 5 × objective magnification.

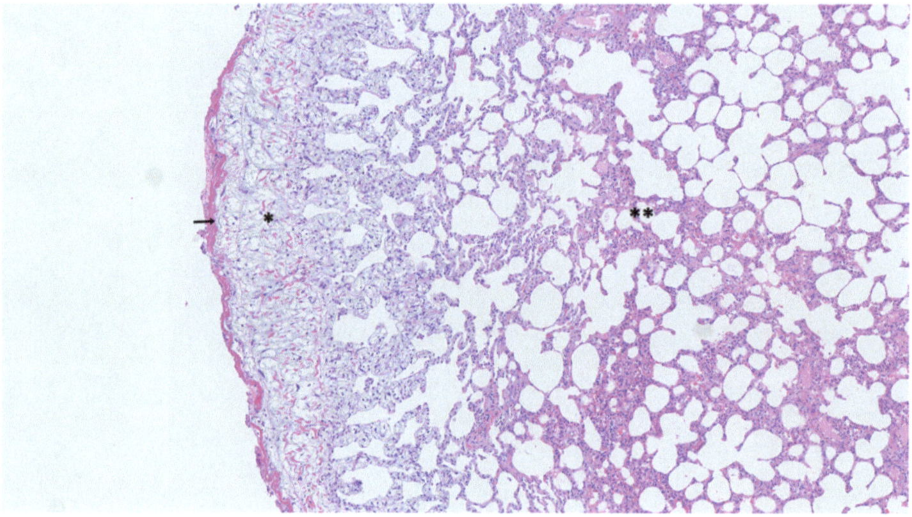

FIGURE 2.8A

Lung specimen from a male rhesus macaque necropsied 31 days after irradiation at 12 Gy. The bright pink band (arrow) on the pleural surface is severely altered pleural collagen, consistent with coagulation-type thermal injury. A broad subpleural zone (*) has thickened alveolar walls that contain numerous myxoid cells. The subpleural zone of thickened alveolar walls blends into the underlying area (**) of microscopically normal alveolar walls. Hematoxylin and eosin stain, 5 × objective magnification.

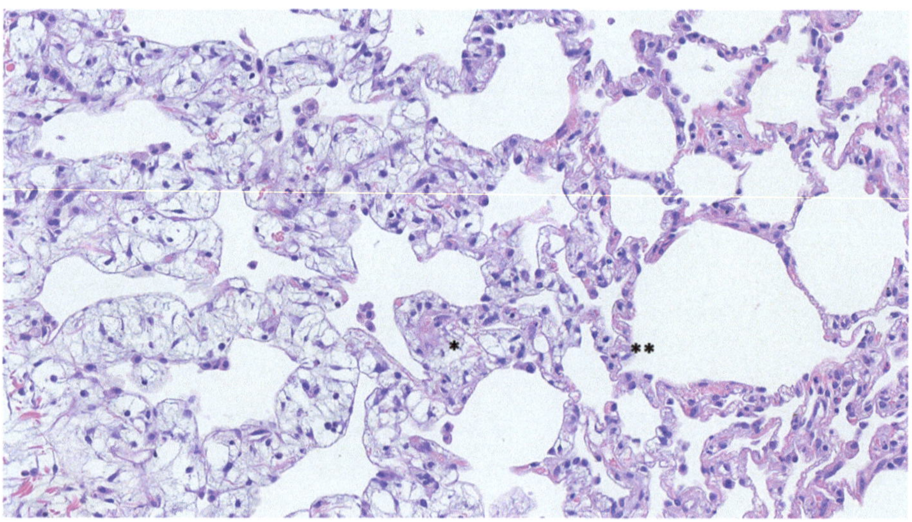

FIGURE 2.8B

Higher magnification of the previous specimen shows details of the thickened alveolar walls with myxoid cells (*) at the junction with normal alveolar tissue (**). Hematoxylin and eosin stain, 20 × objective magnification.

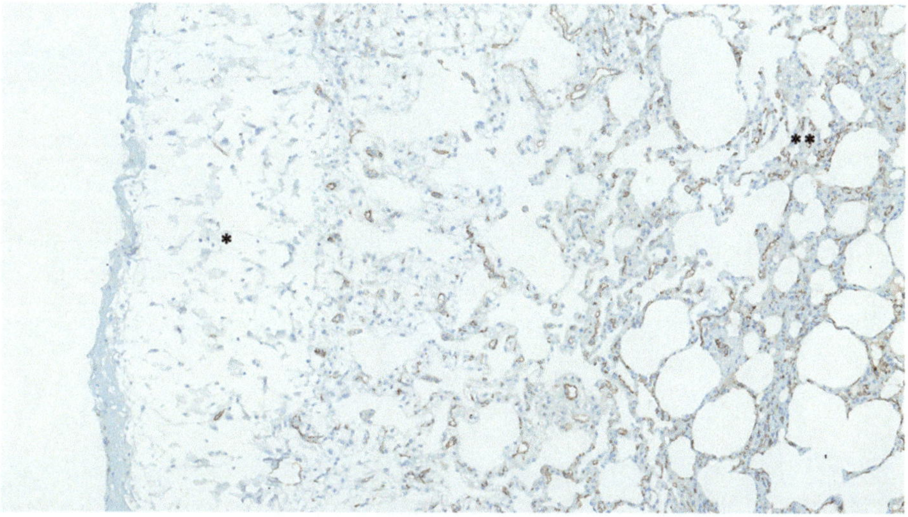

FIGURE 2.8C

Immunohistochemical stain for CD31 on the lung of a male rhesus macaque necropsied 31 days after irradiation at 12 Gy, showing a reduction in microvascular density (brown staining) in the area of pleural and subpleural injury (*). Note the diffuse population of brown stained $CD31^+$ capillary endothelial cells in the normal lung tissue (**) located deep to the area of pleural and subpleural injury. Stain performed with 3,3′-diaminobenzidine chromogen and hematoxylin counterstain, 10 × objective magnification.

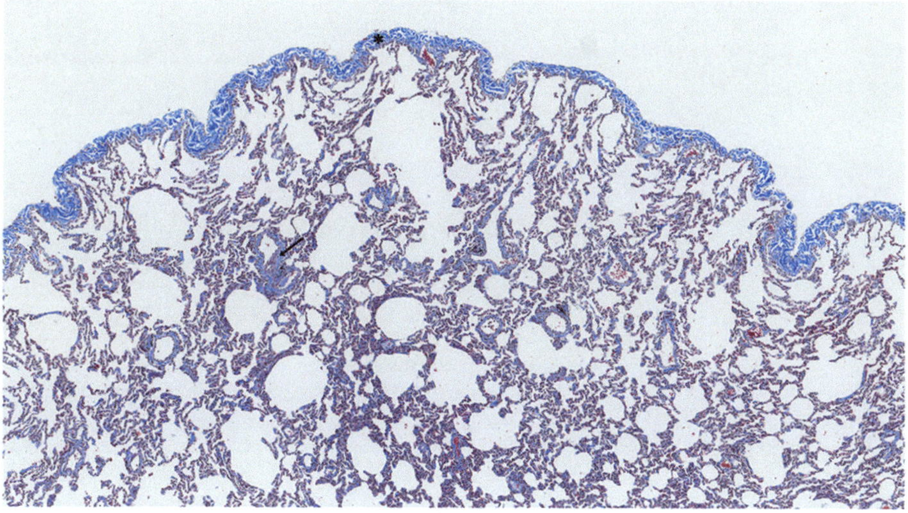

FIGURE 2.9A

Masson's trichrome stain on lung of a naïve male rhesus macaque accentuates the fibrous connective tissue in the visceral pleura (*) and around major pneumovascular tracts (arrow). 5 × objective magnification.

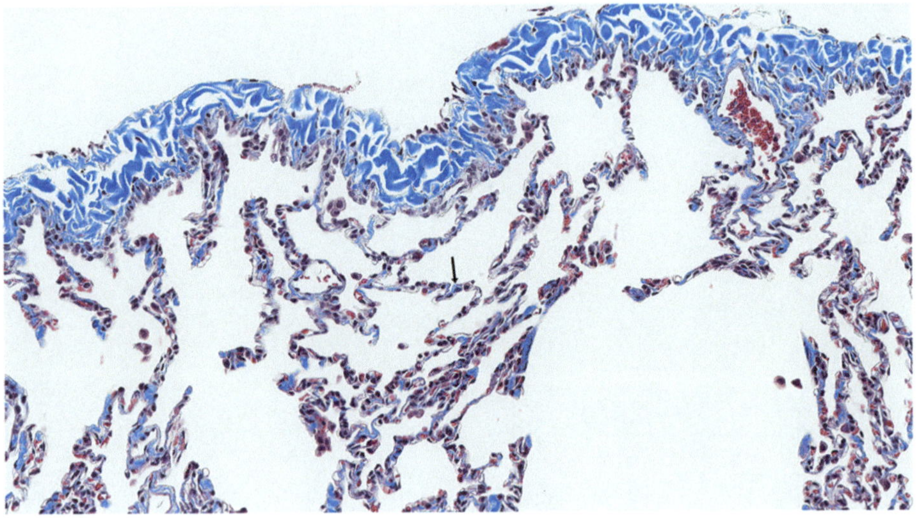

FIGURE 2.9B

A higher magnification of the previous specimen shows the thick fibrous pleural layer (*) and a scant amount of blue-stained fibrous connective tissue associated with alveolar walls (arrow). Masson's trichrome stain, 40× objective magnification.

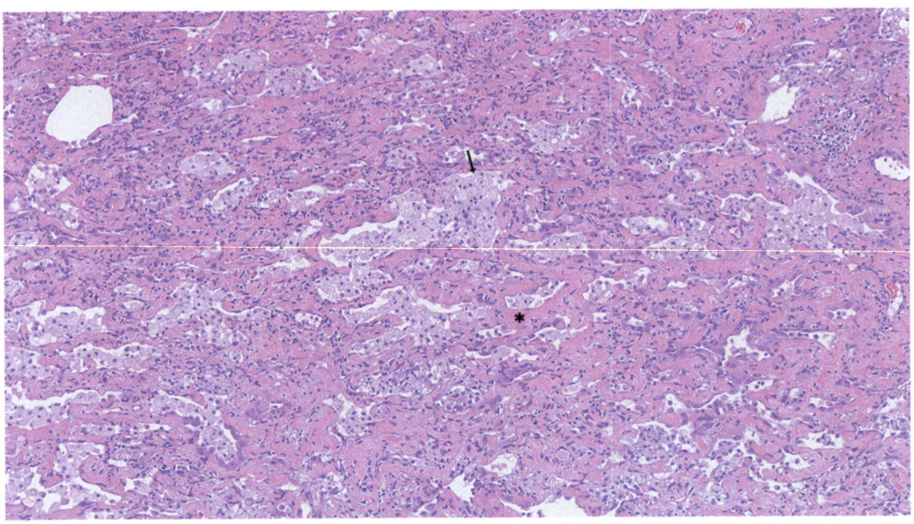

FIGURE 2.10A

Lung of a male rhesus macaque necropsied 177 days after irradiation at 11 Gy. Alveolar walls (*) are thickened by mature fibrous connective tissue, and alveolar lumina contain numerous engorged macrophages (arrow). Hematoxylin and eosin stain, 10× objective magnification.

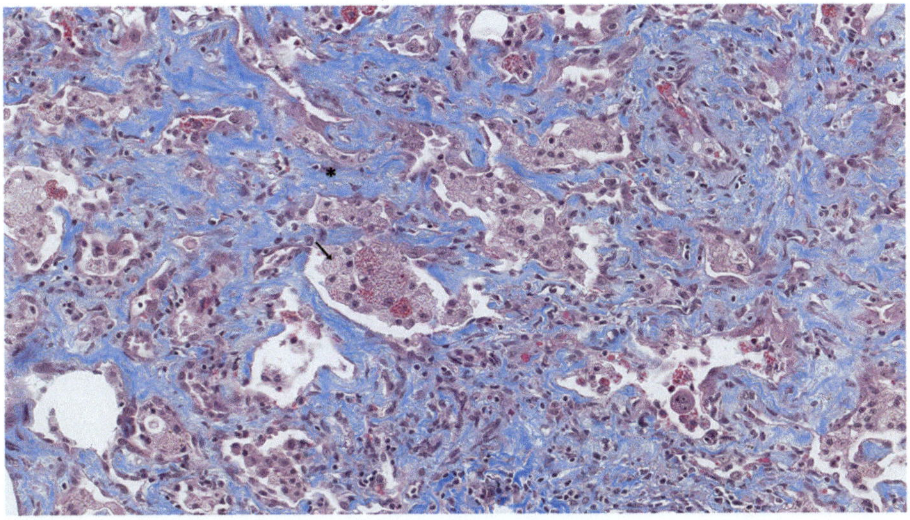

FIGURE 2.10B

Masson's trichrome stain on the same lung specimen shown in the previous image shows alveolar walls (*) are thickened by blue-staining fibrous connective tissue. Alveolar lumina contain numerous alveolar macrophages (arrow). Staining characteristics of the alveolar macrophages are influenced by the nature of the tissue debris that was engulfed by the macrophages. 20 × objective magnification.

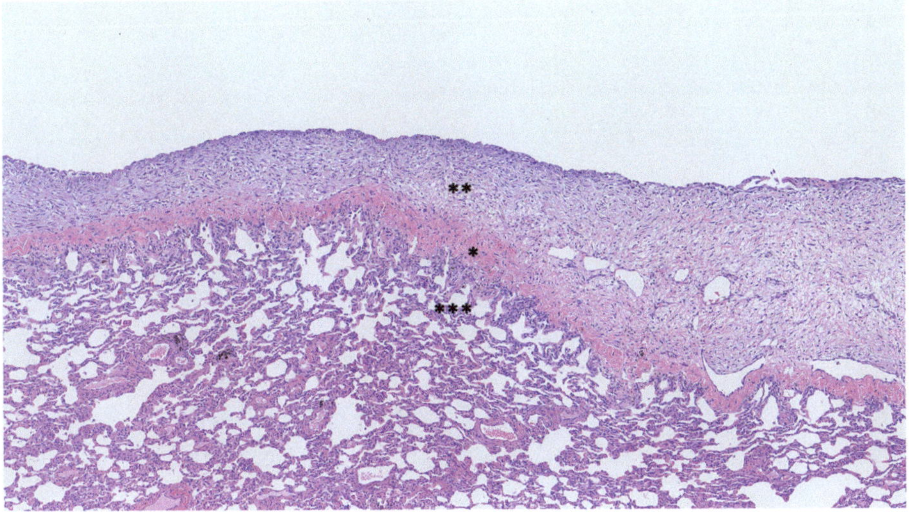

FIGURE 2.11

Lung of a male rhesus macaque necropsied 53 days following irradiation at 11 Gy. The original pleural layer of dense fibrous connective tissue (*) is covered by a thick layer of immature fibrous connective tissue (**). Subpleural alveoli (***) in this image have little evidence of fibrosis. Hematoxylin and eosin stain, 5 × objective magnification.

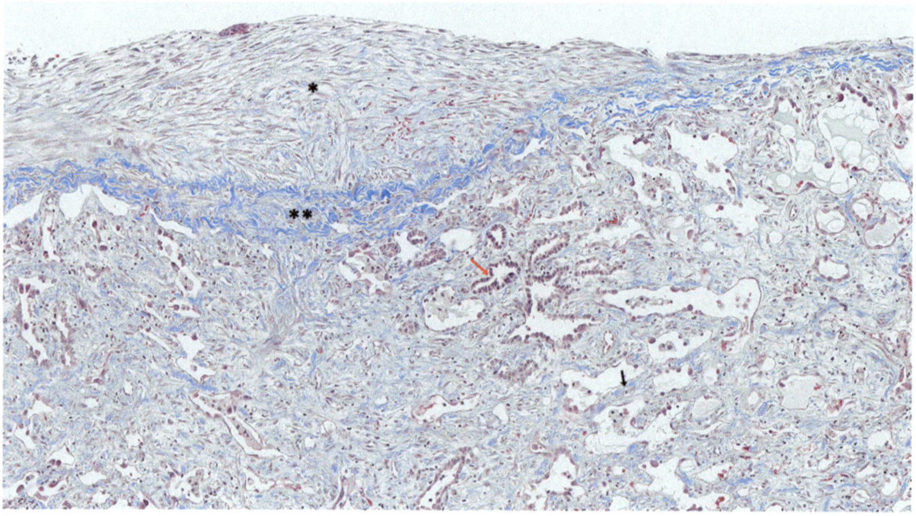

FIGURE 2.12A

Lung of a male rhesus macaque necropsied 53 days following irradiation at 11 Gy. The original pleural layer of dense fibrous connective tissue (*) is covered by a thick layer of immature fibrous connective tissue (**). Subpleural alveoli (***) in this image have little evidence of fibrosis. Hematoxylin and eosin stain, 5 × objective magnification.

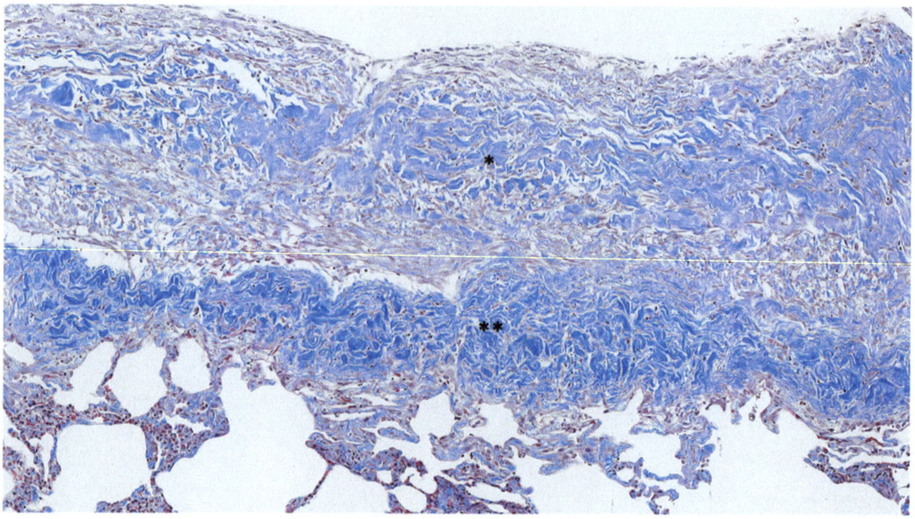

FIGURE 2.12B

The lung of a male rhesus macaque necropsied 105 days after irradiation at 11 Gy has a thick layer of maturing fibrous connective tissue that includes a zone of blue-stained mature connective tissue (*). The original pleural surface is demarcated by the deeply placed band of blue-stained mature connective tissue (**). Masson's trichrome stain, 10 × objective magnification.

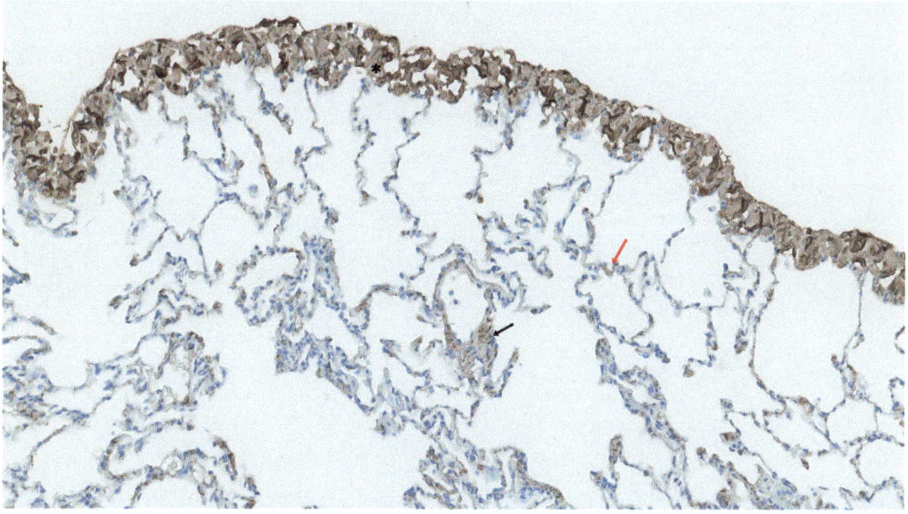

FIGURE 2.13A

Immunohistochemical staining for collagen-1 performed on the lung of a naive control male rhesus macaque. Note the dense collagen-1 staining in the pleura (*) with modest staining associated with larger airways (black arrow) and minimal staining in alveolar walls (red arrow). Stain performed with 3, 3′-diaminobenzidine chromogen and hematoxylin counterstain, 20 × objective magnification.

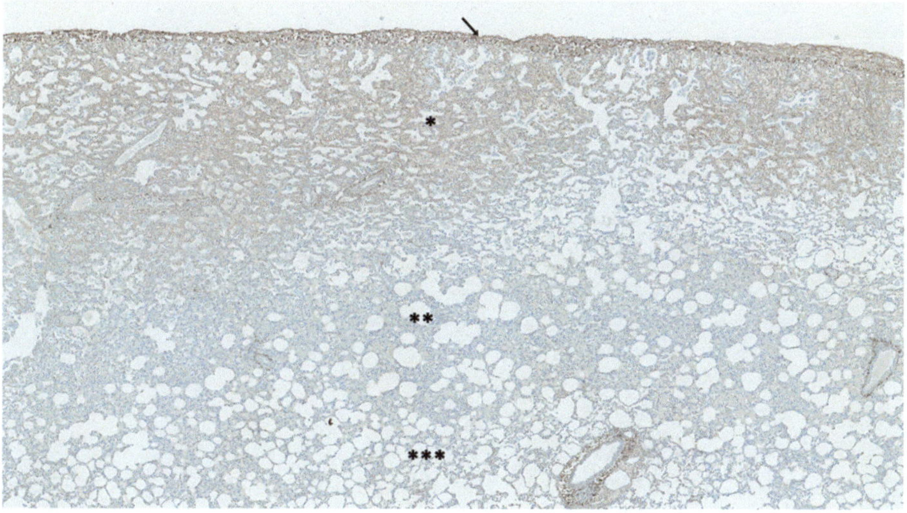

FIGURE 2.13B

Immunohistochemical stain for collagen 1 on lung of a male rhesus macaque collected at day 53 following irradiation at 11 Gy. Immediately subjacent to the pleura (arrow) is a zone of interstitial fibrosis (*) containing brown-stained collagen. Deep to the fibrotic zone is a zone of alveolar edema (**) and relatively sparse inflammatory cell infiltration. The deepest aspect of the specimen (***) has more nearly normal lung tissue. Stain performed with 3,3′-diaminobenzidine chromogen and hematoxylin counterstain, 3.29 × objective magnification.

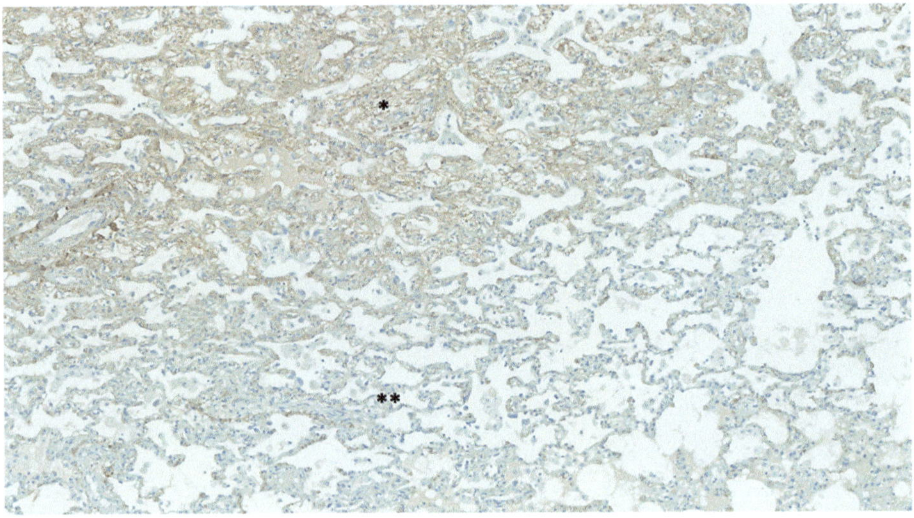

FIGURE 2.13C

Higher magnification of the previous specimen shows the abrupt transition between fibrotic alveolar walls (*) with increased collagen content compared to the normal alveolar walls (**) subjacent to the fibrotic zone. Collagen 1 immunohistochemical stain with 3,3′-diaminobenzidine chromogen and hematoxylin counterstain, 10 × objective magnification.

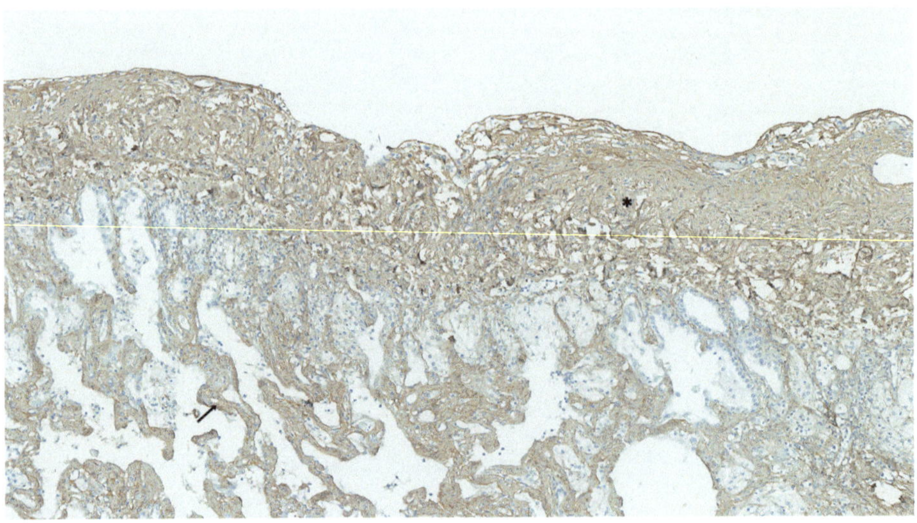

FIGURE 2.14

Immunohistochemical stain for collagen 1 on lung of a male rhesus macaque necropsied 53 days after irradiation at 11 Gy. Note the thick layer of collagen-containing fibrous connective tissue (*) that has replaced the original pleural layer. Also note the thickened, collagen-laden alveolar walls (arrow) in the underlying lung. Stain performed with 3,3′-diaminobenzidine chromogen and hematoxylin counterstain, 10 × objective magnification.

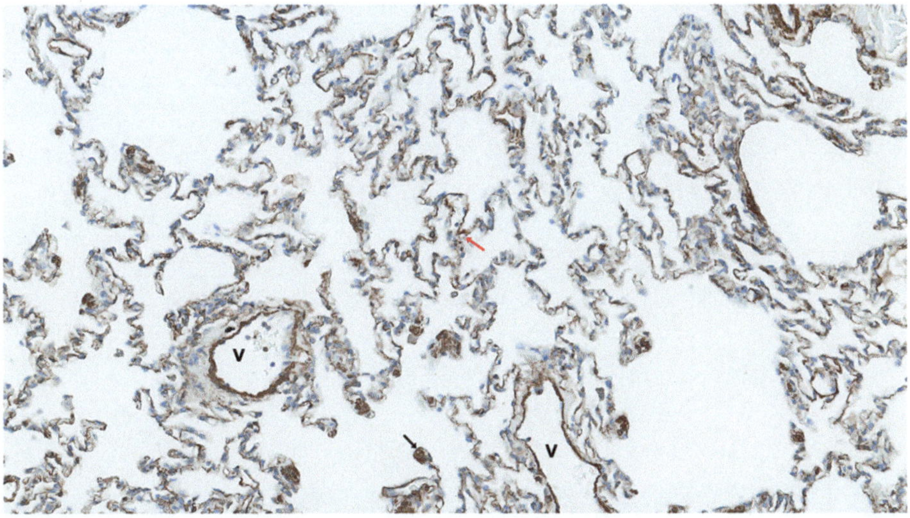

FIGURE 2.15A

Immunohistochemical stain for alpha-smooth muscle actin (αSMA) on the lung of a naïve control male rhesus macaque. Note the prominent staining of blood vessels (v) and smooth muscle aggregates in terminal airways (black arrow), as well as a fine pattern of positive staining associated with microvascular structures in alveolar walls (red arrow). Stain performed with 3,3′-diaminobenzidine chromogen and hematoxylin counterstain, 10 × objective magnification.

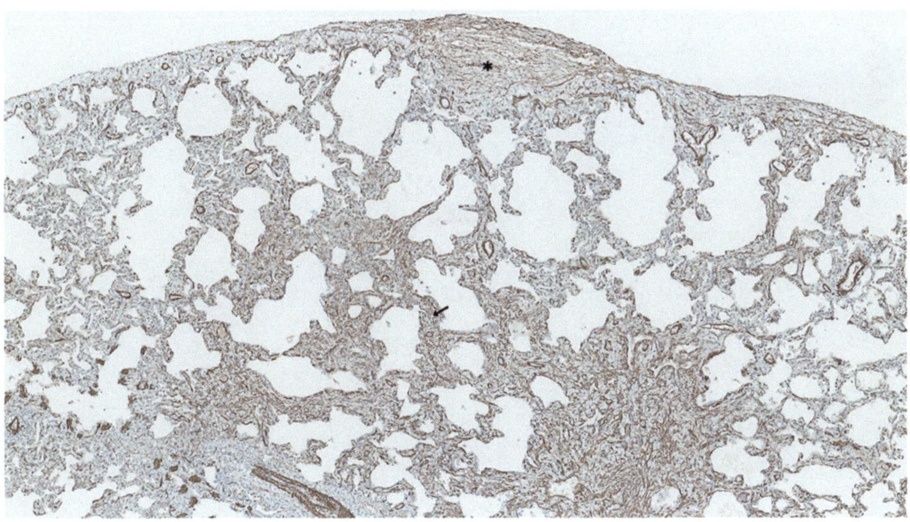

FIGURE 2.15B

Immunohistochemical stain for alpha smooth muscle actin (αSMA) performed on the lung from a male rhesus macaque necropsied 182 days after irradiation at 10 Gy. Note the increased amount of αSMA staining in the thickened alveolar walls (arrow) and in the plaque-like area of immature fibrous connective tissue (*) on the pleural surface. Stain performed with 3,3′-diaminobenzidine chromogen and hematoxylin counterstain, 5 × objective magnification.

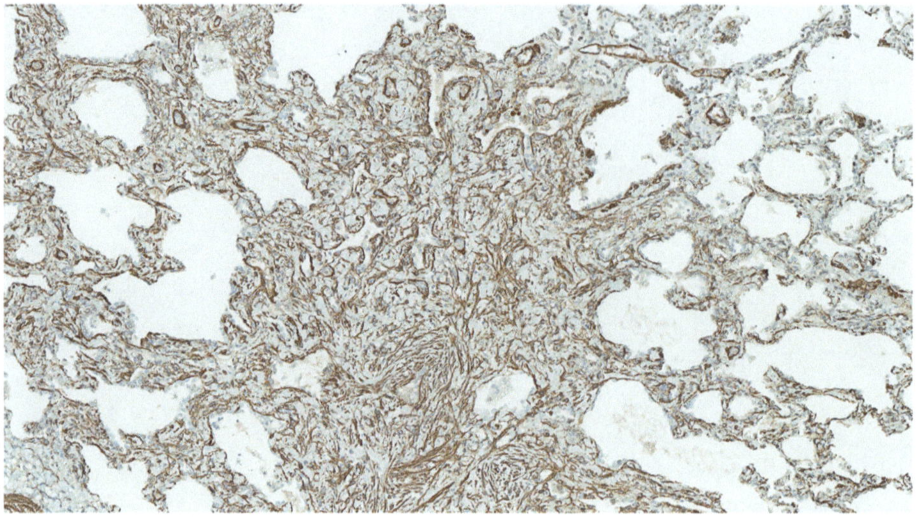

FIGURE 2.15C

Higher magnification of the previous specimen showing the fibrillar pattern of αSMA staining in the thickened alveolar walls. Alpha smooth muscle actin immunohistochemical stain with 3,3′-diaminobenzidine chromogen and hematoxylin counterstain, 10 × objective magnification.

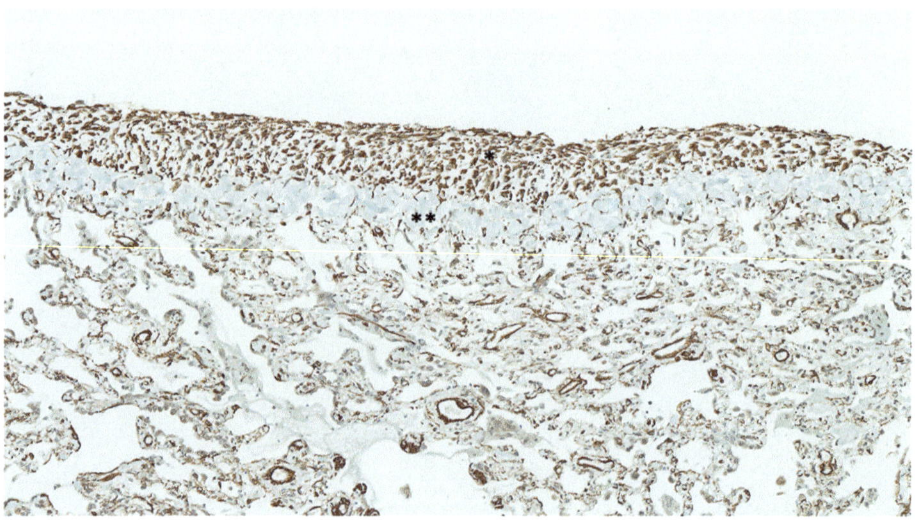

FIGURE 2.15D

Immunohistochemical stain for alpha smooth muscle actin (αSMA) performed on the lung from a male rhesus macaque necropsied 44 days after irradiation at 10 Gy. Note the pronounced increase in αSMA staining in a superficial cellular zone (*) overlying the original fibrous connective tissue layer of the pleura (**). Stain performed with 3,3′-diaminobenzidine chromogen and hematoxylin counterstain, 10 × objective magnification.

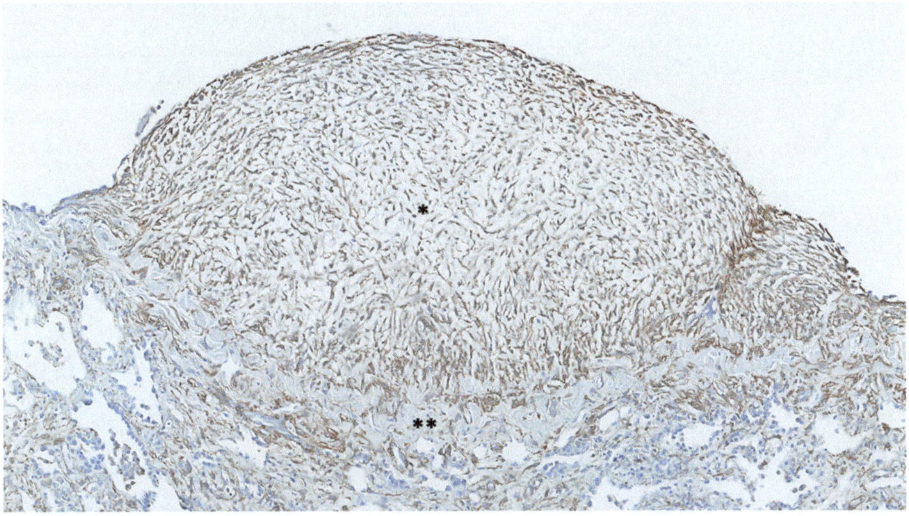

FIGURE 2.15E

Immunohistochemical stain for alpha smooth muscle actin (αSMA) performed on the lung from a male rhesus macaque necropsied 53 days after irradiation at 11 Gy. Note the nodular proliferation of αSMA$^+$ cells in the superficial cellular zone (*) overlying the original fibrous connective tissue layer of the pleura (**). Stain performed with 3,3′-diaminobenzidine chromogen and hematoxylin counterstain, 10 × objective magnification.

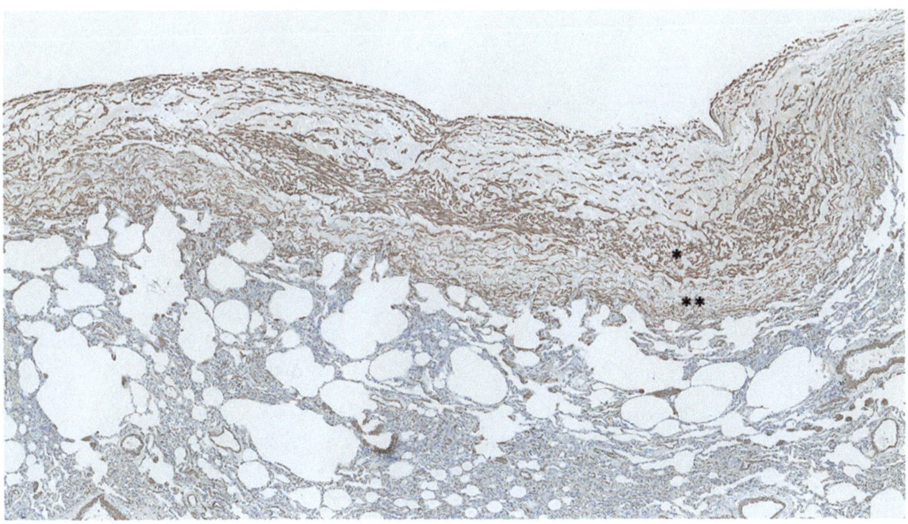

FIGURE 2.15F

Immunohistochemical stain for alpha smooth muscle actin (αSMA) performed on the lung from a male rhesus macaque necropsied 105 days after irradiation at 11 Gy. Note the thick layer of αSMA$^+$ cells in the superficial cellular zone (*) that is closely adherent to the original fibrous connective tissue layer of the pleura (**). Stain performed with 3,3′-diaminobenzidine chromogen and hematoxylin counterstain, 5 × objective magnification.

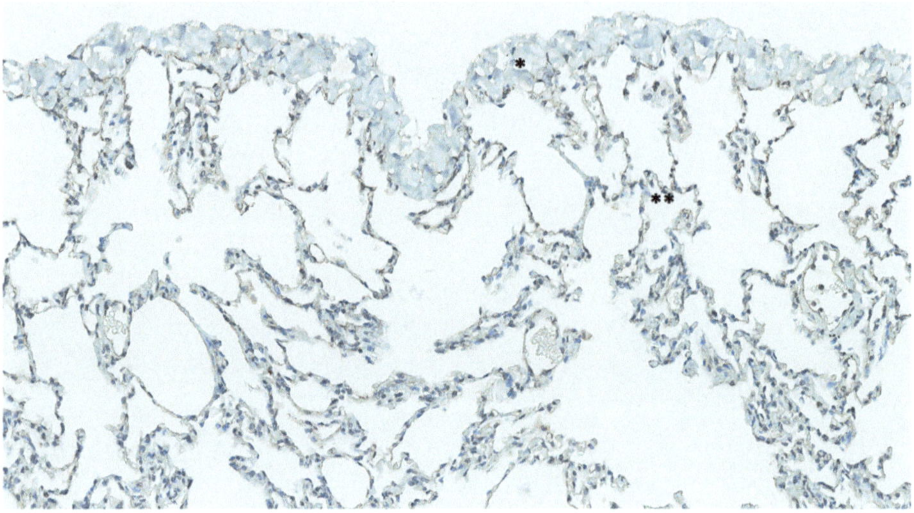

FIGURE 2.16A

Immunohistochemical staining for transforming growth factor-beta (TGF-β) performed on the lung of a naive control male rhesus macaque. Note the virtual absence of staining in the pleura (*) and alveolar tissue (**). Stain performed with 3, 3′-diaminobenzidine chromogen and hematoxylin counterstain, 20 × objective magnification.

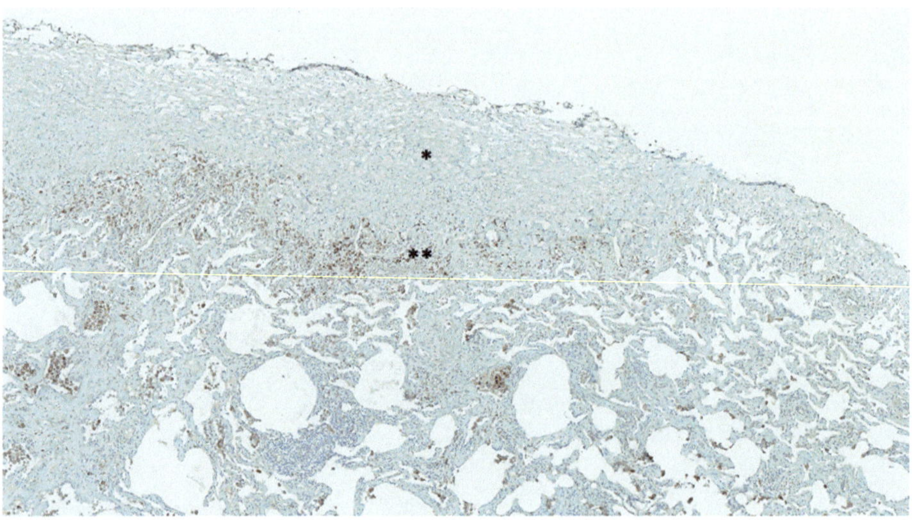

FIGURE 2.16B

Immunohistochemical stain for transforming growth factor β (TGF-β) on lung of a male rhesus macaque collected at day 183 following irradiation at 11 Gy. A zone of fibrosis (*) is present on the pleural surface and a zone of brown-stained TGF-β^+ cells (**) is present immediately deep to the fibrotic pleura. Evidence of abundant TGFβ production by inflammatory cells, largely alveolar macrophages, at this late stage following irradiation suggests traditional inflammatory mediators may be involved in perpetuation and extension of the radiation-associated pulmonary injury. Stain performed with 3, 3′-diaminobenzidine chromogen and hematoxylin counterstain, 5 × objective magnification.

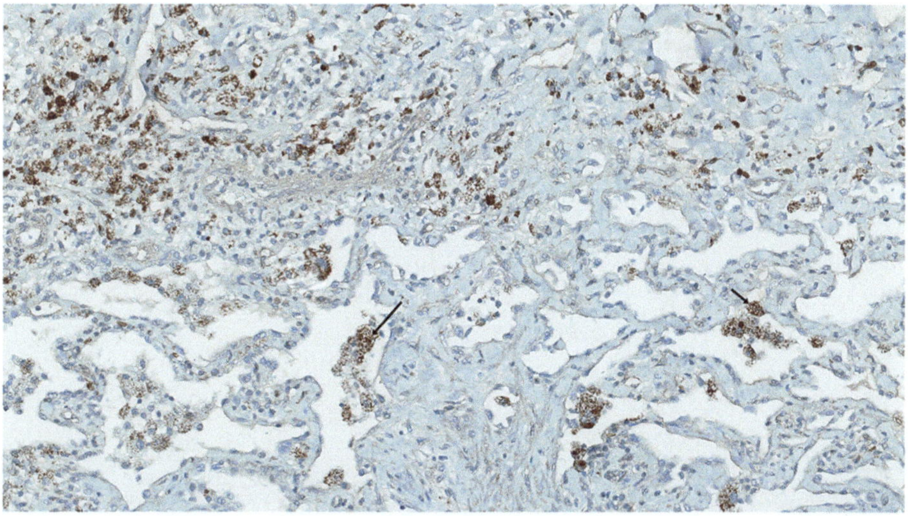

FIGURE 2.16C

Higher magnification of the previous specimen. Note the TGF-β^{+} alveolar macrophages (arrow) in the lumina of alveoli that have thickened, fibrotic walls. Stain performed with 3,3′-diaminobenzidine chromogen and hematoxylin counterstain, 20 × objective magnification.

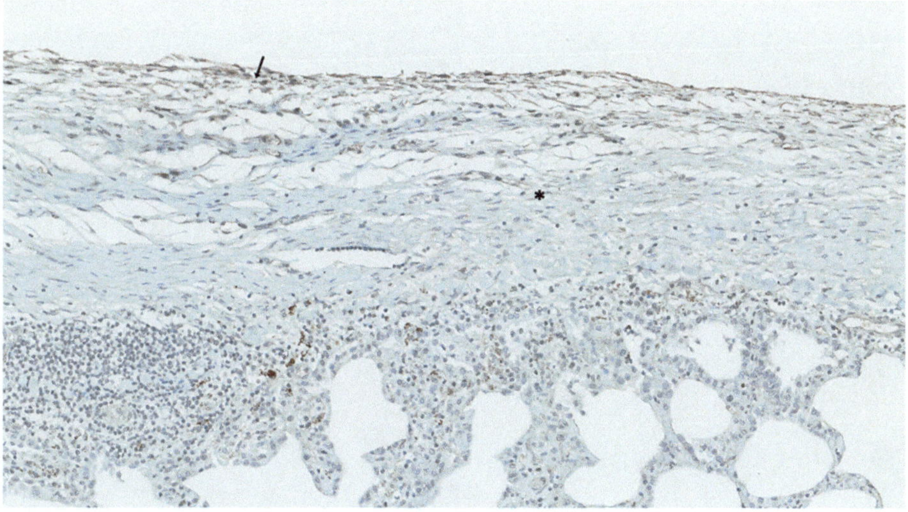

FIGURE 2.16D

Immunohistochemical stain for transforming growth factor β (TGF-β) on lung of a male rhesus macaque collected at day 183 following irradiation at 11 Gy. A zone of TGF-β staining (arrow) is present on the surface of the thickened, fibrotic pleural surface (*), suggesting TGFβ signaling may be involved in perpetuation and amplification of the radiation-associated superficial pleural reaction. Stain performed with 3,3′-diaminobenzidine chromogen and hematoxylin counterstain, 15 × objective magnification.

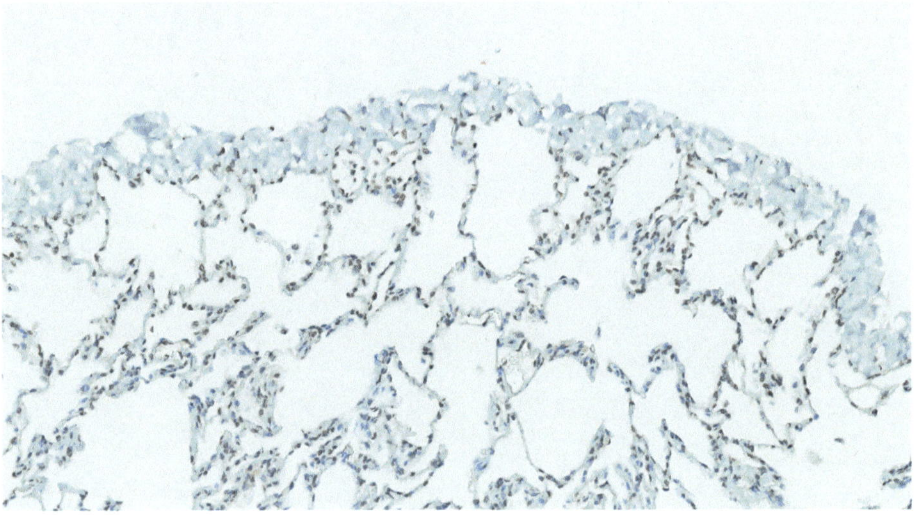

FIGURE 2.17A

Immunohistochemical staining for connective tissue growth factor (CTGF) performed on the lung of a naive control male rhesus macaque. Note the virtual absence of CTGF staining. Stain performed with 3, 3′-diaminobenzidine chromogen and hematoxylin counterstain, 20 × objective magnification.

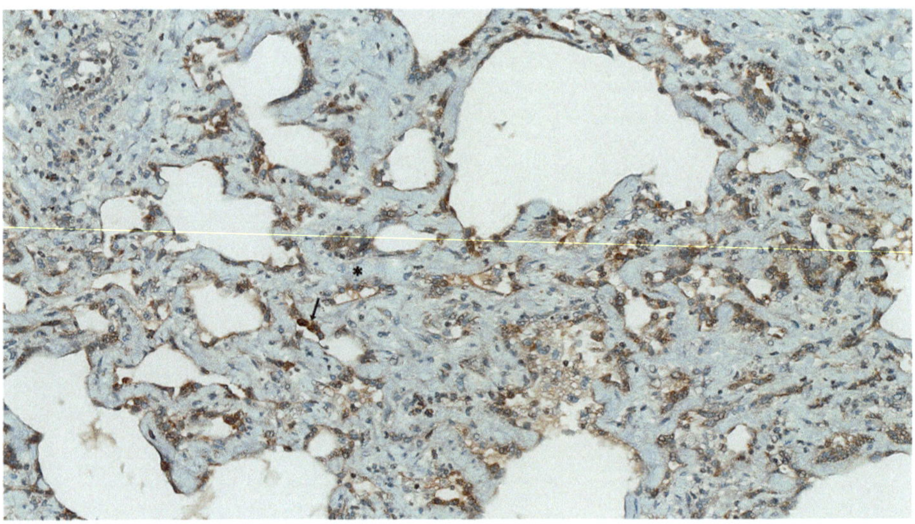

FIGURE 2.17B

Immunohistochemical stain for connective tissue growth factor (CTGF) on lung of a male rhesus macaque collected at day 183 following irradiation at 10 Gy. Note the brown staining representing expression of CTGF in proliferative type II pneumocytes (arrow) on the surface of thickened, fibrotic alveolar walls (*). Stain performed with 3,3′-diaminobenzidine chromogen and hematoxylin counterstain, 20 × objective magnification.

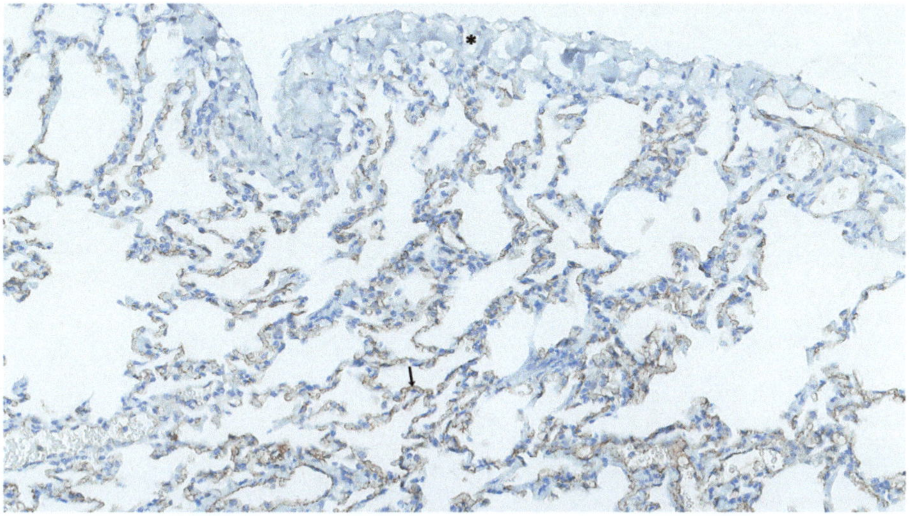

FIGURE 2.18A

Immunohistochemical stain for CD31 on the lung of a naïve control male rhesus macaque. Note the lacework pattern of CD31 staining in alveolar walls (arrow), and absence of staining in fibrous pleura (*). Stain performed with 3,3′-diaminobenzidine chromogen and hematoxylin counterstain, 20 × objective magnification.

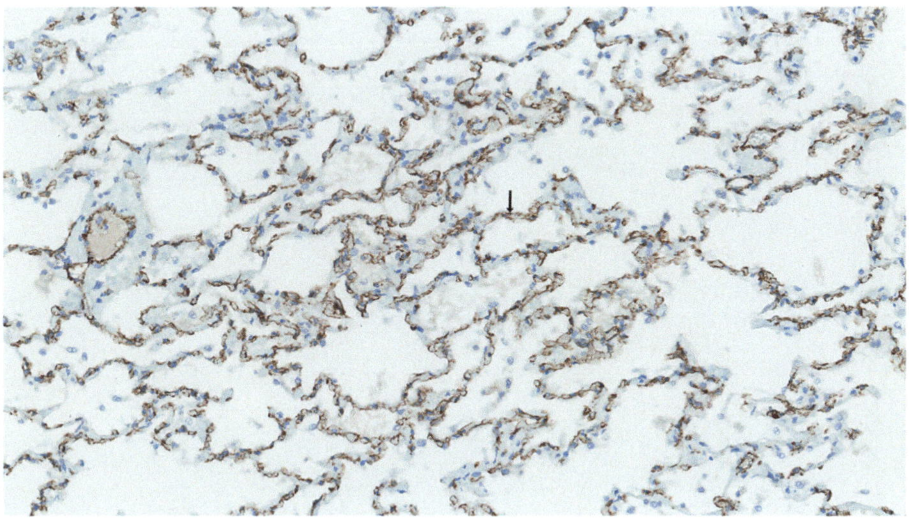

FIGURE 2.18B

Immunohistochemical stain for CD31 on non-fibrotic region lung from a male rhesus macaque necropsied 183 days following irradiation at 10 Gy. Note the diffuse population of brown stained $CD31^+$ endothelial cells (arrow) within normal alveolar walls throughout the image. Stain performed with 3,3′-diaminobenzidine chromogen and hematoxylin counterstain, 20 × objective magnification.

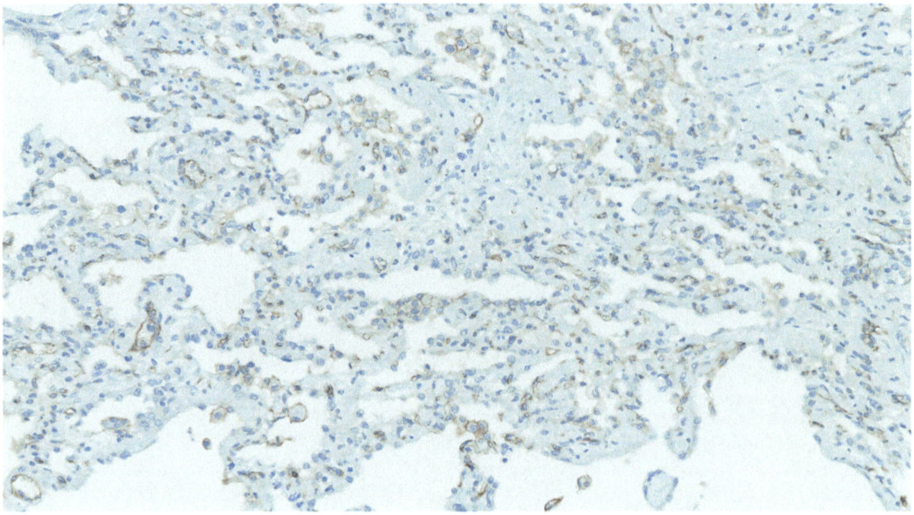

FIGURE 2.18C

Immunohistochemical stain for CD31 on fibrotic region of the lung of the male rhesus macaque presented in the previous specimen. Note the relatively sparse population of brown stained CD31^{+} endothelial cells within thickened alveolar walls. Stain performed with 3,3′-diaminobenzidine chromogen and hematoxylin counterstain, 20× objective magnification.

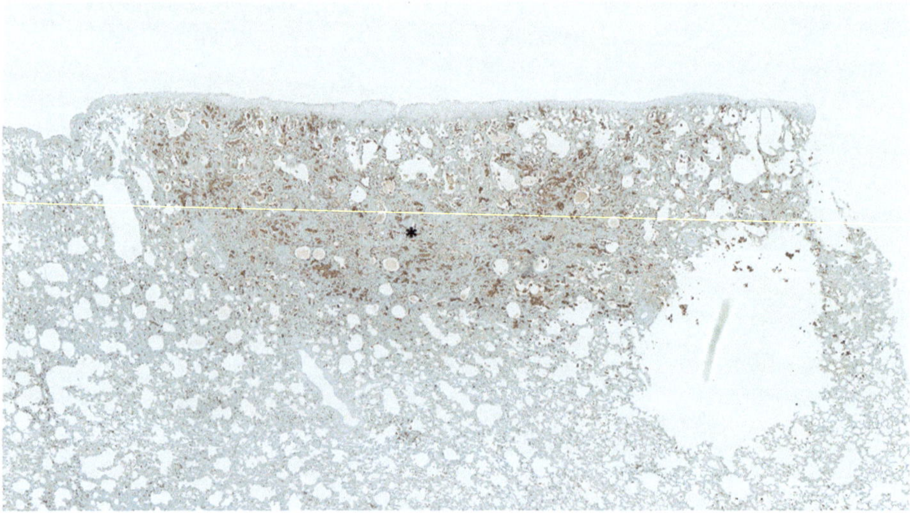

FIGURE 2.19A

Immunohistochemical staining for galectin-3 on the lung of a male rhesus macaque necropsied 183 days after irradiation at 10 Gy. Note the zone of intense galectin-3 staining in the subpleural fibrotic zone (*). Stain performed with 3, 3′-diaminobenzidine chromogen and hematoxylin counterstain, 1.65× objective magnification.

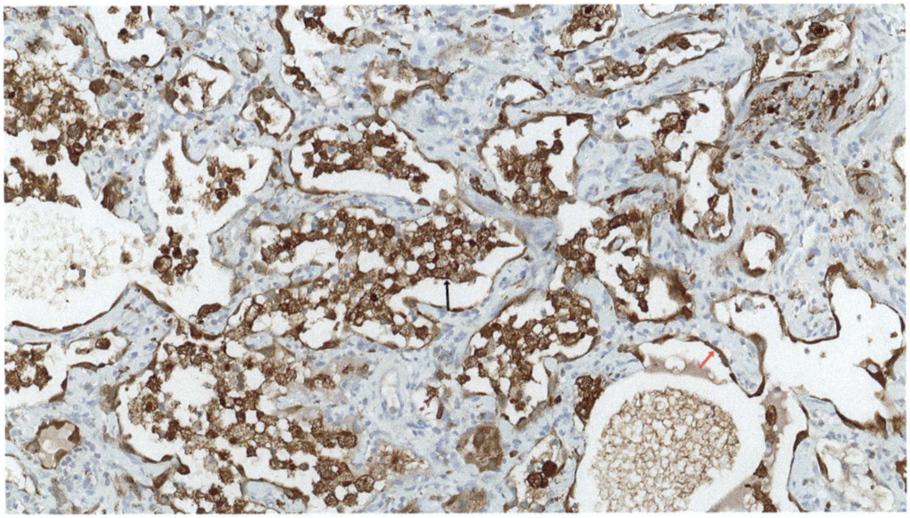

FIGURE 2.19B

Higher magnification of the previous specimen showing the galectin-3 staining is concentrated in alveolar macrophages (black arrow) and epithelial cells lining alveoli (red arrow). Stain performed with 3, 3′-diaminobenzidine chromogen and hematoxylin counterstain, 20 × objective magnification.

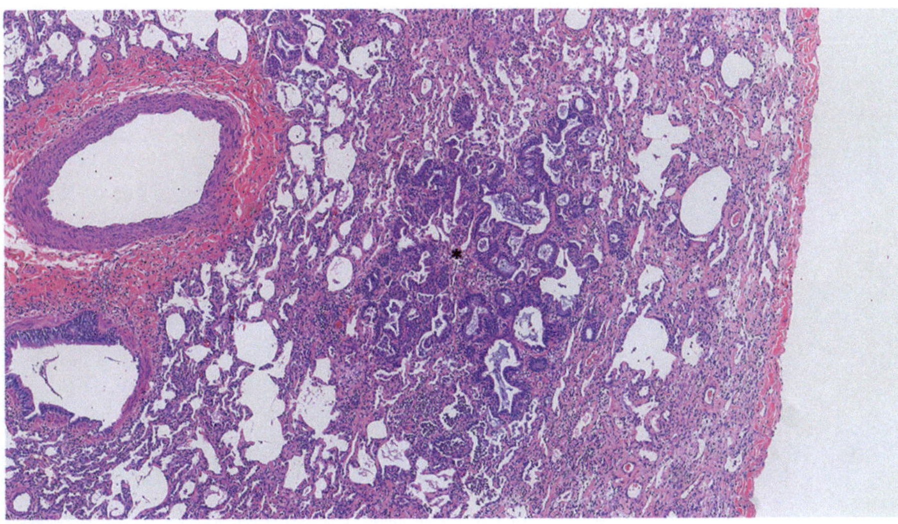

FIGURE 2.20A

The lung of a male rhesus macaque irradiated at 10 Gy and necropsied on Day 154 post-irradiation has a deeply basophilic area (*) of alveolar-bronchiolar hyperplasia. Hematoxylin and eosin stain, 5 × objective magnification.

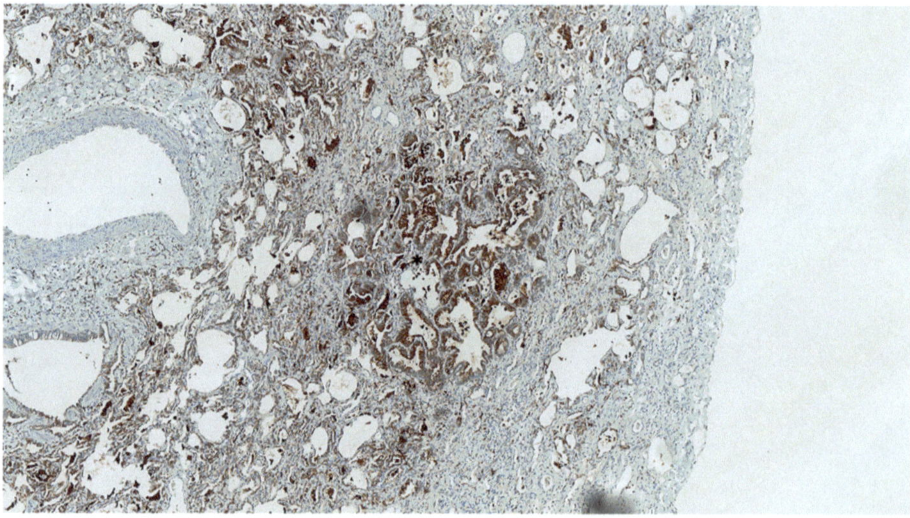

FIGURE 2.20B

Immunohistochemical staining for galectin-3 performed on the same lesion shown in the previous specimen. Note the area of alveolar-bronchiolar hyperplasia (circle) has intense galectin-3 expression (*). Stain performed with 3, 3′-diaminobenzidine chromogen and hematoxylin counterstain, 5 × objective magnification.

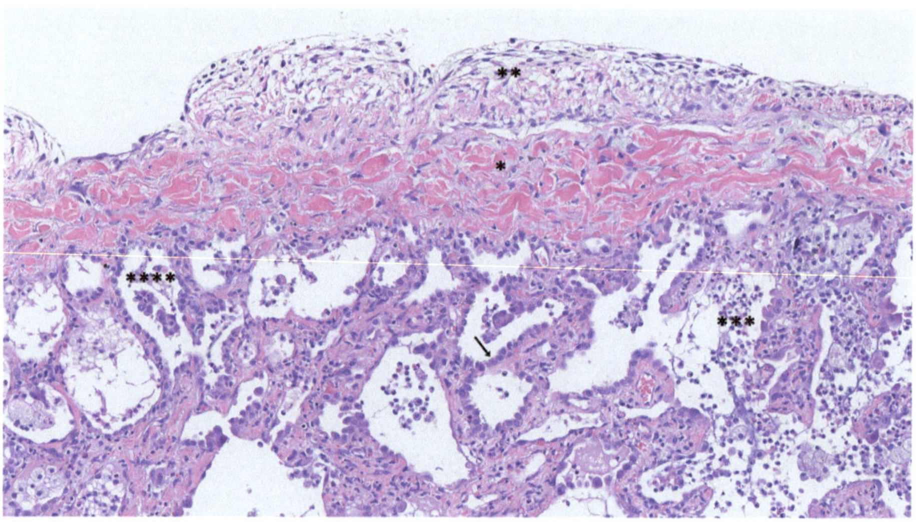

FIGURE 2.21

Lung of a male rhesus macaque necropsied 119 days after irradiation at 10 Gy. Alveolar walls beneath the fibrous pleura (*) are covered by plump, proliferative type II pneumocytes (arrow). Note the layer of immature fibrous connective tissue on the pleural surface (**). Mixed (lymphocytic/neutrophilic) inflammatory cell infiltration is present in one region of the image (***), while alveolar macrophage accumulation predominates in other regions (****). Hematoxylin and eosin stain, 14.7 × objective magnification.

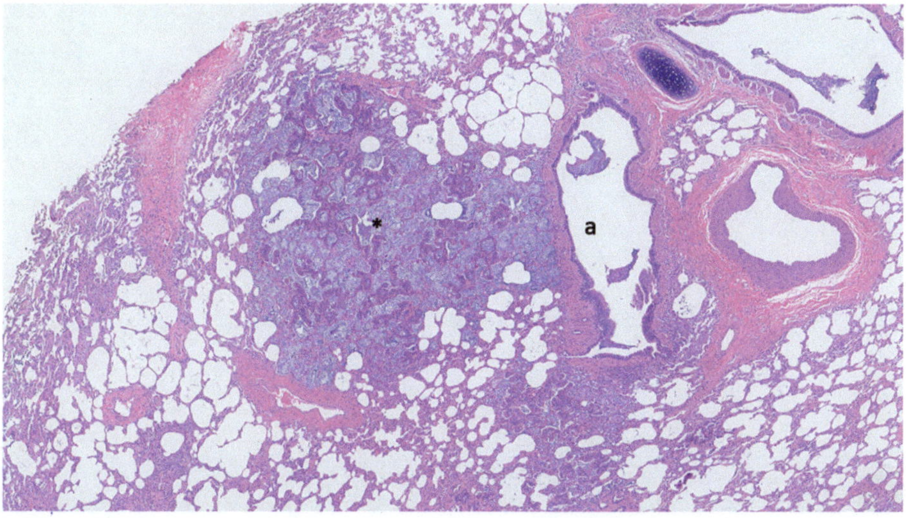

FIGURE 2.22A

Lung of a male rhesus macaque necropsied 141 days after irradiation at 11 Gy. Note the focal alveolar-bronchiolar hyperplasia (*) composed of epithelial cords and fronds surrounded by amphophilic secretory material. The proliferative lesion is located adjacent to a major airway (a). Hematoxylin and eosin stain, 2.5 × objective magnification.

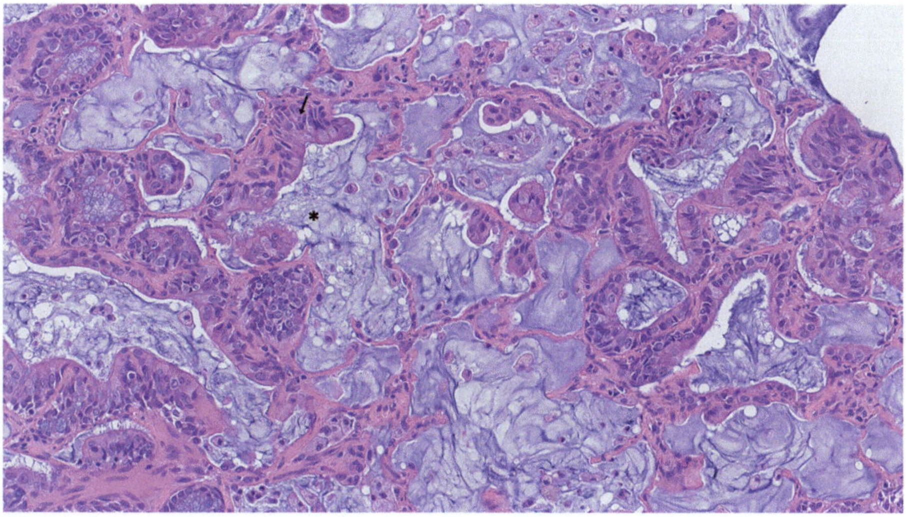

FIGURE 2.22B

Higher magnification of the lung specimen shown in the previous image shows alveoli and bronchioles are lined by thick layers of epithelial cells (arrow) which have few mitotic figures and exhibit a mild to moderate degree of cellular atypism. Alveolar lumina contain mucinous secretory material (*) produced by mucin-secreting cells within the proliferative epithelial cell population. Hematoxylin and eosin stain, 20 × objective magnification.

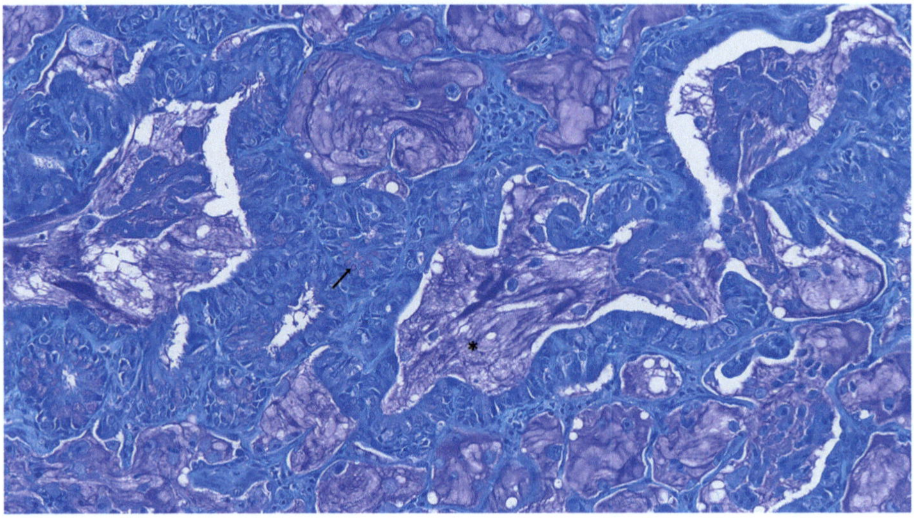

FIGURE 2.22C

Toluidine blue stain of the proliferative lung lesion in a male rhesus macaque necropsied at day 141 after irradiation at 11 Gy. Lumina within the area of alveolar-bronchiolar hyperplasia contain metachromatic (purple) material (*), indicating carbohydrate content of the mucinous material produced by the proliferating epithelial cells. Note the mucin-containing cytoplasmic vacuoles (arrow) in proliferating epithelial cells. 20 × objective magnification.

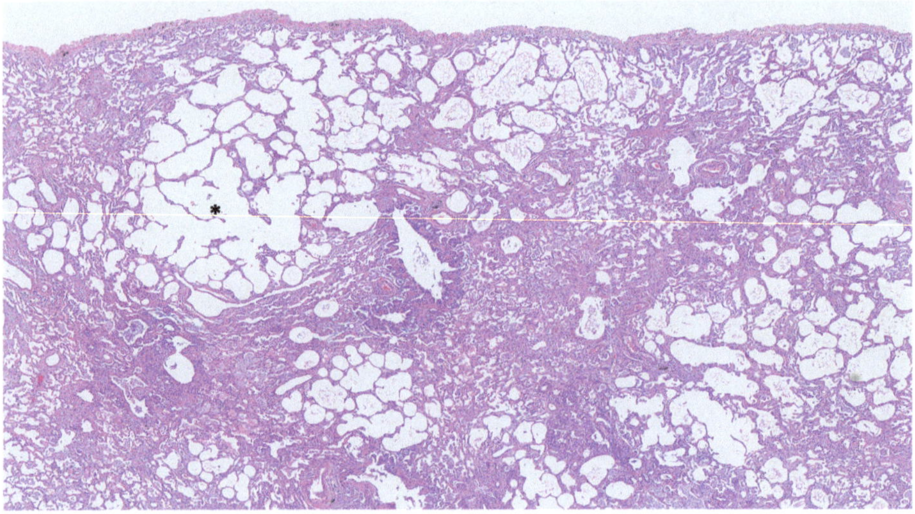

FIGURE 2.23

The lung specimen of a male rhesus macaque necropsied 111 days after irradiation at 11 Gy has multifocal alveolar emphysema (*), consisting of areas of pronounced airway and alveolar dilation. Breakdown of alveolar walls within emphysematous areas reduces the surface area available for oxygen exchange, thus contributing to hypoxemia. Hematoxylin and eosin stain, 2.55 × objective magnification.

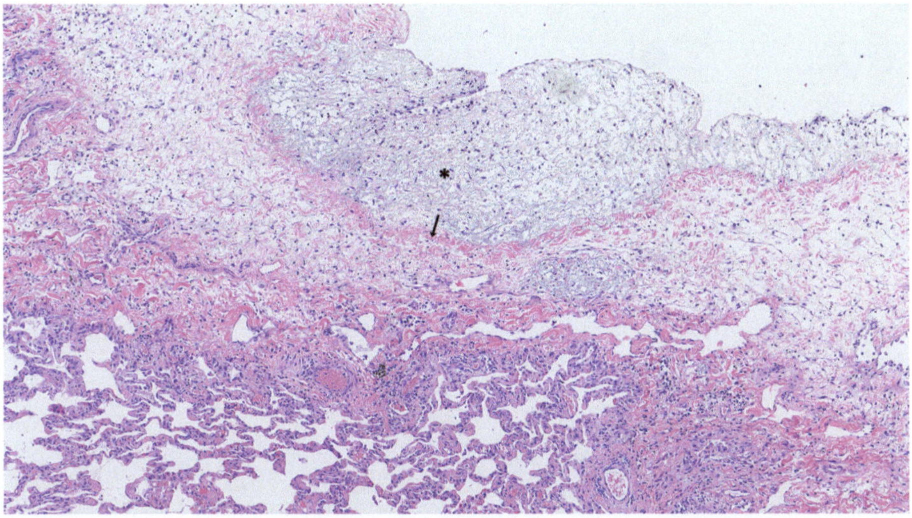

FIGURE 2.24

The lung specimen from a male rhesus macaque necropsied 177 days after irradiation at 11 Gy has a thick layer of loosely woven immature fibrous connective tissue and inflammatory cells (*) on the pleural surface. The original pleural surface is apparent as a band of dense fibrous connective tissue (arrow). Hematoxylin and eosin stain, 6.7 × objective magnification.

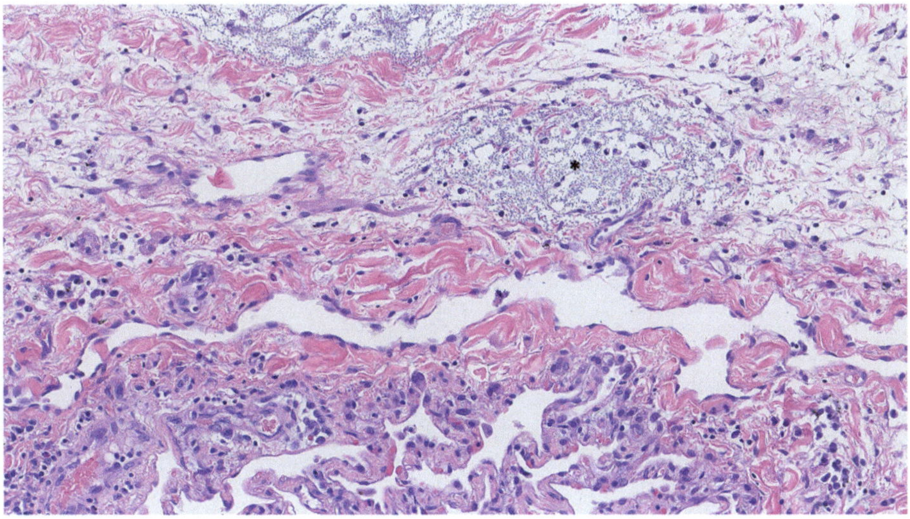

FIGURE 2.25

Higher magnification of the previous specimen reveals an extensive bacterial population (*) within the thickened pleural surface. Hematoxylin and eosin stain, 20 × objective magnification.

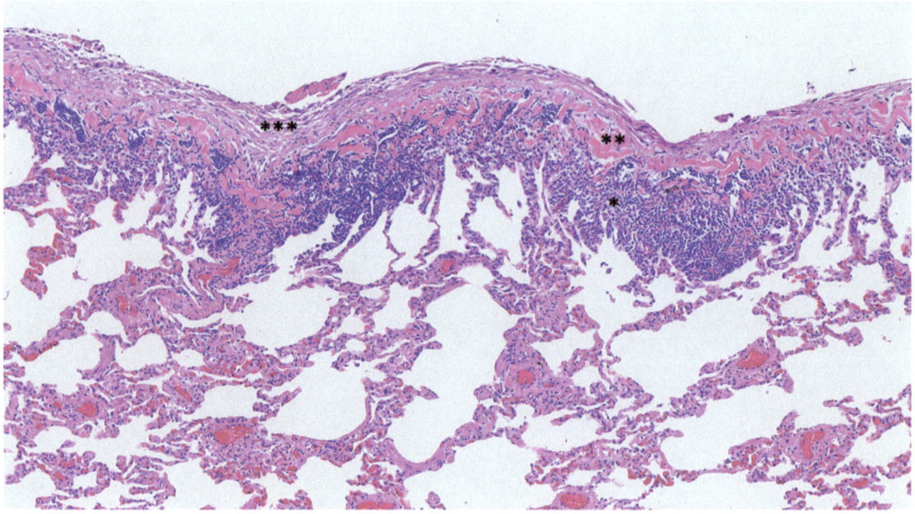

FIGURE 2.26A

The lung of a male rhesus macaque necropsied 85 days after irradiation at 11 Gy has a prominent band of densely stained inflammatory cells (*) located immediately beneath the pleura (**). A zone of immature fibrous connective tissue (***) is present on the pleural surface. Hematoxylin and eosin stain, 10× objective magnification.

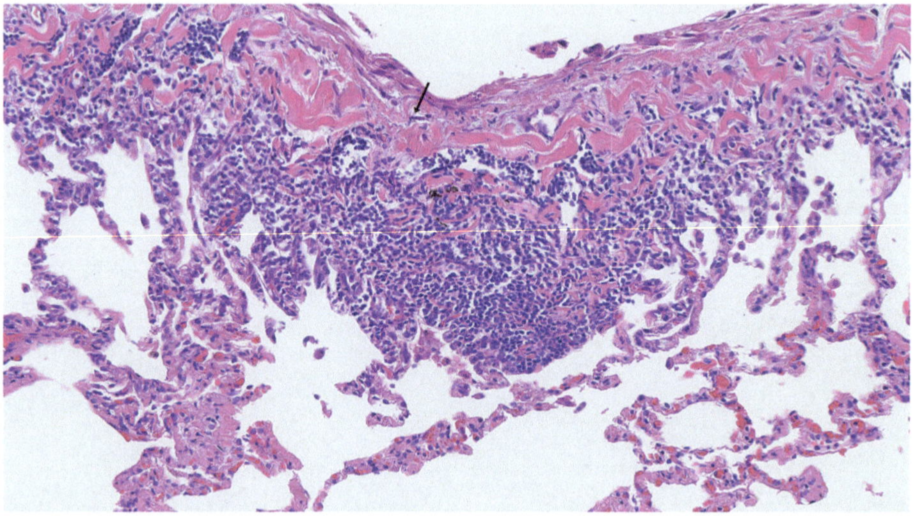

FIGURE 2.26B

Higher magnification of the previous specimen reveals numerous small particulate structures (arrow), consistent with bacteria, in the pleura. The subpleural inflammatory cell infiltration consists primarily of lymphocytes and macrophages. Hematoxylin and eosin stain, 20× objective magnification.

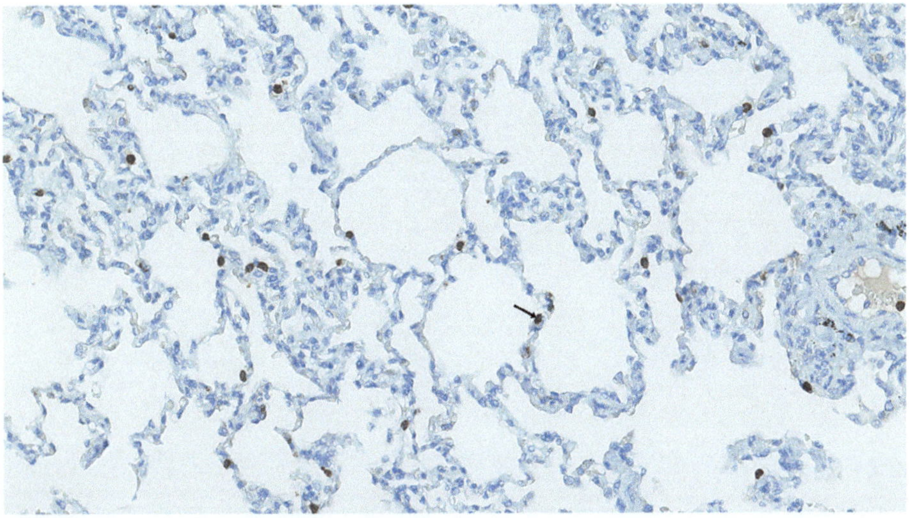

FIGURE 2.27A

Immunohistochemical staining for myeloperoxidase (MPO) performed on the lung of a naive control male rhesus macaque. Note the sparse population of MPO^+ large mononuclear cells, consistent with macrophages, located primarily within alveolar walls rather than alveolar lumina. Stain performed with 3, 3′-diaminobenzidine chromogen and hematoxylin counterstain, 29 × objective magnification.

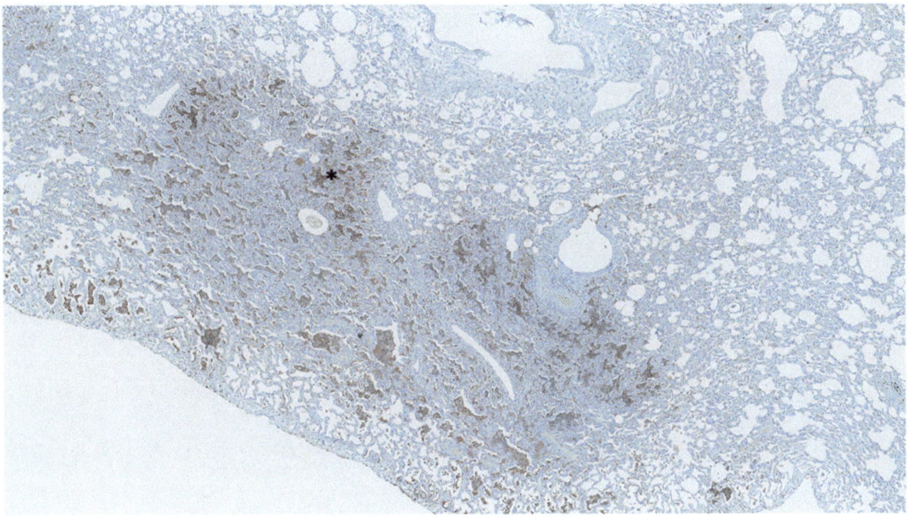

FIGURE 2.27B

Immunohistochemical stain for myeloperoxidase (MPO) on lung of a male rhesus macaque necropsied 180 days following irradiation at 10 Gy. A zone of MPO^+ macrophages (outlined) is associated with an area of interstitial fibrosis. Stain performed with 3,3′-diaminobenzidine chromogen and hematoxylin counterstain, 2.5 × objective magnification.

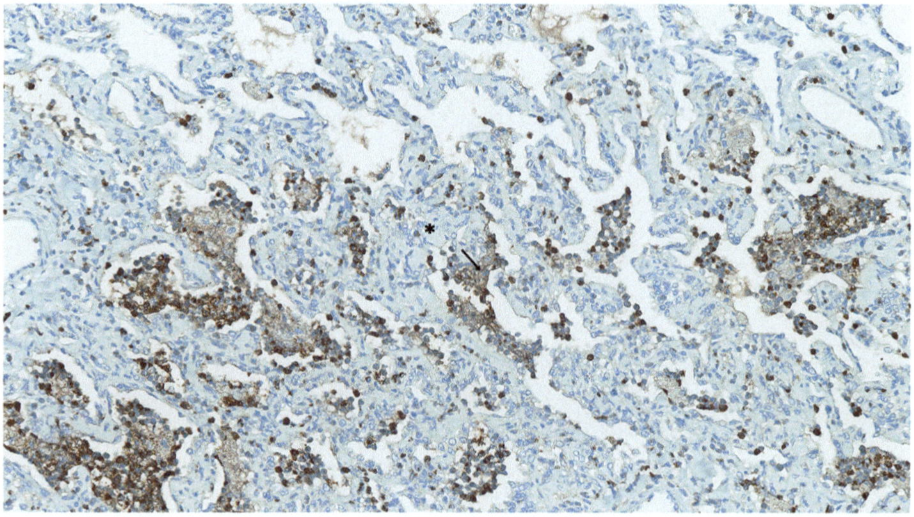

FIGURE 2.27C

Higher magnification of the specimen shown in the previous figure demonstrates entrapment of MPO^+ macrophages (arrow) in the lumen of alveoli that have thickened, fibrotic alveolar walls (*). Presence of MPO^+ macrophages with attendant inflammatory cytokine release probably contributes to the maintenance and progression of the radiation-associated pulmonary lesion complex, but absence of a significant macrophage population in areas of substantial interstitial fibrosis indicates the fibrotic process was not entirely dependent on macrophage mediation. Immunohistochemical stain for myeloperoxidase with 3,3′-diaminobenzidine chromogen and hematoxylin counterstain, 20 × objective magnification.

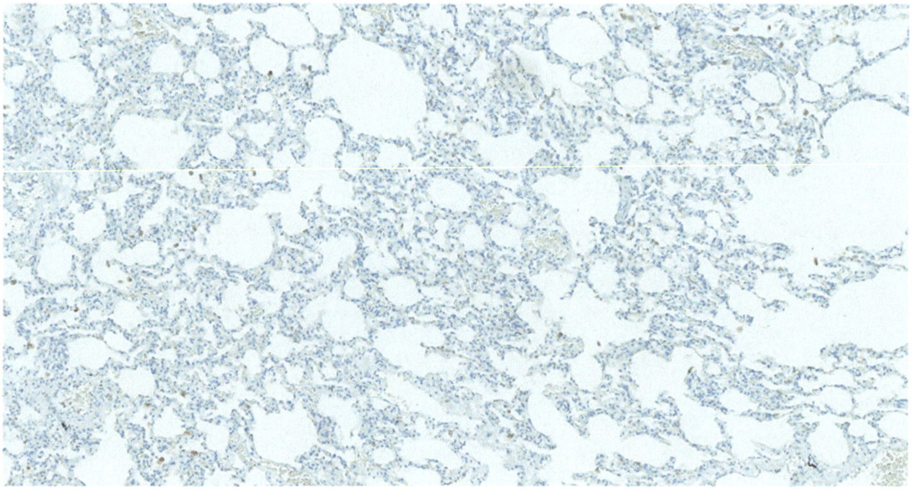

FIGURE 2.28A

Immunohistochemical stain for CD163 on the lung of a naïve control male rhesus macaque. Note the sparse population of positively stained cells. Stain performed with 3,3′-diaminobenzidine chromogen and hematoxylin counterstain, 10 × objective magnification.

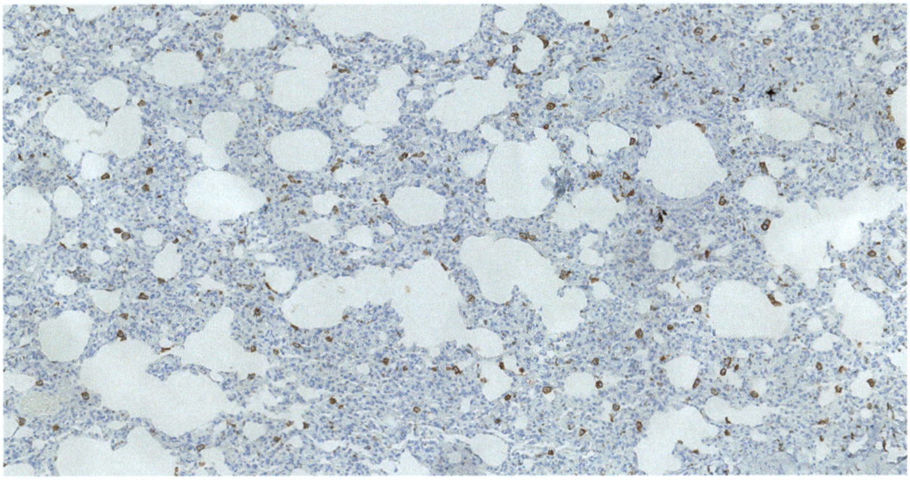

FIGURE 2.28B

Immunohistochemical stain for CD163 on the lung of a male rhesus macaque necropsied 24 days after irradiation at 11 Gy. Note the population of brown stained CD163$^+$ cells within alveolar walls, with no evidence of the intraluminal aggregates of CD163$^+$ cells seen in the lungs of animals necropsied after a prolonged post-irradiation interval (shown below). Stain performed with 3,3′-diaminobenzidine chromogen and hematoxylin counterstain, 10 × objective magnification.

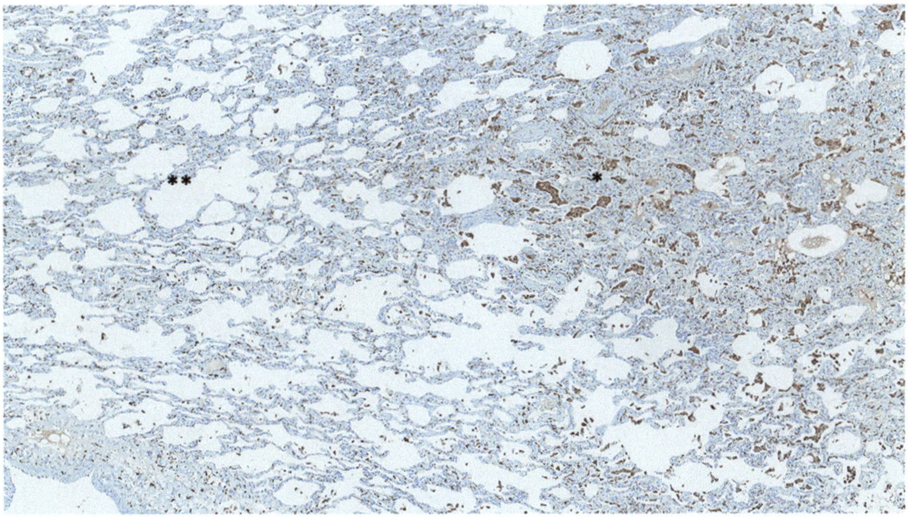

FIGURE 2.28C

Immunohistochemical stain for CD163 on the lung of a male rhesus macaque necropsied 177 days following irradiation at 11 Gy. Note the population of brown stained CD163$^+$ cells in the fibrotic region (*) as compared to the non-fibrotic region (**). While it is possible the presence of fibrosis hinders the normal process of alveolar macrophage exiting up the mucociliary escalator, it seems likely that presence of an increased population of activated macrophages could contribute to the general inflammatory milieu in areas of pulmonary fibrosis, thus serving to perpetuate the pathological process. Stain performed with 3,3′-diaminobenzidine chromogen and hematoxylin counterstain, 5 × objective magnification.

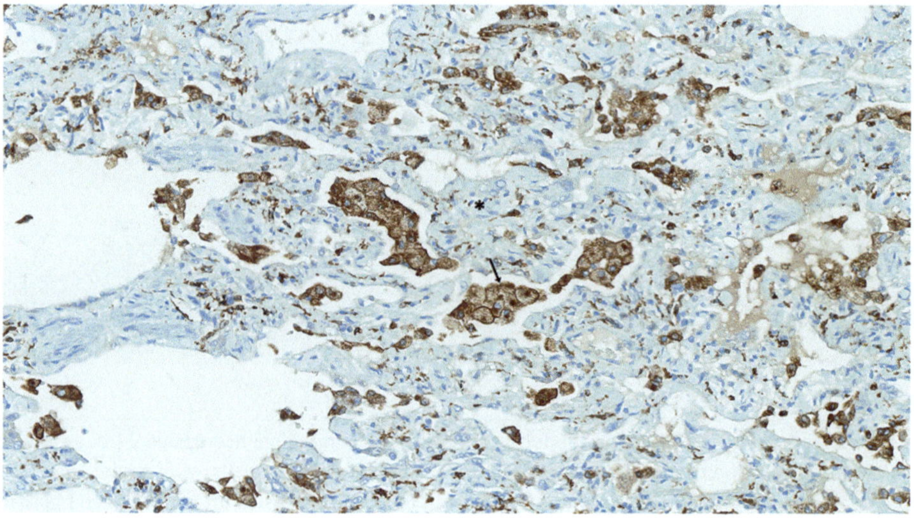

FIGURE 2.28D

Higher magnification of the previous specimen. Note the $CD163^+$ alveolar macrophages (arrow) within lumina of alveoli that have thickened walls (*). Stain performed with 3,3′-diaminobenzidine chromogen and hematoxylin counterstain, 20 × objective magnification.

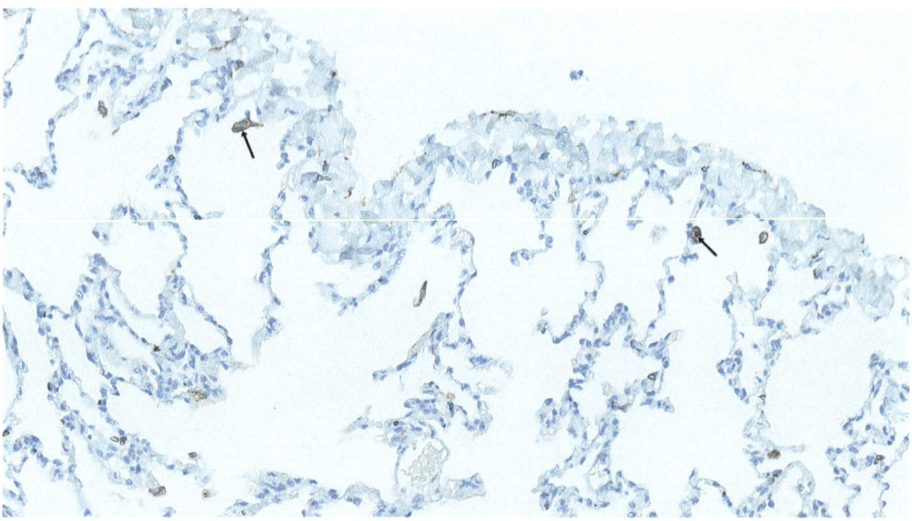

FIGURE 2.29A

Immunohistochemical stain for CD206 on the lung of a naïve control male rhesus macaque. Note the sparse population of positively stained cells (arrows). Stain performed with 3,3′-diaminobenzidine chromogen and hematoxylin counterstain, 25.4 × objective magnification.

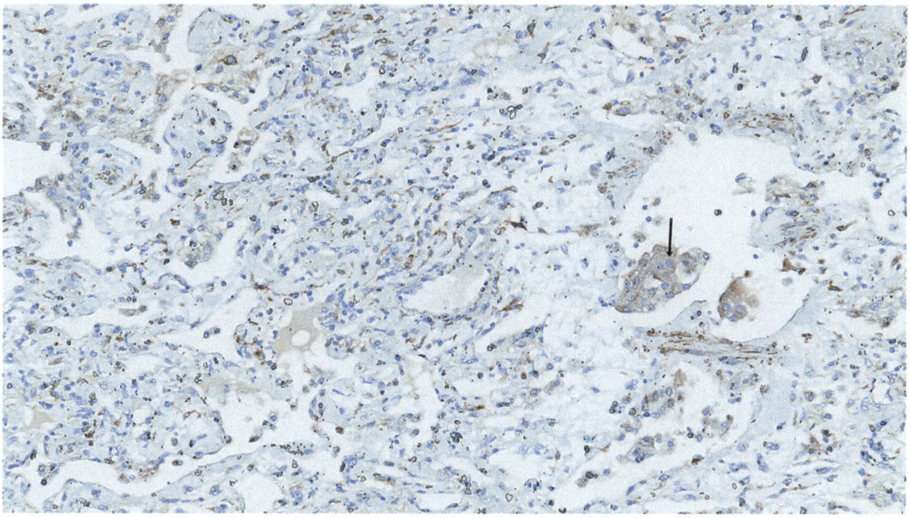

FIGURE 2.29B

Immunohistochemical stain for CD206 on lung of the same male rhesus macaque used for CD163 staining, necropsied 177 days following irradiation at 11 Gy. Note the $CD206^{+}$ alveolar macrophages (arrow) within lumina of airways. The difference in intensity of brown staining reflects the binding characteristics of the antibodies and technical procedures used in IHC staining and does not necessarily indicate a difference in the level of expression of CD206 versus CD163. However, the relative numerical predominance of $CD206^{+}$ versus $CD163^{+}$ cells is not influenced by technical considerations. Stain performed with 3,3′-diaminobenzidine chromogen and hematoxylin counterstain, 20 × objective magnification.

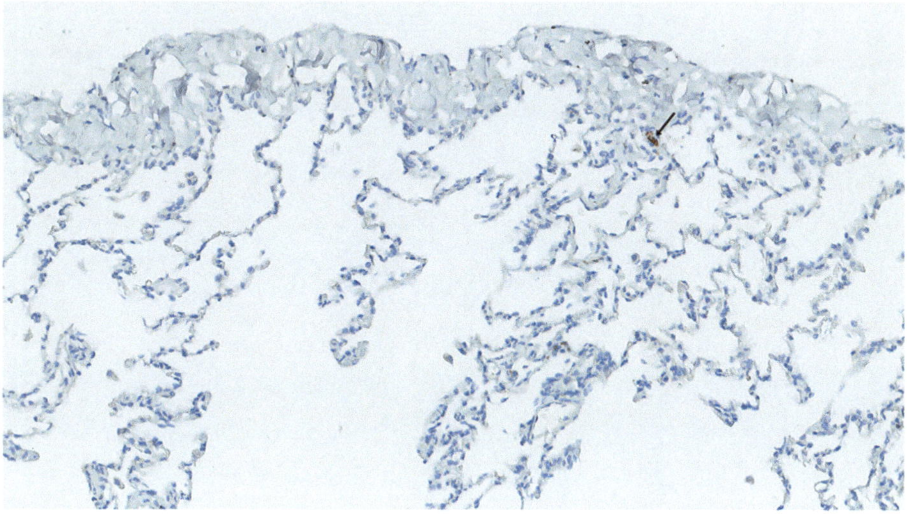

FIGURE 2.30A

Immunohistochemical staining for tryptase in mast cell granules performed on the lung of a naive control male rhesus macaque. Note the minimal staining of mast cells (arrow). Stain performed with 3, 3′-diaminobenzidine chromogen and hematoxylin counterstain, 20 × objective magnification.

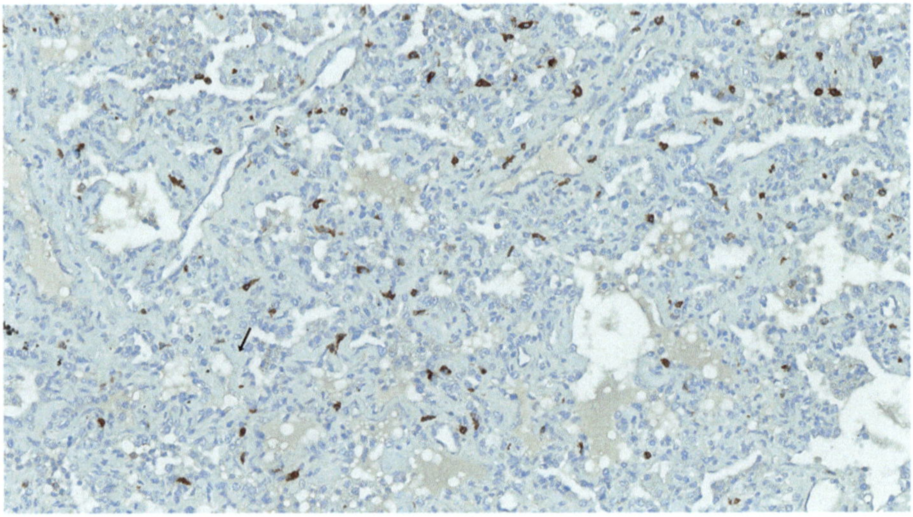

FIGURE 2.30B

Immunohistochemical stain for tryptase on fibrotic lung region of a male rhesus macaque necropsied 182 days after irradiation at 10 Gy. Note the diffuse population of brown-stained mast cells in this fibrotic region of lung that has thickened alveolar walls (arrow). While it is possible the mast cell population has contributed to the fibrosis, it is equally possible the increased mast cell population in this region of the lung represents the known tendency of mast cells to congregate in tissues that contain collagenous tissue. Stain performed with 3,3′-diaminobenzidine chromogen and hematoxylin counterstain, 20 × objective magnification.

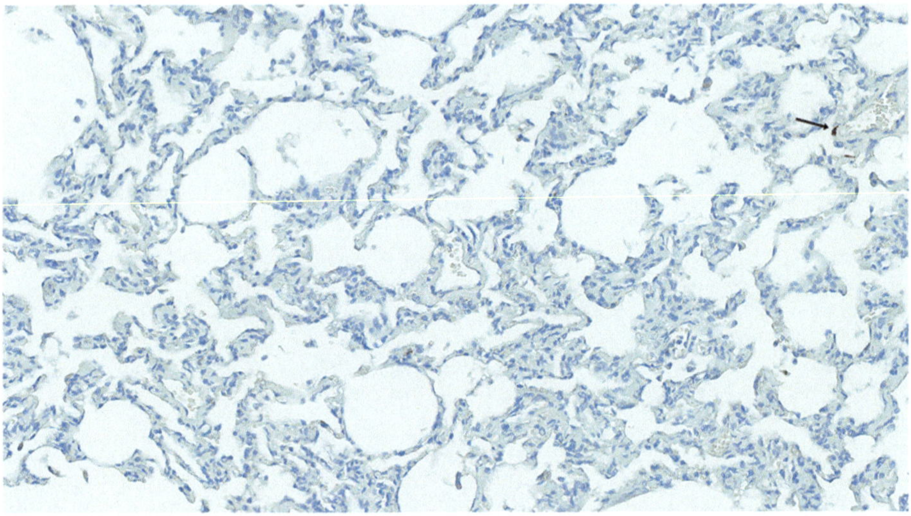

FIGURE 2.30C

Immunohistochemical stain for tryptase on non-fibrotic region of lung from the same male rhesus macaque as presented in the previous figure. Note the sparse population of brown-stained mast cells (arrow) in this region of lung that has histologically normal alveolar walls. Stain performed with 3,3′-diaminobenzidine chromogen and hematoxylin counterstain, 20 × objective magnification.

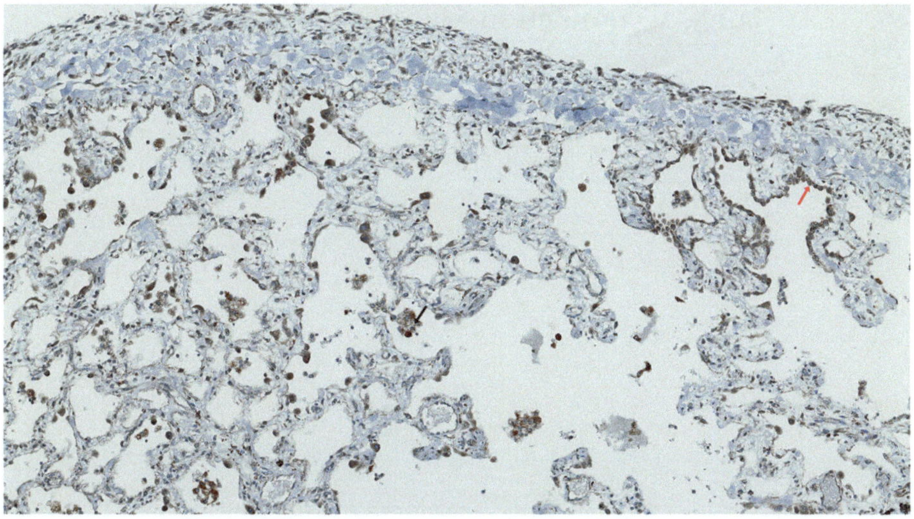

FIGURE 2.31

Immunohistochemical stain for caspase 3 performed on the lung from a male rhesus macaque necropsied 44 days after irradiation at 10 Gy. Note the caspase 3^+ alveolar macrophages (black arrow) and proliferating type II pneumocytes (red arrow). Stain performed with 3,3′-diaminobenzidine chromogen and hematoxylin counterstain, 10× objective magnification.

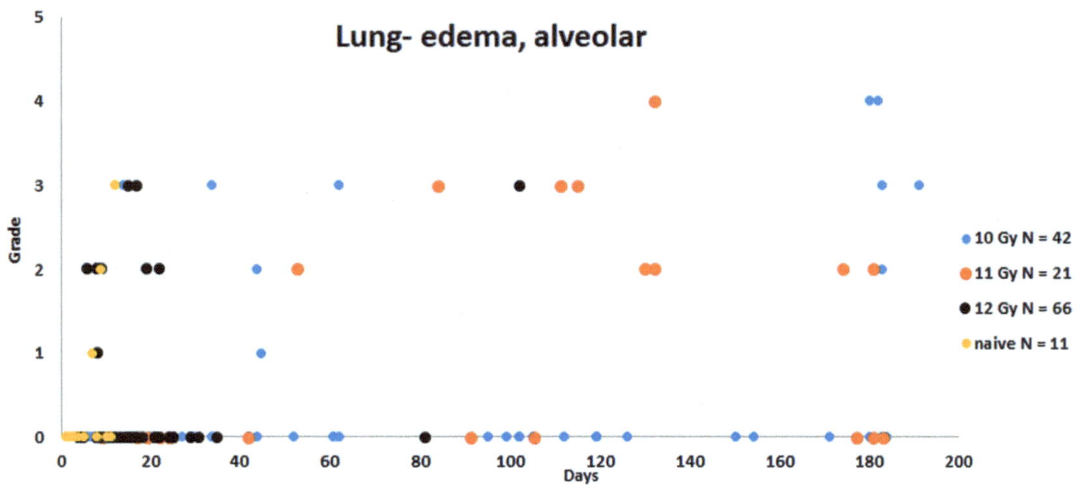

FIGURE 2.32

Scatter plot: Lung—edema, alveolar. Observations based on light microscopic examination of paraffin-embedded specimens stained with hematoxylin and eosin.

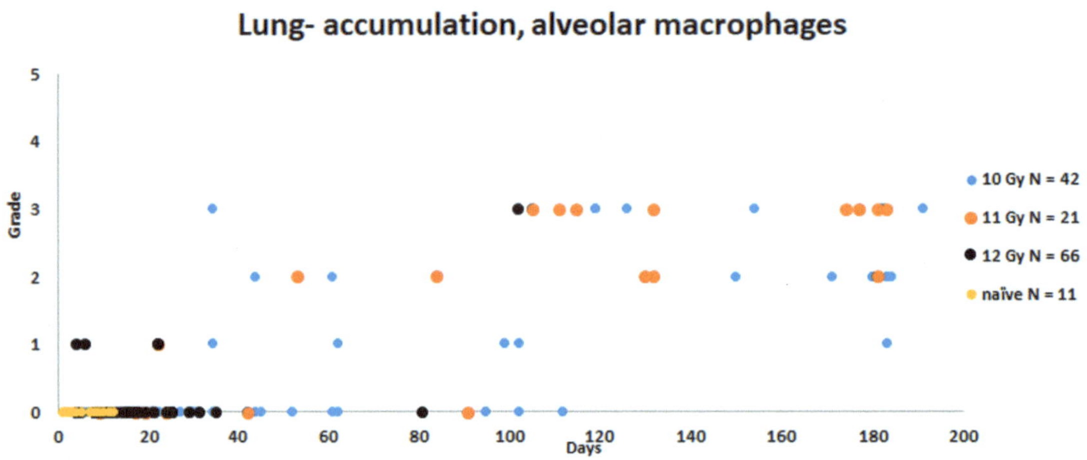

FIGURE 2.33

Scatter plot: Lung—accumulation, alveolar macrophages. Observations based on light microscopic examination of paraffin-embedded specimens stained with hematoxylin and eosin.

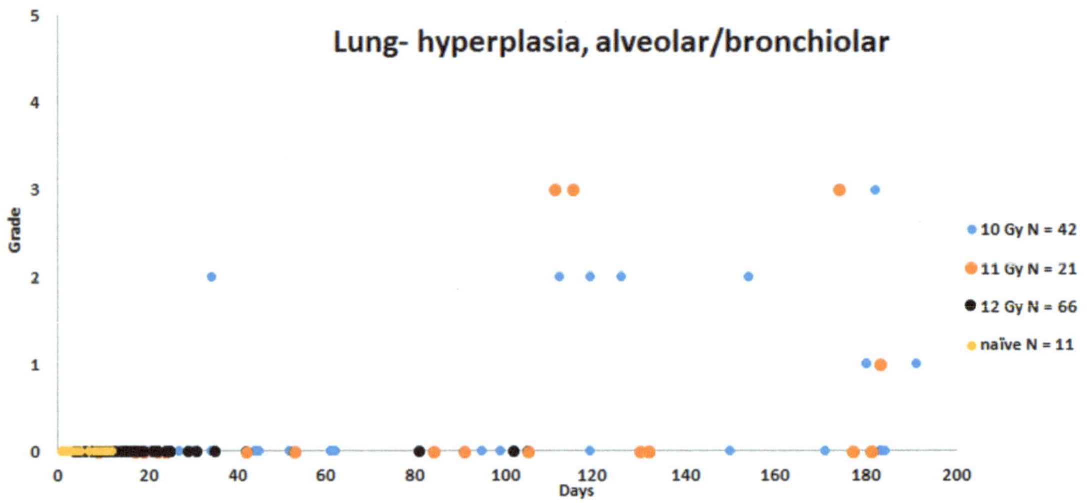

FIGURE 2.34

Scatter plot: Lung—hyperplasia, alveolar/bronchiolar. Observations based on light microscopic examination of paraffin-embedded specimens stained with hematoxylin and eosin.

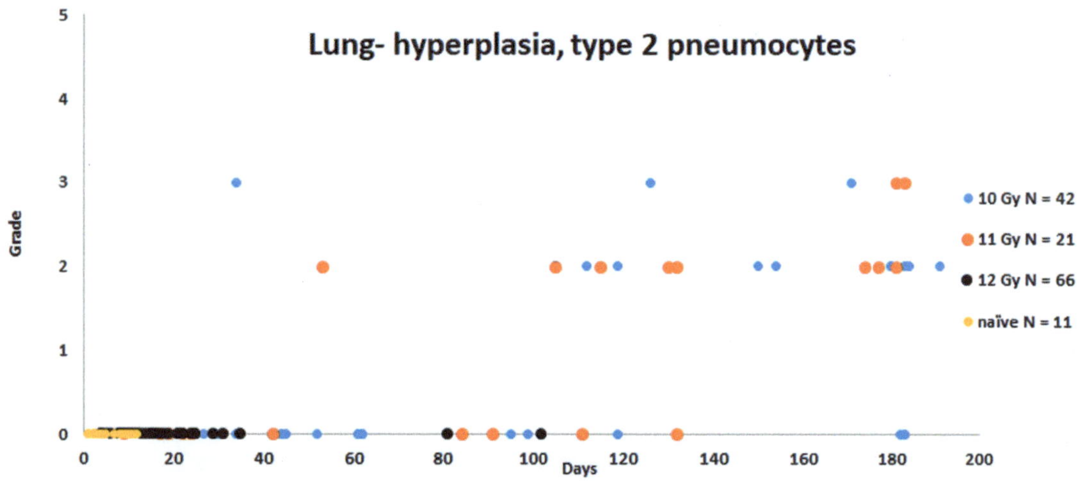

FIGURE 2.35

Scatter plot: Lung—hyperplasia, type 2 pneumocytes. Observations based on light microscopic examination of paraffin-embedded specimens stained with hematoxylin and eosin.

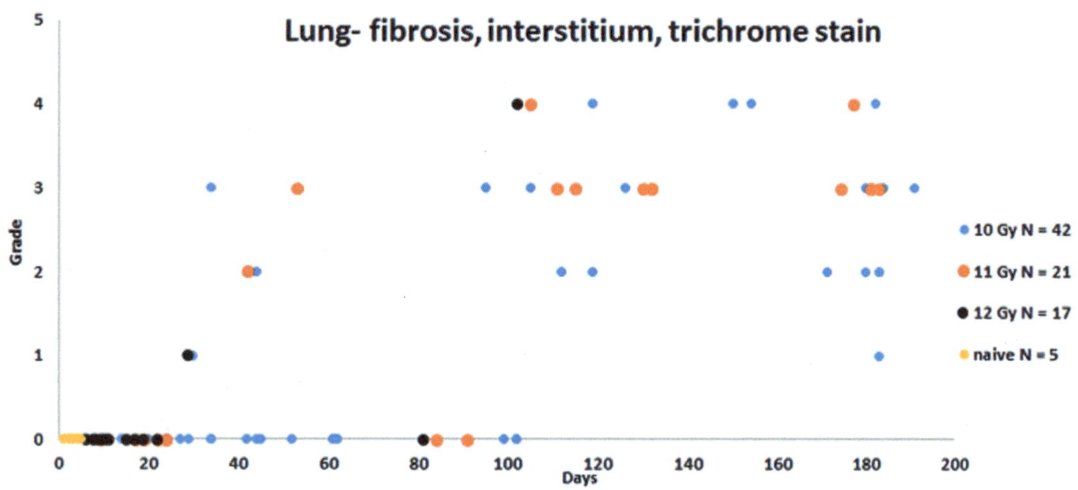

FIGURE 2.36

Scatter plot: Lung—fibrosis, interstitium. Observations based on light microscopic examination of paraffin-embedded specimens stained with Masson's trichrome stain.

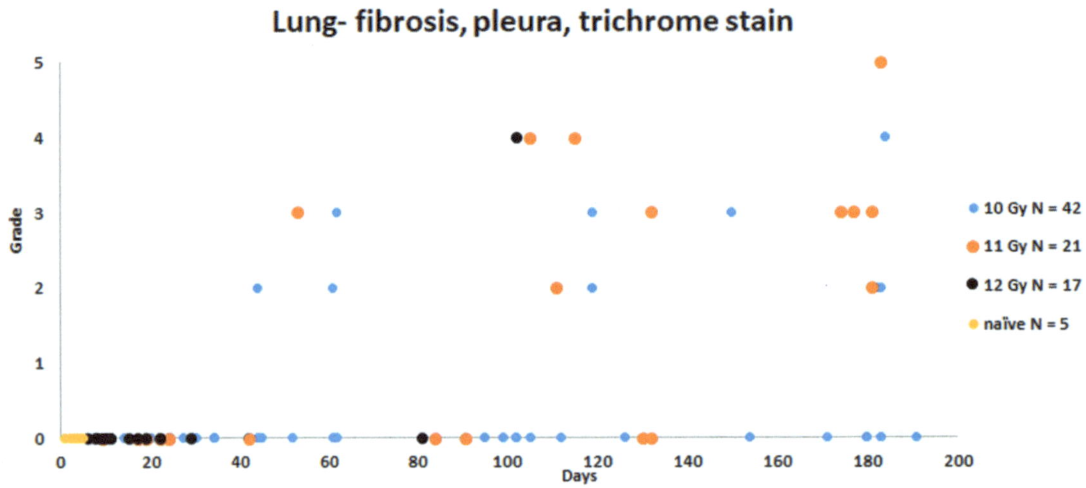

FIGURE 2.37

Scatter plot: Lung—fibrosis, pleura. Observations based on light microscopic examination of paraffin-embedded specimens stained with Masson's trichrome stain.

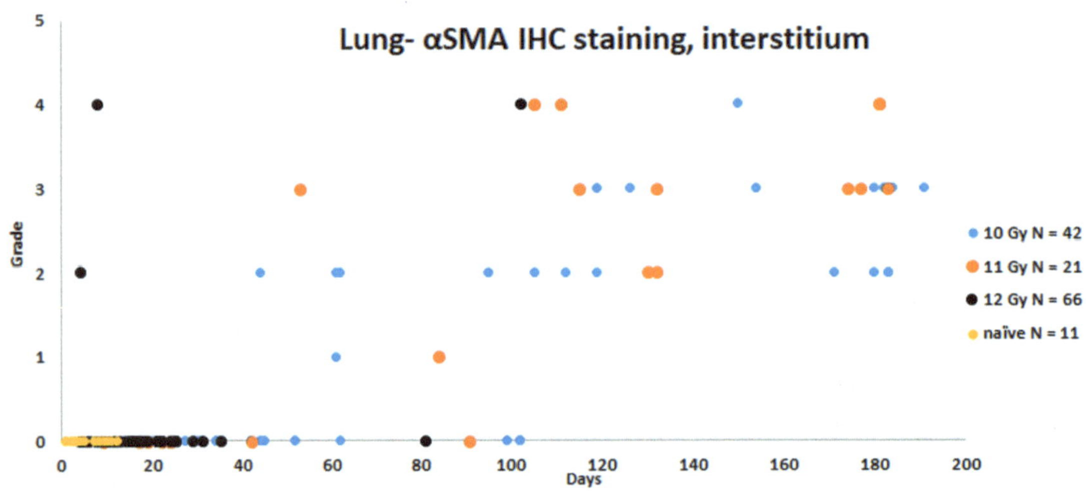

FIGURE 2.38

Scatter plot: Lung—alpha smooth muscle actin (αSMA) staining, interstitium. Observations based on light microscopic examination of paraffin-embedded specimens stained by immunohistochemical procedures for αSMA with 3,3′-diaminobenzidine chromogen and hematoxylin counterstain.

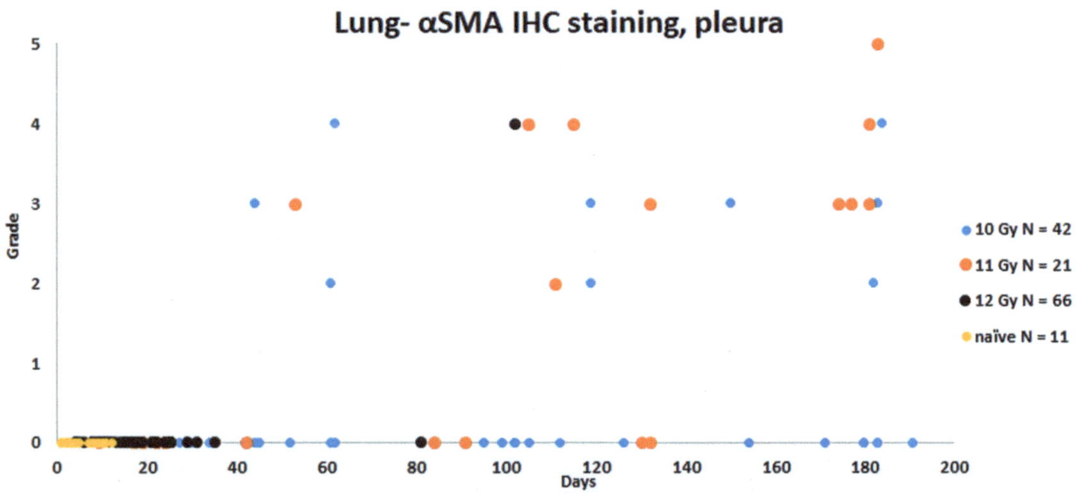

FIGURE 2.39

Scatter plot: Lung—alpha smooth muscle actin (αSMA) staining, pleura. Observations based on light microscopic examination of paraffin-embedded specimens stained by immunohistochemical procedures for αSMA with 3,3′-diaminobenzidine chromogen and hematoxylin counterstain.

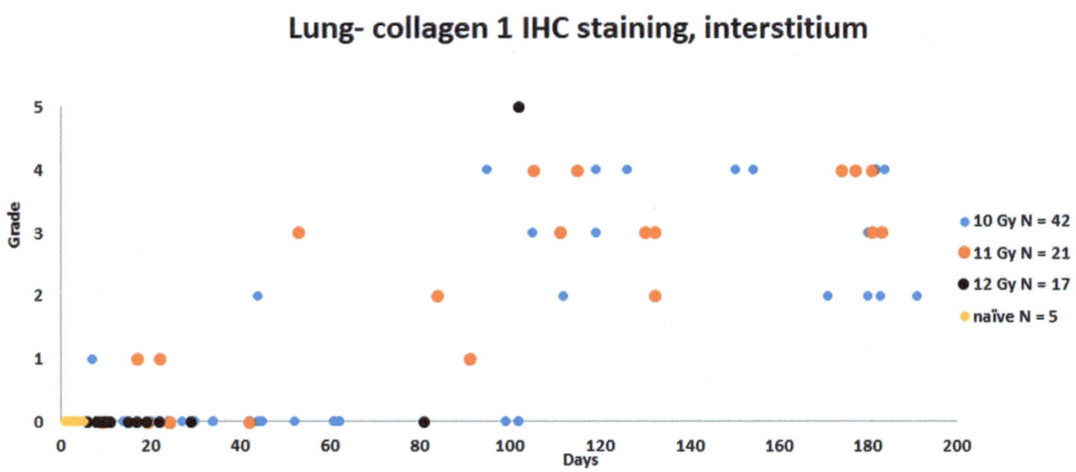

FIGURE 2.40

Scatter plot: Lung—collagen 1 staining, interstitium. Observations based on light microscopic examination of paraffin-embedded specimens stained by immunohistochemical procedures for collagen 1 with 3,3′-diaminobenzidine chromogen and hematoxylin counterstain.

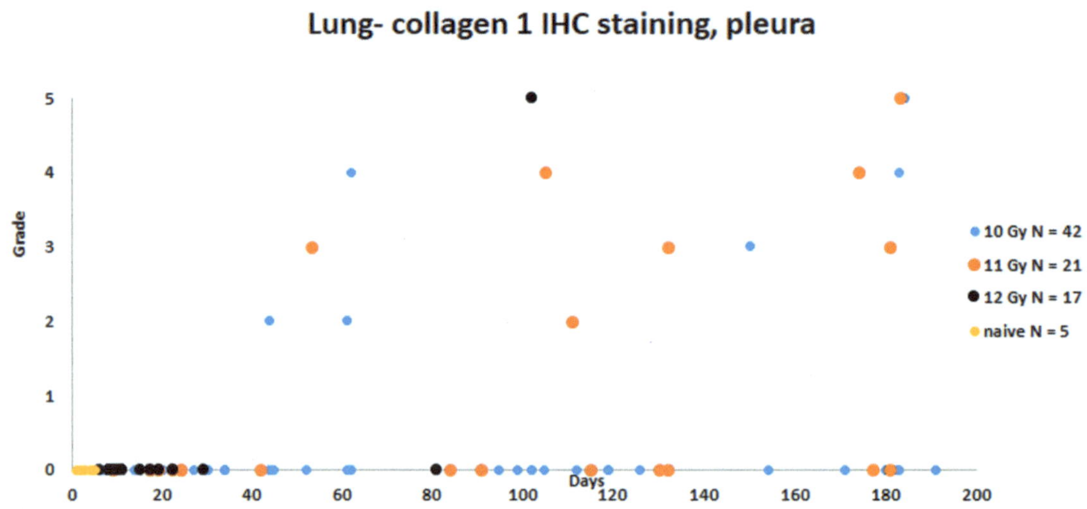

FIGURE 2.41

Scatter plot: Lung—collagen 1 staining, pleura. Observations based on light microscopic examination of paraffin-embedded specimens stained by immunohistochemical procedures for collagen 1 with 3,3′-diaminobenzidine chromogen and hematoxylin counterstain.

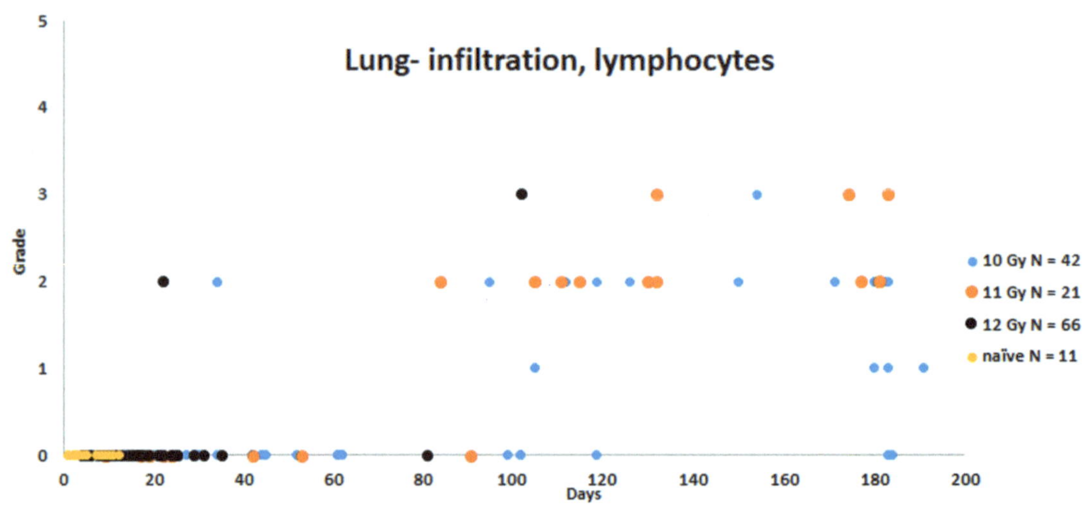

FIGURE 2.42

Scatter plot: Lung—infiltration, lymphocytes. Observations based on light microscopic examination of paraffin-embedded specimens stained with hematoxylin and eosin.

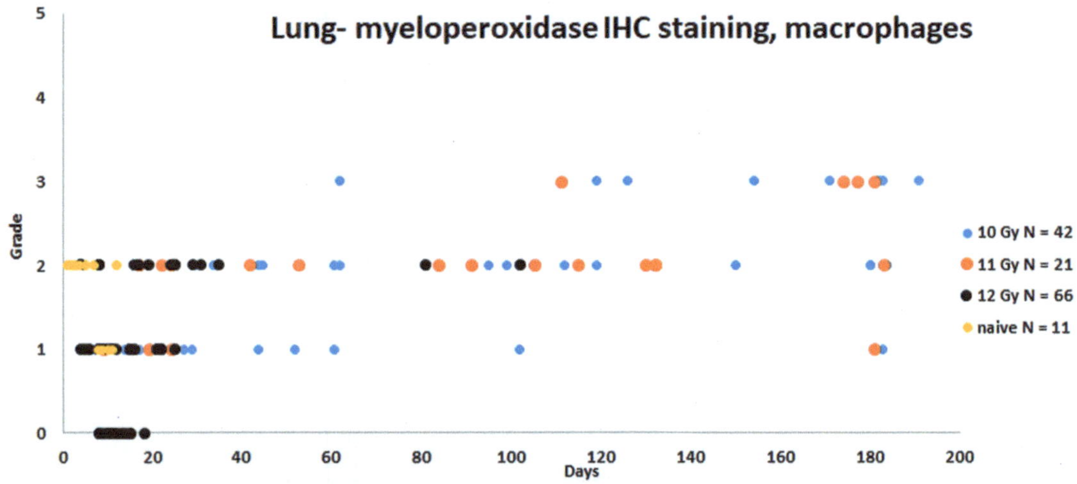

FIGURE 2.43

Scatter plot: Lung—myeloperoxidase staining, macrophages. Observations based on light microscopic examination of paraffin-embedded specimens stained by immunohistochemical procedures for myeloperoxidase with 3,3′-diaminobenzidine chromogen and hematoxylin counterstain.

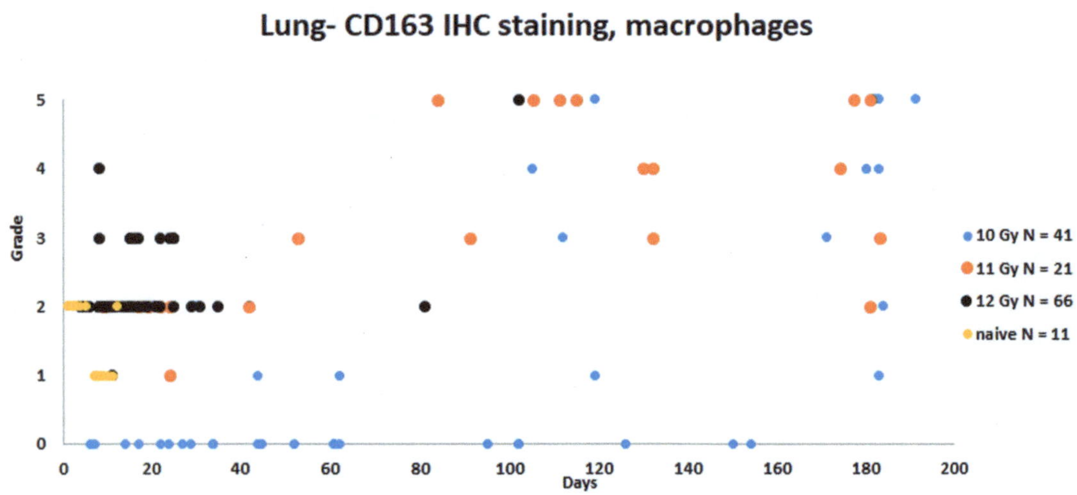

FIGURE 2.44

Scatter plot: Lung—CD163 staining, macrophages. Observations based on light microscopic examination of paraffin-embedded specimens stained by immunohistochemical procedures for CD163 with 3,3′-diaminobenzidine chromogen and hematoxylin counterstain.

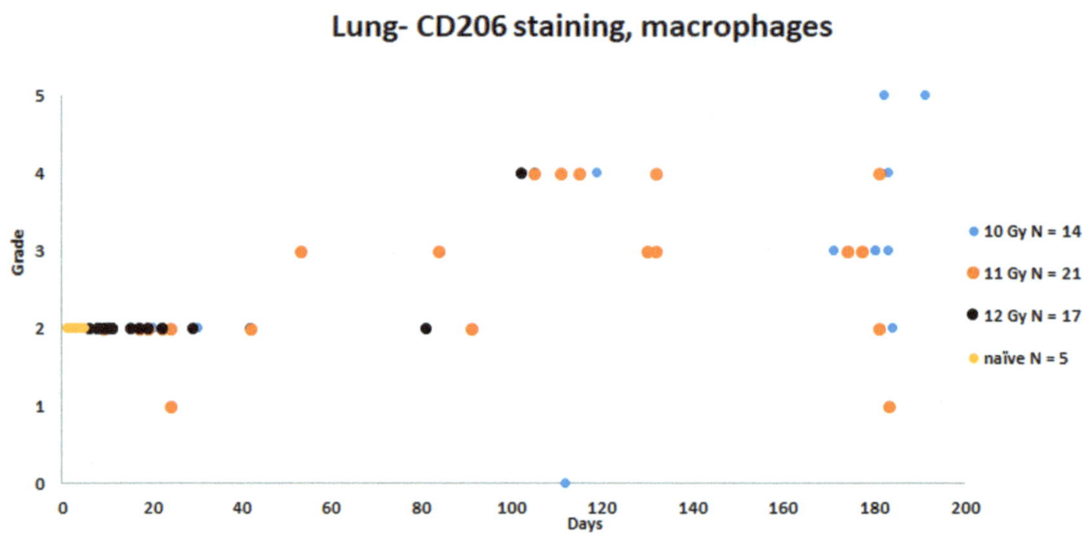

FIGURE 2.45

Scatter plot: Lung—CD206 staining, macrophages. Observations based on light microscopic examination of paraffin-embedded specimens stained by immunohistochemical procedures for CD206 with 3,3′-diaminobenzidine chromogen and hematoxylin counterstain.

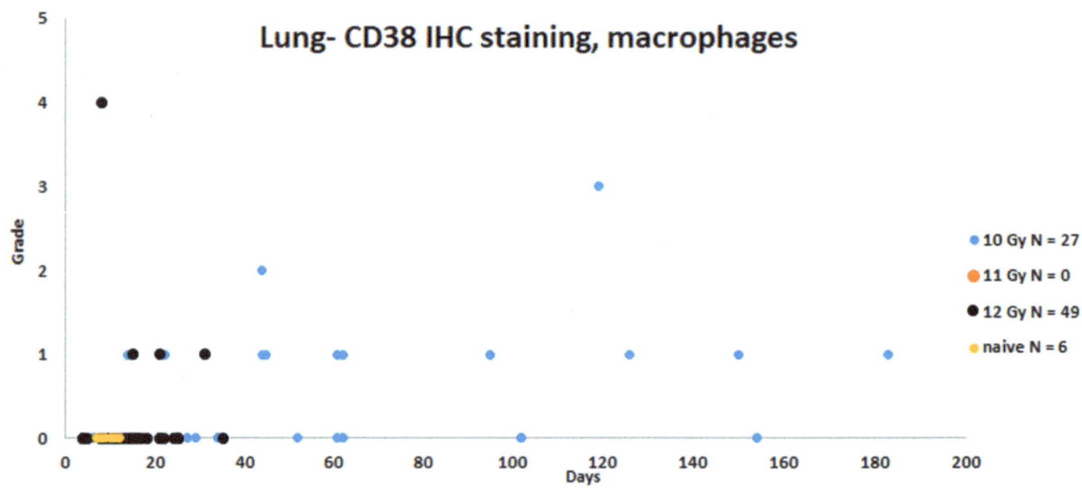

FIGURE 2.46

Scatter plot: Lung—CD38 staining, macrophages. Observations based on light microscopic examination of paraffin-embedded specimens stained by immunohistochemical procedures for CD38 with 3,3′-diaminobenzidine chromogen and hematoxylin counterstain.

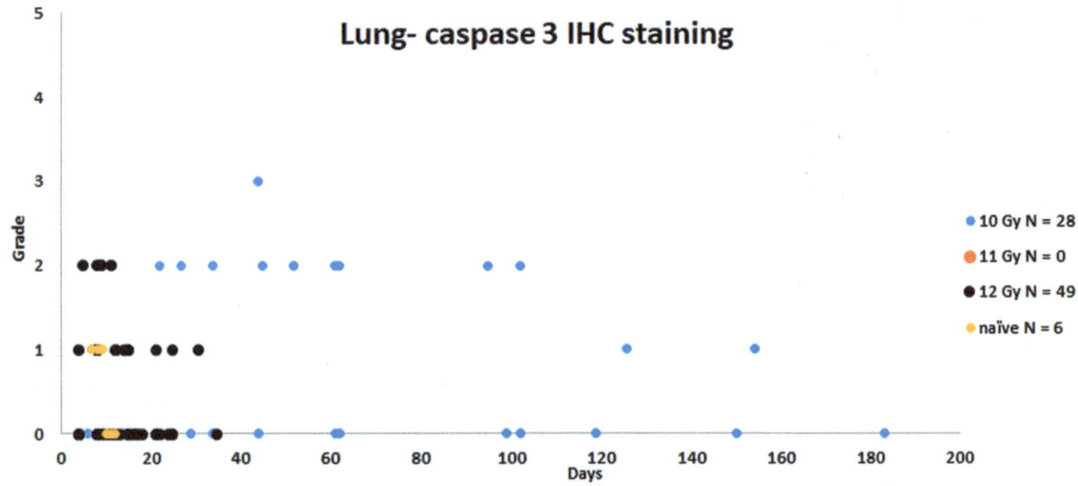

FIGURE 2.47

Scatter plot: Lung—caspase 3, macrophages. Observations based on light microscopic examination of paraffin-embedded specimens stained by immunohistochemical procedures for caspase 3 with 3,3′-diaminobenzidine chromogen and hematoxylin counterstain.

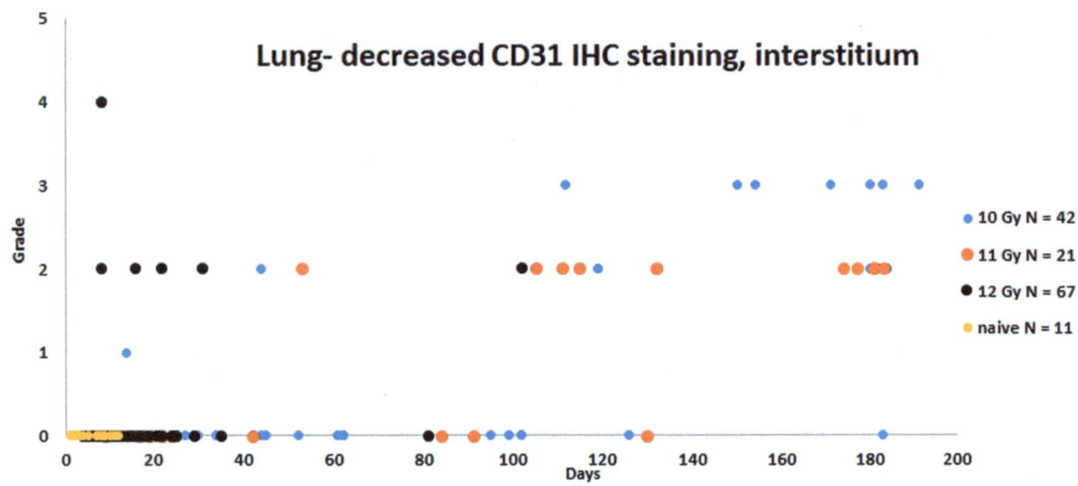

FIGURE 2.48

Scatter plot: Lung—decreased CD31 staining. Observations based on light microscopic examination of paraffin-embedded specimens stained by immunohistochemical procedures for CD31 with 3,3′-diaminobenzidine chromogen and hematoxylin counterstain.

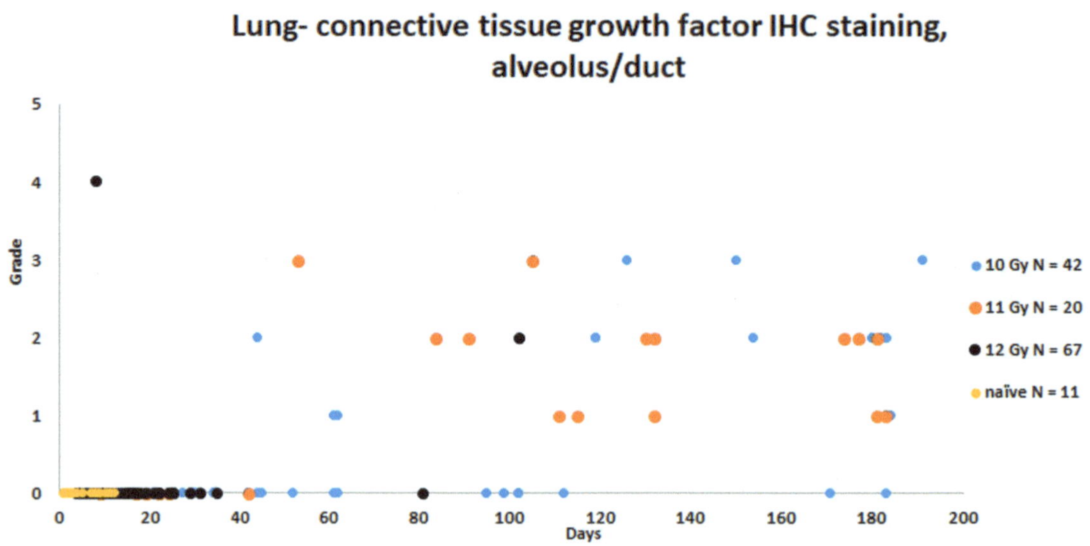

FIGURE 2.49

Scatter plot: Lung—connective tissue growth factor (CTGF) staining, alveolar/duct. Observations based on light microscopic examination of paraffin-embedded specimens stained by immunohistochemical procedures for CTGF with 3,3′-diaminobenzidine chromogen and hematoxylin counterstain.

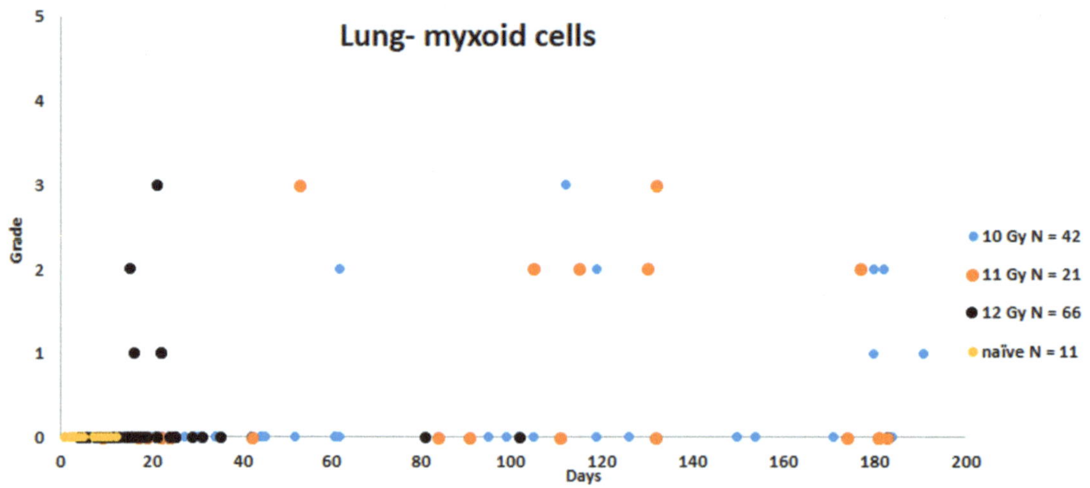

FIGURE 2.50

Scatter plot: Lung—myxoid cells. Observations based on light microscopic examination of paraffin-embedded specimens stained with hematoxylin and eosin.

CHAPTER

Radiation-associated changes in the heart of rhesus macaques

3

Overview

Myocardial fibrosis, which was more readily discernible in the trichrome-stained sections than in the H&E-stained sections, consisted of an increase in the amount of endomysial fibrous connective tissue in one or more distinct areas of the myocardium. The myocardial fibrosis was not histologically typical of spontaneous disease or degeneration and thus was considered to be associated with irradiation. Inflammatory cell infiltration, when present, typically consisted of minimal to mild mononuclear cell infiltration. Low-level mononuclear cell infiltration is commonly encountered in the heart of nonhuman primates; thus it is not possible to conclude that the mononuclear cell infiltration was definitively linked to the myocardial fibrosis.

Immunohistochemical staining for collagen 1, alpha smooth muscle actin (α-SMA) and connective tissue growth factor (CTGF) revealed increased staining for these target molecules in areas of myocardial fibrosis. Immunohistochemical staining for galectin-3 was most pronounced in defined areas of myocardial degeneration.

Animals irradiated at 12 Gy had an unusual caspase 3 immunohistochemical staining pattern which consisted of caspase 3^+ particles located within myofiber cytoplasm, with accentuated accumulation at the nuclear poles of myofibers.

One animal irradiated at 10 Gy and necropsied 29 days after irradiation had a large myocardial abscess with extensive bacterial colonization. That animal also had a large mural thrombus attached to the endocardium in the region of the right atrioventricular valve. Another animal irradiated at 10 Gy and necropsied 87 days after irradiation had a focally severe area of myocardial degeneration, hemorrhage, and mononuclear inflammatory cell infiltration located in an atrial wall. These unusual cardiac lesions were considered to be associated with irradiation, though the exact pathogenesis was not apparent from microscopic examination.

Specimen preparation varied between animals but typically included only one or two specimens from unidentified regions of the heart. The true incidence of radiation-associated myocardial injury may not have been revealed by the suboptimal heart sampling procedures. It was not possible to determine whether the myocardial changes were preferentially located in a specific region of the heart.

Stains and related figures

Routine histology stain

Hematoxylin & eosin (H&E)—Figs. 3.1, 3.2A, 3.2B, 3.3A, 3.3B, 3.3C, 3.4A, 3.4B, 3.4C, 3.6A, 3.6B

Histopathology Atlas of Acute Radiation Syndrome and Delayed Effects in Rhesus Macaques. DOI: https://doi.org/10.1016/B978-0-323-91393-5.00003-7

Histochemical stains

Masson's trichrome—Figs. 3.5A, 3.5B, 3.7A, 3.7Bstains

Immunohistochemical stains

Caspase 3—Figs. 3.10A, 3.10B, 3.10C, 3.10D, 3.10E
Galectin-3—Figs. 3.8A, 3.8B
α-SMA—Figs. 3.9A, 3.9B, 3.9C, 3.9D

Graphical representations (scatter plots) of individual histologic observations in the heart are shown in the following figures:

Fig. 3.11. Heart—degeneration, myocardium.
Fig. 3.12. Heart—fibrosis, myocardium
Fig. 3.13. Heart—alpha smooth muscle actin staining
Fig. 3.14. Heart—mast cells, toluidine blue stain
Fig. 3.15. Heart—mast cells, tryptase immunohistochemical staining
Fig. 3.16. Heart—caspase 3 staining

Detailed microscopic observations

Toluidine blue stain

The toluidine blue stain revealed a minimal population of mast cells in the myocardium. The mast cell population had no discernible relationship to the areas of myocardial fibrosis, and there was no discernible alteration in mast cell populations as related to the radiation dose.

Tryptase immunohistochemical stain

The tryptase immunohistochemical stain revealed a low-level population of mast cells in the myocardium that was slightly increased above that seen with toluidine blue staining, thus the tryptase IHC staining was considered to be a more sensitive indicator of mast cell populations. The mast cell population had no discernible relationship to the areas of myocardial fibrosis, and there was no discernible alteration in mast cell populations as related to the radiation dose.

Alpha smooth muscle actin immunohistochemical stain

Minimal to mild α-SMA staining was typically, but not invariably, associated with areas of myocardial fibrosis. Areas of myocardial degeneration had a reduction in the α-SMA that is normally present in cardiac myocytes and an increase in endomysial staining for α-SMA.

Galectin-3 immunohistochemical staining

Galectin-3 expression was most pronounced in the areas of myocardial degeneration.

Caspase 3 immunohistochemical staining

Immunohistochemical staining for caspase 3 revealed unusual staining patterns in the myocardium of animals irradiated at 12 Gy. The caspase 3 staining consisted of multimorphous intracytoplasmic caspase 3^+ structures or particles of uncertain identity, as opposed to the whole-cell caspase 3 staining that is typically seen with apoptotic cells. Prominent caspase 3^+ globules were commonly located near the nuclear poles of myofibers, which is the typical location of lipofuscin "aging pigment" in the myocardium of aged macaques. The overall significance of the myocardial observation in this study is uncertain but is intriguing relative to previously reported indications of increased rate of senescence in various tissues of irradiated animals.

Additional details of cardiac observations have been presented previously by Parker et al. (2019).

References

Parker GA, Li N, Takayama K, Farese AM, Macvittie TJ: Lung and heart injury in a nonhuman primate model of partial-body irradiation with minimal bone marrow sparing: histopathological evidence of lung and heart injury, *Health Phys* 116:383–400, 2019.

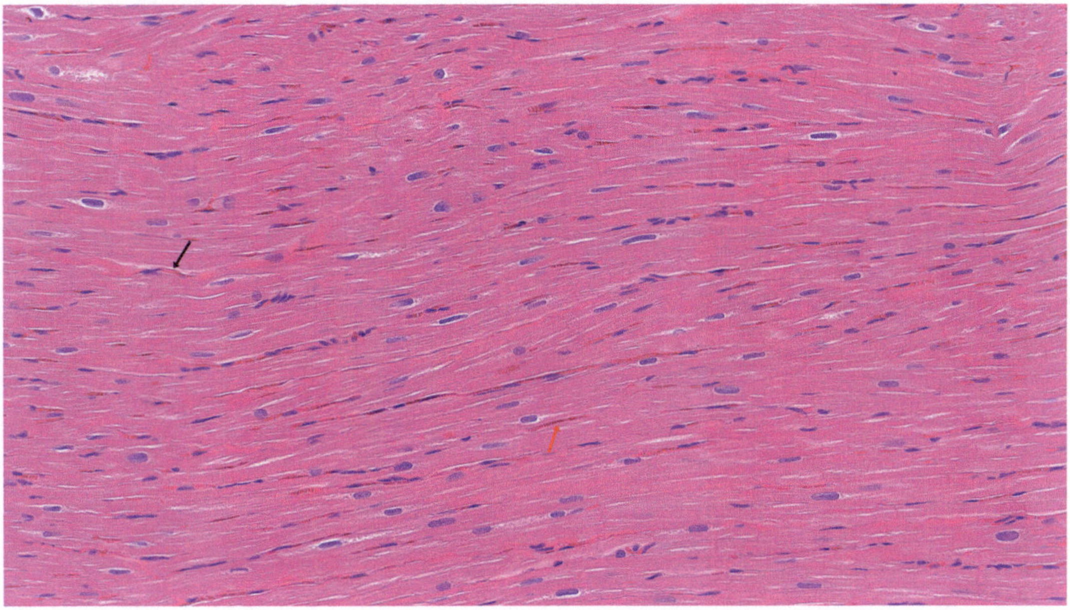

FIGURE 3.1

Myocardium of a naïve male rhesus macaque showing the orderly linear arrays of myocytes with interspersed blood-filled capillaries (*red arrow*) and minimal endomysial connective tissue (*black arrow*). Variation in nuclear size is a species characteristic of rhesus macaques. Hematoxylin and eosin stain, 20 × objective magnification.

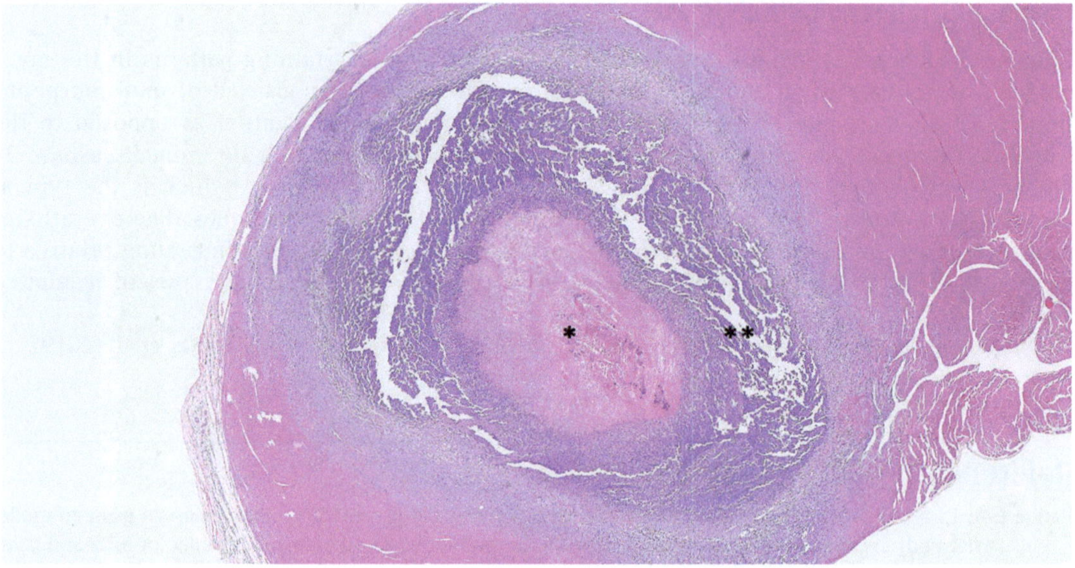

FIGURE 3.2A

Left ventricular free wall of a male rhesus macaque necropsied 29 days after irradiation at 10 Gy. Note the large abscess consisting of a central zone of necrotic debris (*) surrounded by a broad zone (**) of infiltrating inflammatory cells. Hematoxylin and eosin stain, 1.25 × objective magnification.

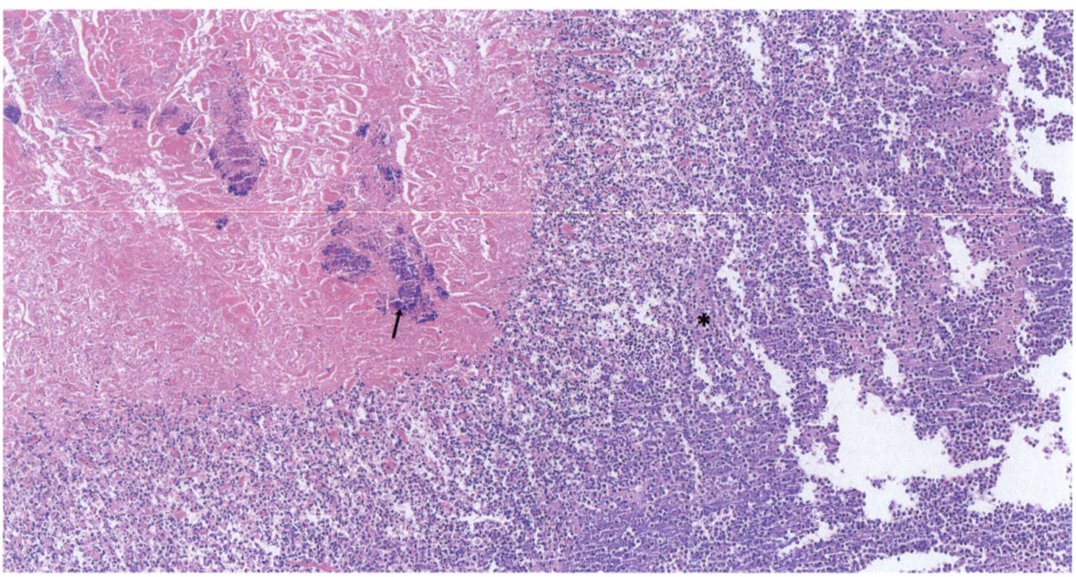

FIGURE 3.2B

Higher magnification of the specimen shows basophilic bacterial colonies (*arrow*) surrounded by a broad zone of intense neutrophilic infiltration (*). Hematoxylin and eosin stain, 10 × objective magnification.

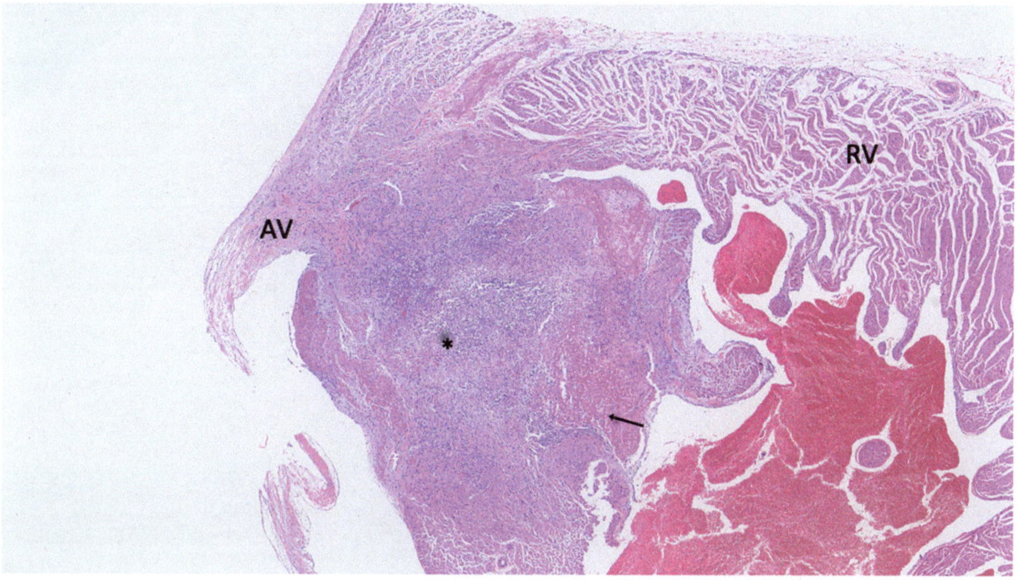

FIGURE 3.3A

The right ventricle of the same animal shown in the previous images has a large mural thrombus (*) consisting of inflammatory cell debris with a superficial layer of coagulated blood (*arrow*). *RV*, right ventricular free wall; *AV*, right atrioventricular valve. Hematoxylin and eosin stain, 2.5 × objective magnification.

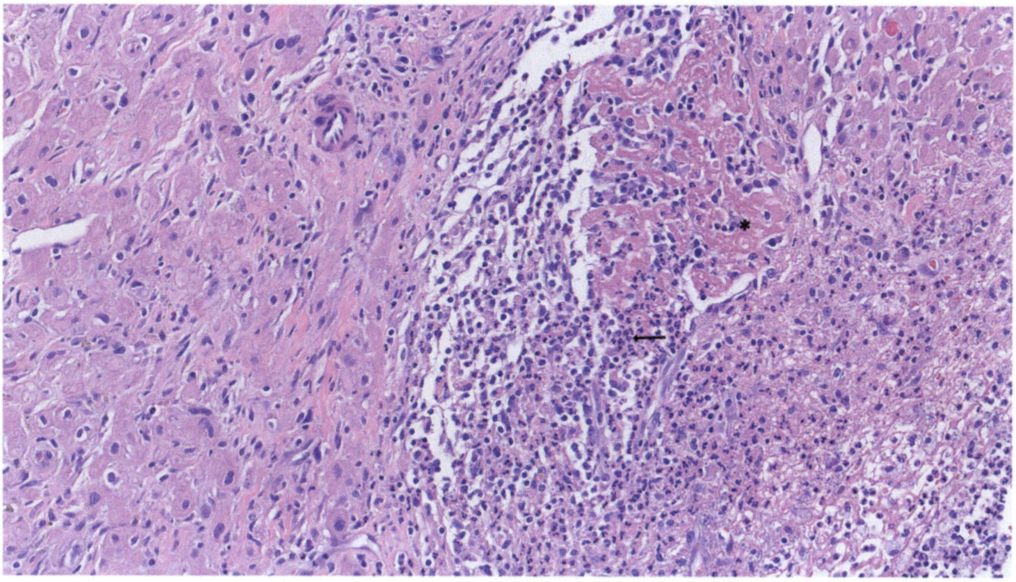

FIGURE 3.3B

Higher magnification of the previous lesion showing neutrophilic infiltration (*arrow*) and tissue debris (*) within the mural thrombus. Hematoxylin and eosin stain, 20 × objective magnification.

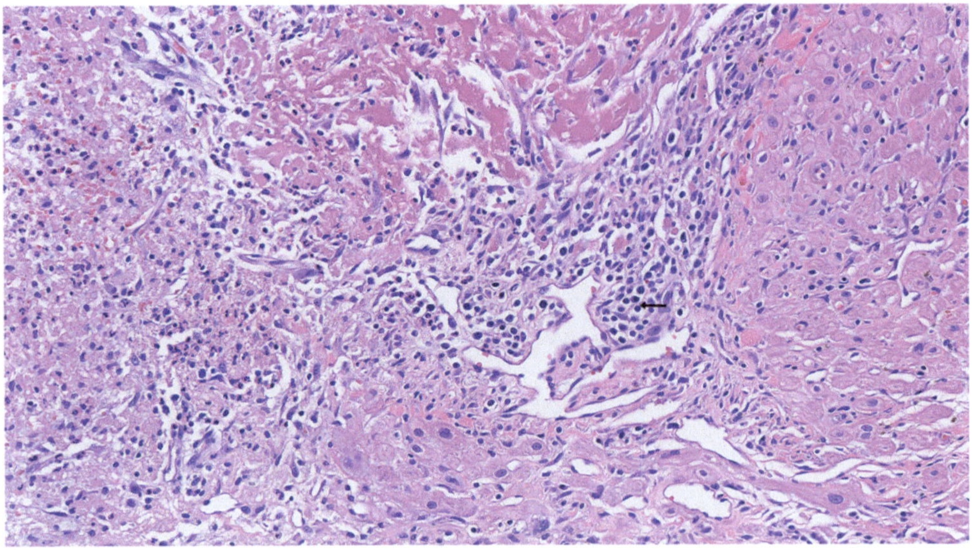

FIGURE 3.3C

An additional area with the mural thrombus has a mononuclear inflammatory cell infiltration (*arrow*) that consists primarily of lymphocytes and plasma cells. Hematoxylin and eosin stain, 20 × objective magnification.

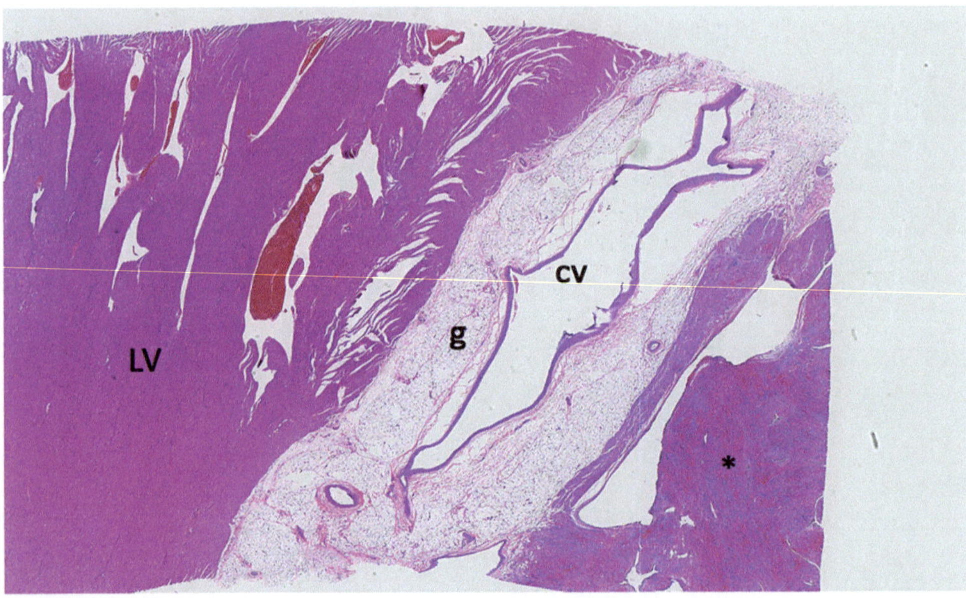

FIGURE 3.4A

Heart specimen from a male rhesus macaque necropsied 87 days after irradiation at 10 Gy has a large area (*) of myocardial disruption and hemorrhage located in an atrial wall. *LV*, left ventricular wall; *CV*, coronary vein contained within a fat-filled coronary groove (*g*). Hematoxylin and eosin stain, 0.65 × objective magnification.

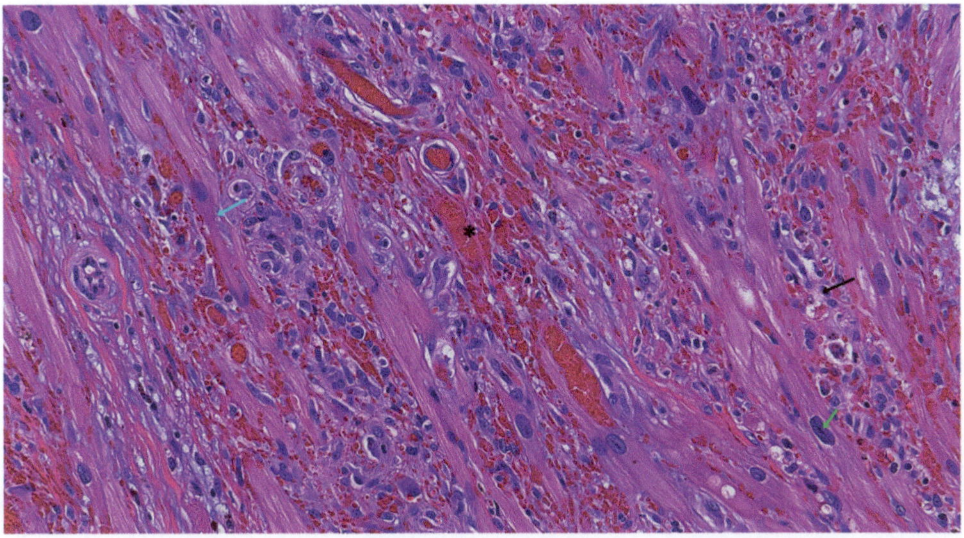

FIGURE 3.4B

Higher magnification of the myocardial lesion shown in the previous image shows degeneration of myofibers (*blue arrow*), extravasated blood (*), and mononuclear inflammatory cell infiltration (*black arrow*). Enlarged myofiber nuclei (*green arrow*) are a species characteristic of rhesus macaques and are unrelated to the pathologic process. Hematoxylin and eosin stain, 20 × objective magnification.

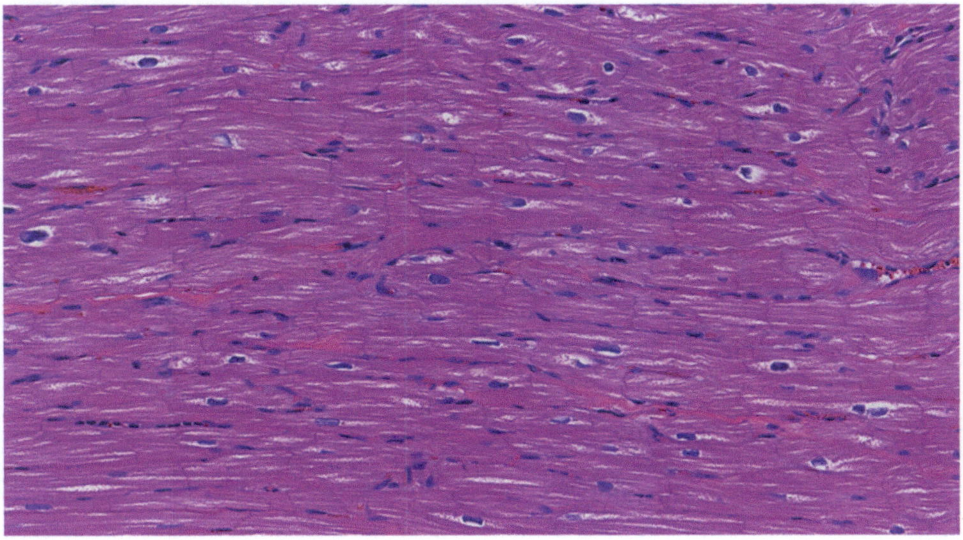

FIGURE 3.4C

An area of unaffected myocardium from the same animal shown in the previous figure has relatively normal histologic features. Accumulation of uncharacterized material is commonly present near the poles of ovoid myofiber nuclei in normal rhesus macaques, but the particulate accumulations shown in this image are unusually prominent. Compare this image to the caspase 3 immunohistochemical stain. Hematoxylin and eosin stain, 20 × objective magnification.

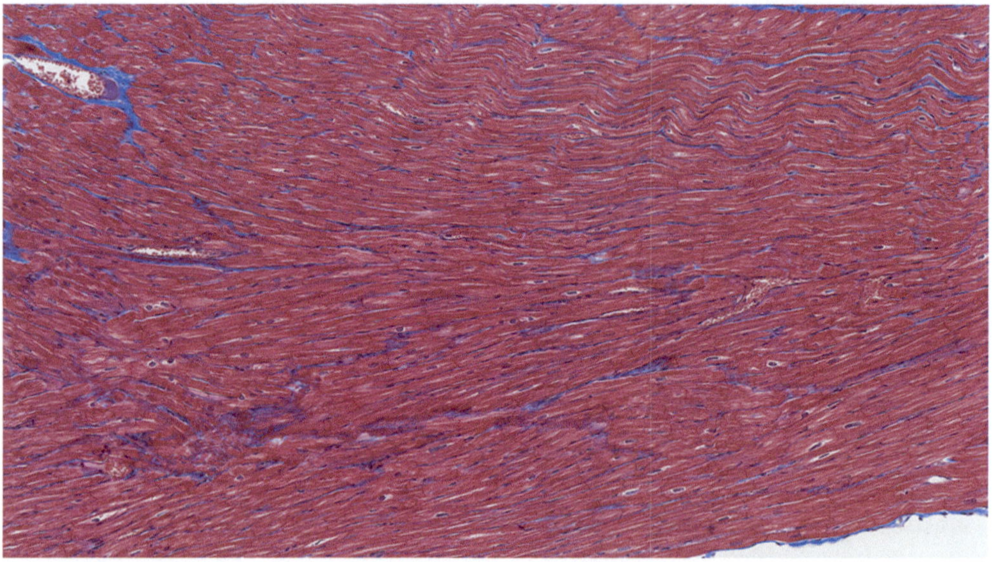

FIGURE 3.5A

Masson's trichrome stain performed on myocardium of a naïve male rhesus macaque shows a small amount of blue-stained endomysial fibrous connective tissue, 20 × objective magnification.

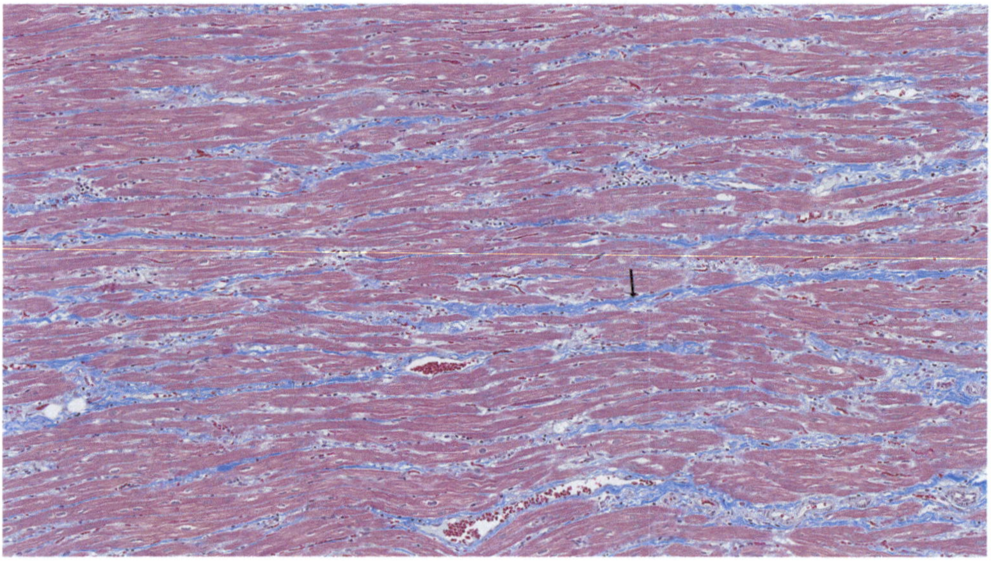

FIGURE 3.5B

A Masson's trichrome stain performed on the myocardium of a male rhesus macaque necropsied 84 days after irradiation at 11 Gy has a substantial amount of blue-stained fibrous connective (*arrow*) in the endomysium surrounding myofibers. Masson's trichrome stain, 10 × objective magnification.

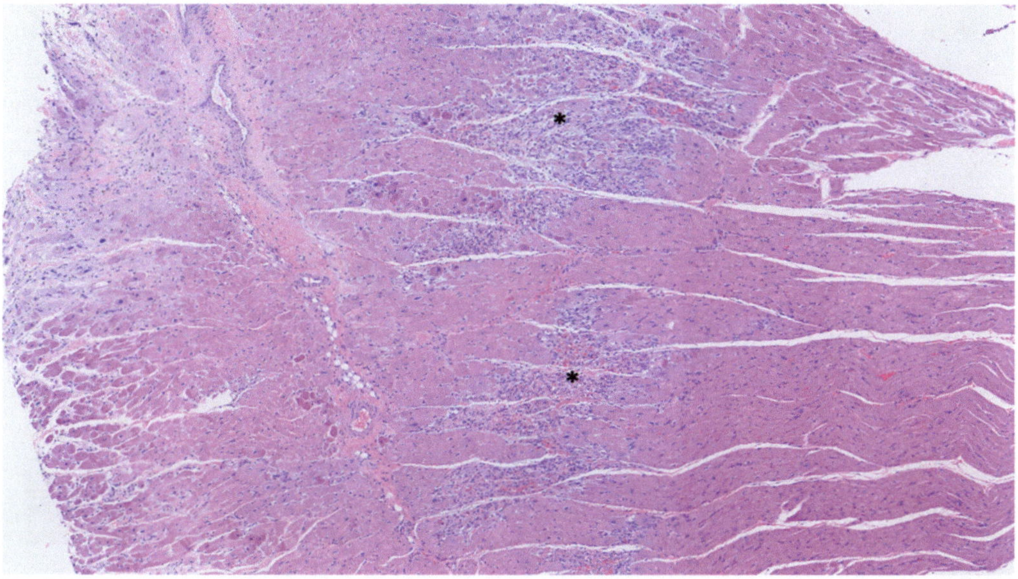

FIGURE 3.6A

The myocardium of a male rhesus macaque necropsied 44 days after irradiation at 10 Gy has multifocal areas of myofiber degeneration (*). Hematoxylin and eosin stain, 20 × objective magnification.

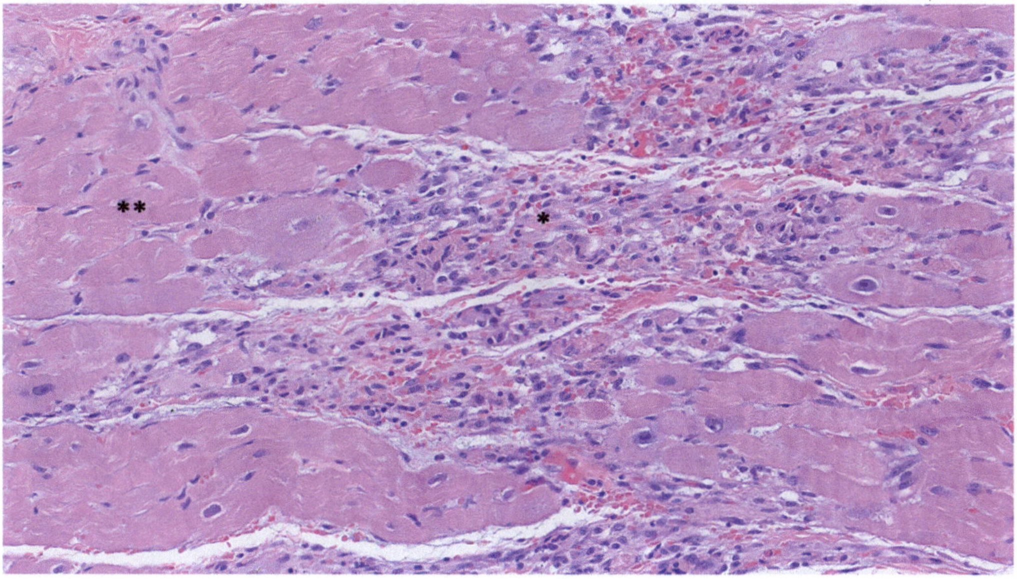

FIGURE 3.6B

Higher magnification of the previous specimen reveals near-total loss of myofibers within the damaged area (*), with relative sparing of adjacent areas of myocardium (**). Hematoxylin and eosin stain, 20 × objective magnification.

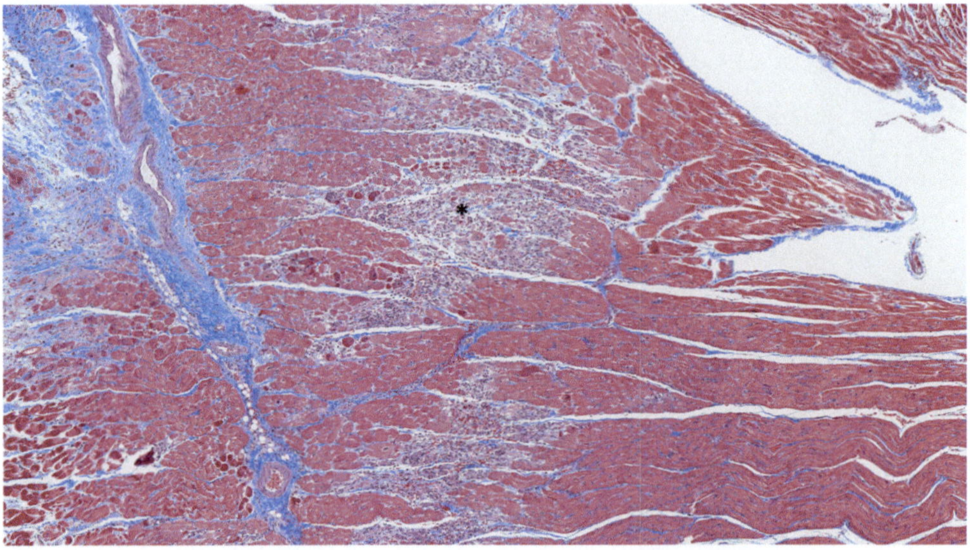

FIGURE 3.7A

Masson's trichrome stain performed on the same lesion complex shown in the previous images. Note the distinct borders of the multifocal myocardial injury (*). 5 × objective magnification.

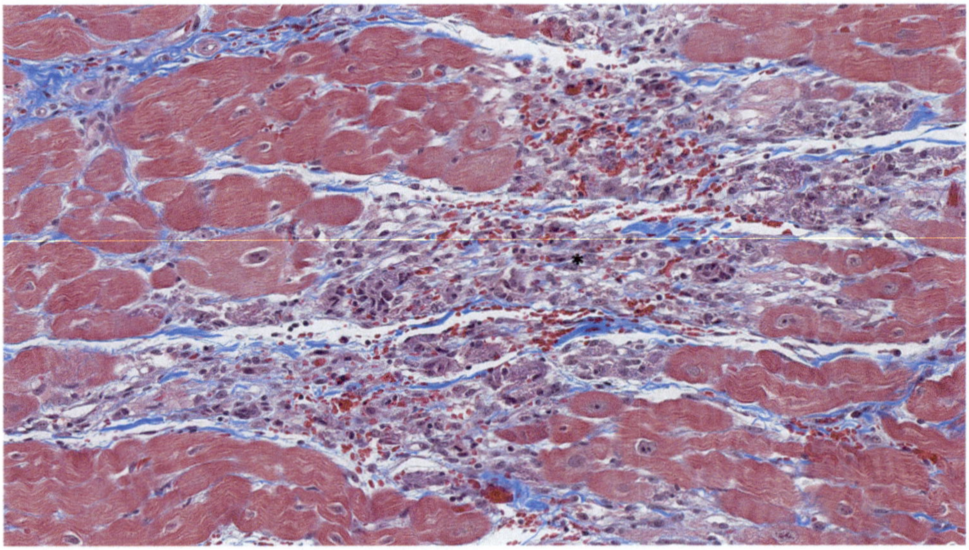

FIGURE 3.7B

Higher magnification of the previous specimen shows collapse of myofibers with persistence of blue-stained endomysial connective tissue within the area of radiation injury (*). There is no evidence of increased endomysial fibrous connective tissue in the adjacent normal myocardium (**). Masson's trichrome stain, 20 × objective magnification.

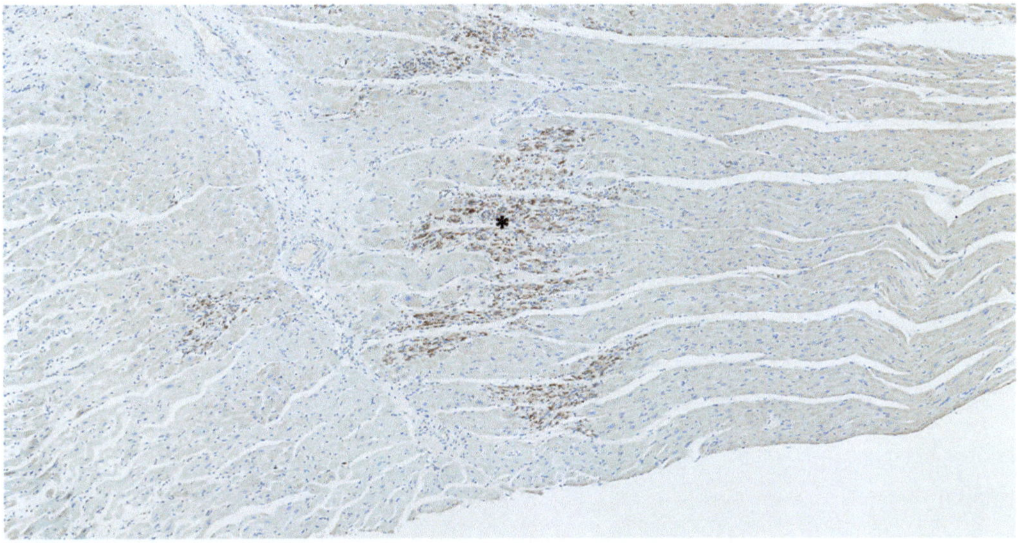

FIGURE 3.8A

Immunohistochemical (IHC) staining for galectin-3 performed on the same specimen shown in the previous images. Note the galectin-3 expression in the areas of radiation injury (*) in the myocardium. Stain performed with 3, 3′-diaminobenzidine chromogen and hematoxylin counterstain, 5 × objective magnification.

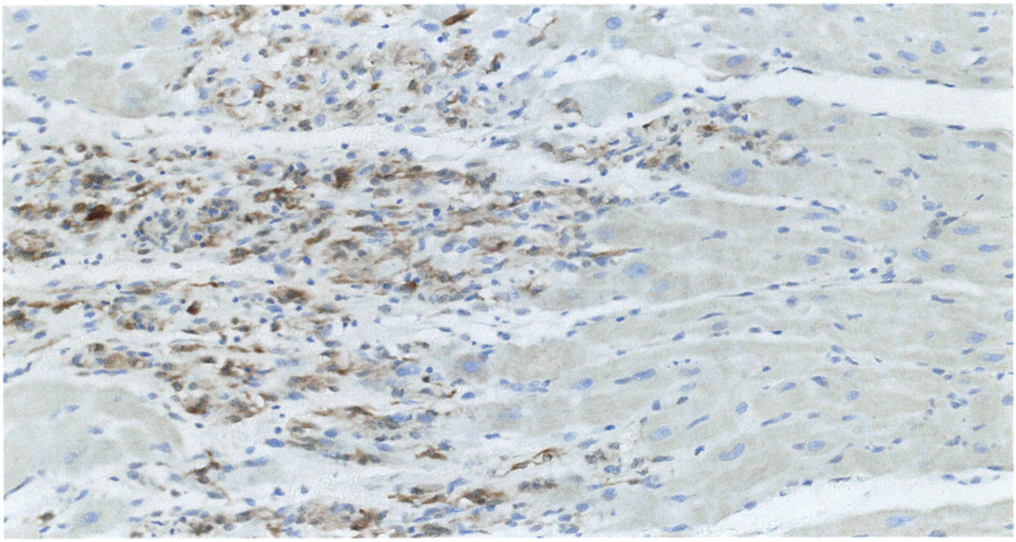

FIGURE 3.8B

Higher magnification of the previous specimen shows galectin-3 expression concentrated in a population of unidentified large mononuclear cells. Immunohistochemical (IHC) stain for galectin-3 with 3, 3′-diaminobenzidine chromogen and hematoxylin counterstain, 20 × objective magnification.

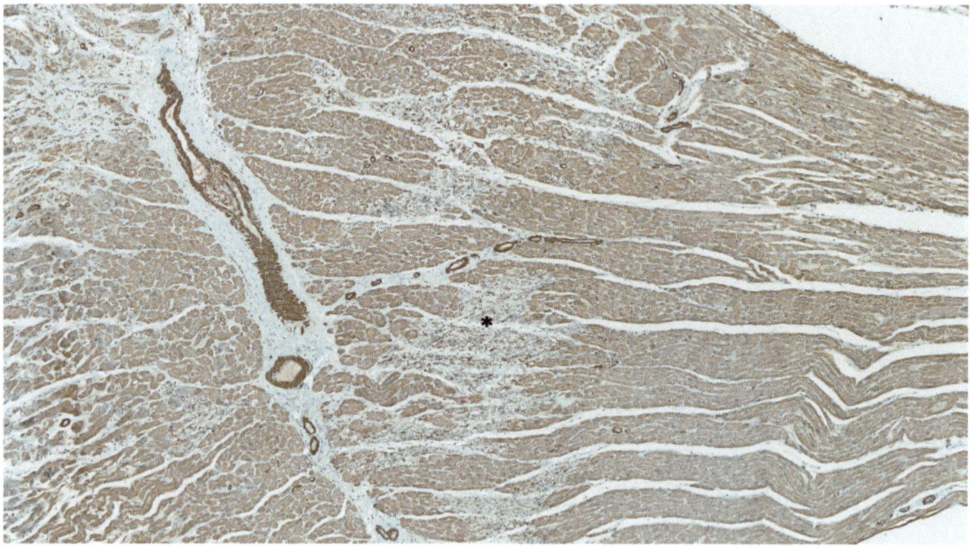

FIGURE 3.9A

Immunohistochemical (IHC) staining for alpha smooth muscle actin (α-SMA) performed on the same specimen shown in the previous images. Note the α-SMA expression in the areas of radiation injury (*) in the myocardium. Stain performed with 3, 3′-diaminobenzidine chromogen and hematoxylin counterstain, 5 × objective magnification.

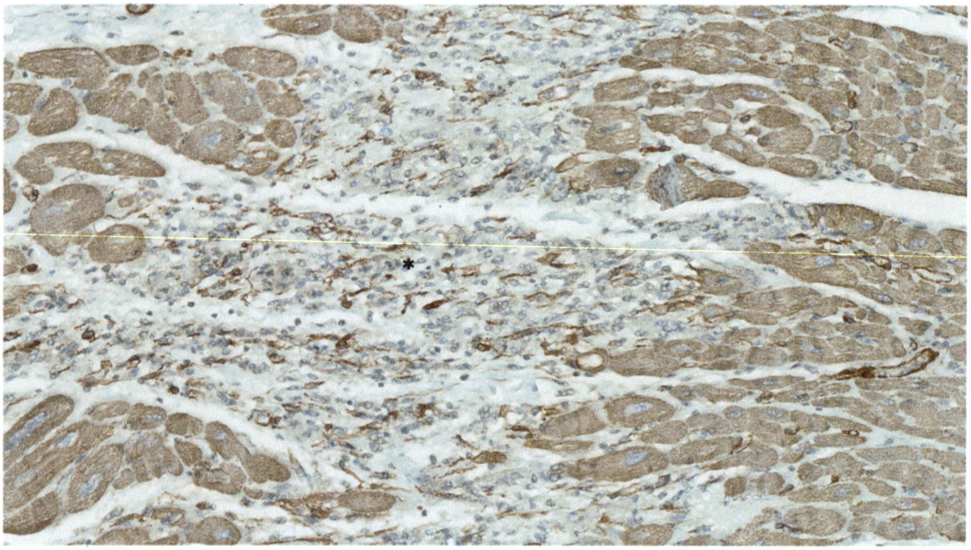

FIGURE 3.9B

Immunohistochemical (IHC) staining for alpha smooth muscle actin (α-SMA) reveals the lesion area (*) has marked reduction in the normal level of α-SMA expression in myofibers and a slight increase in endomysial α-SMA expression. Stain performed with DAB chromogen and hematoxylin counterstain, 20 × objective magnification.

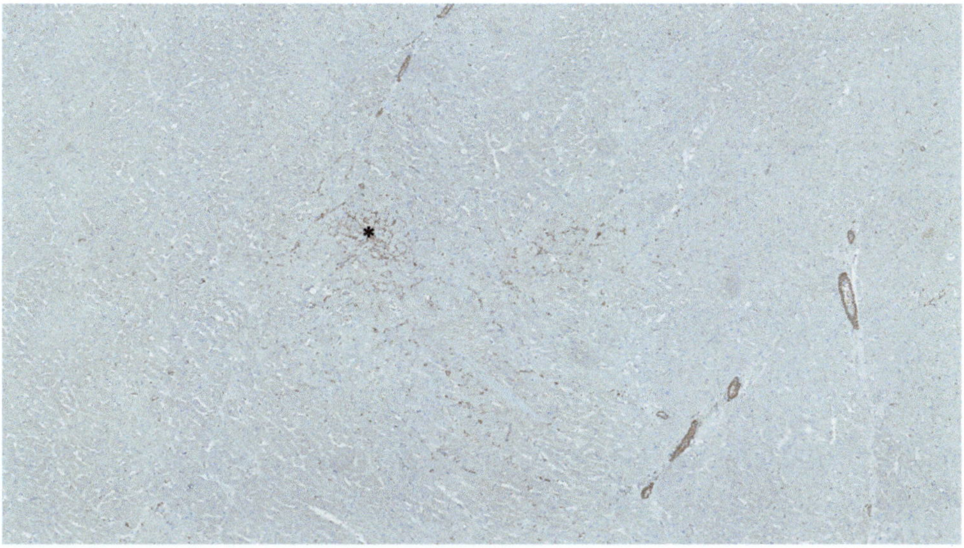

FIGURE 3.9C

Immunohistochemical (IHC) staining for alpha smooth muscle actin (α-SMA) performed on the myocardium of a male rhesus macaque necropsied 115 after irradiation at 11 Gy. Note the focal area (*) of α-SMA expression in the myocardium. This tissue alteration was not discernible in sections stained with hematoxylin and eosin or Masson's trichrome stain. Stain performed with 3, 3′-diaminobenzidine chromogen and hematoxylin counterstain, 5 × objective magnification.

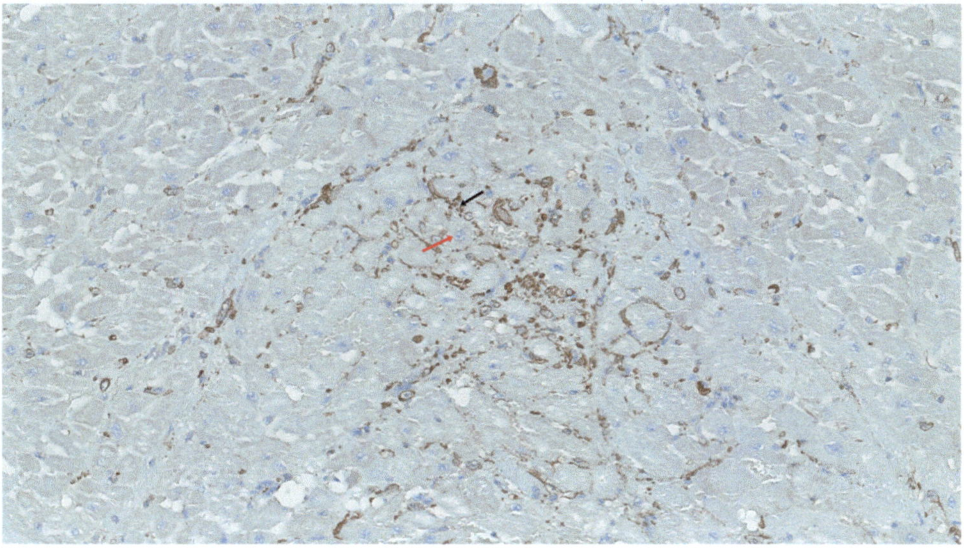

FIGURE 3.9D

Higher magnification of the previous specimen reveals the alpha smooth muscle actin expression is in endomysial cell populations (*black arrow*) as opposed to cardiac myocytes (*red arrow*). Stain performed with 3, 3′-diaminobenzidine chromogen and hematoxylin counterstain, 20 × objective magnification.

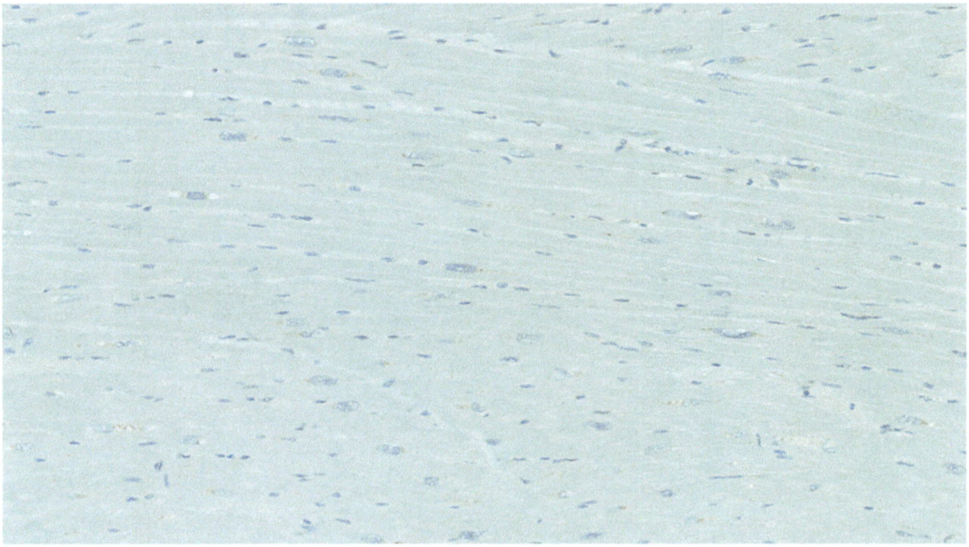

FIGURE 3.10A

Immunohistochemical (IHC) staining for caspase 3 performed on the myocardium of a naïve male rhesus macaque reveals essentially no caspase 3 staining. Stain performed with 3, 3′-diaminobenzidine chromogen, and hematoxylin counterstain, 20 × objective magnification.

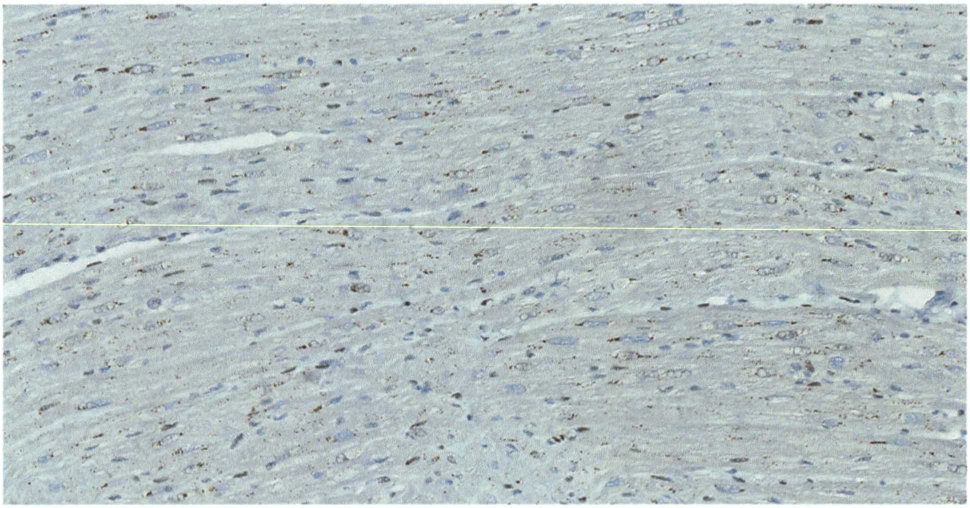

FIGURE 3.10B

Immunohistochemical (IHC) staining on the myocardium of a male rhesus macaque necropsied 9 days after irradiation at 12 Gy reveals substantial caspase 3 staining consisting of brown-stained particulate structures in the myocardium. Caspase immunohistochemical stain with 3, 3′-diaminobenzidine chromogen and hematoxylin counterstain, 20 × objective magnification.

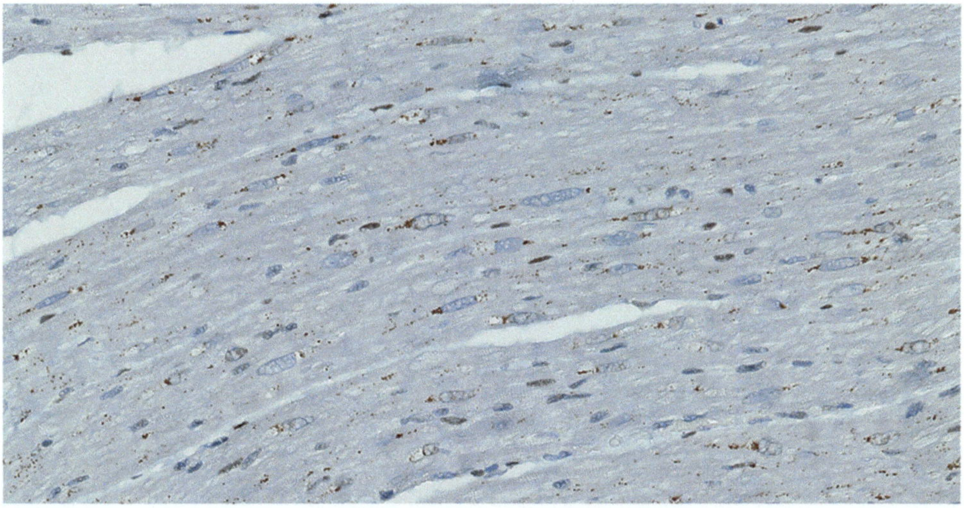

FIGURE 3.10C

Higher magnification of the previous specimen reveals caspase 3^+ internal structures within myofibers, but no evidence of the whole-cell apoptosis that is typically seen with caspase 3 expression in cellular apoptosis. The caspase 3^+ material is most pronounced in cytoplasmic regions at the poles of ovoid myofiber nuclei, similar to the location of the lipofuscin "aging pigment" seen in the myocardium of aged animals. Caspase immunohistochemical stain with 3, 3′-diaminobenzidine chromogen and hematoxylin counterstain, 30 × objective magnification.

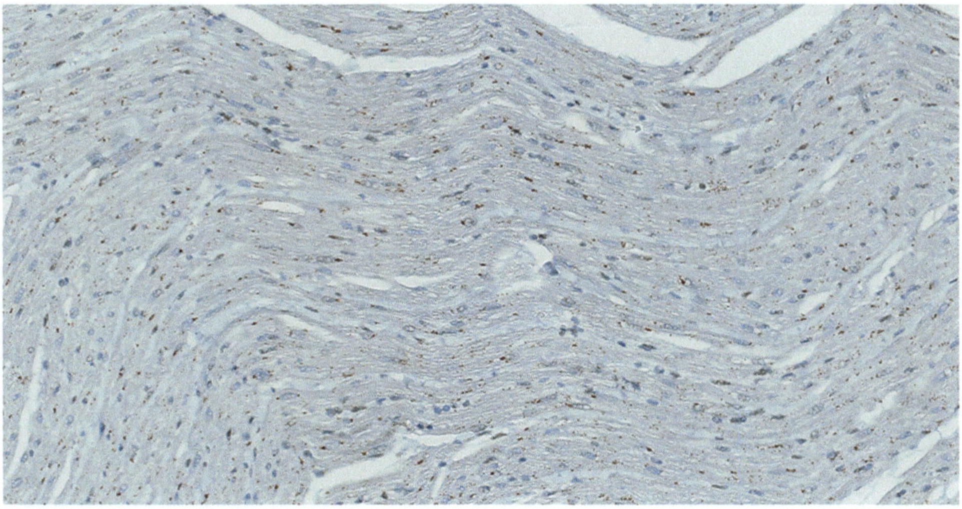

FIGURE 3.10D

Immunohistochemical (IHC) staining on the myocardium of a male rhesus macaque necropsied 16 days after irradiation at 12 Gy reveals substantial caspase 3 staining in the myocardium, similar to that seen in an animal necropsied 9 days after irradiation. Caspase immunohistochemical stain with 3, 3′-diaminobenzidine chromogen and hematoxylin counterstain, 20 × objective magnification.

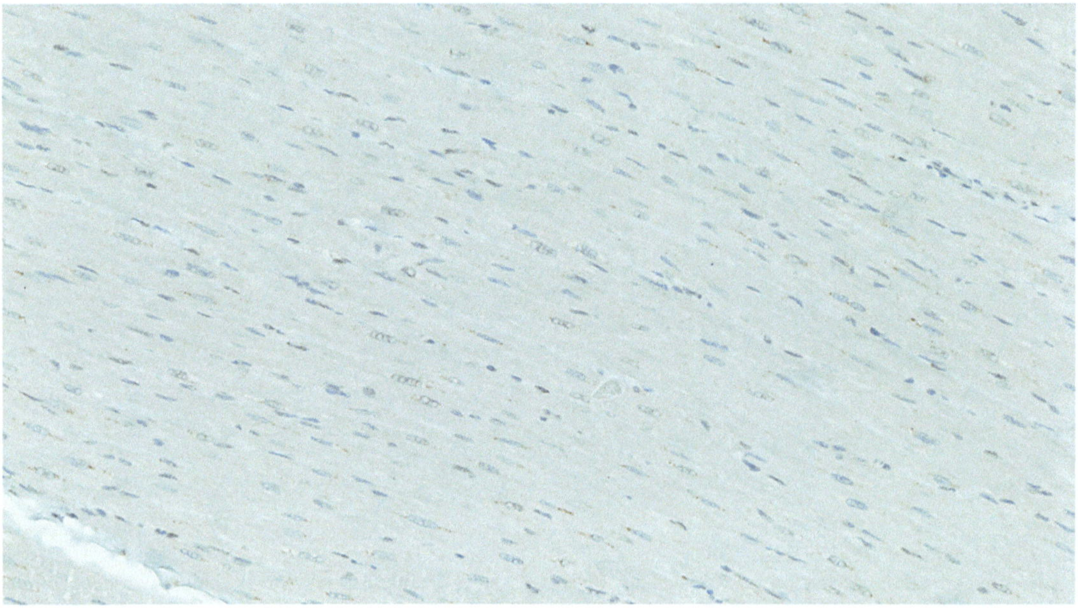

FIGURE 3.10E

Immunohistochemical (IHC) staining on the myocardium of a male rhesus macaque necropsied 25 days after irradiation at 12 Gy reveals no caspase 3 staining in the myocardium. Caspase 3 immunohistochemical stain with 3, 3′-diaminobenzidine chromogen and hematoxylin counterstain, 20 × objective magnification.

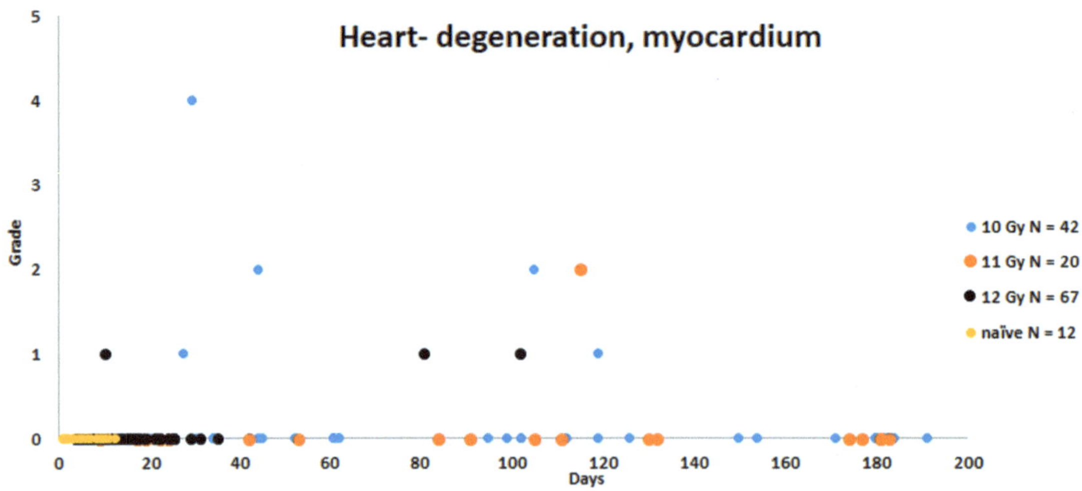

FIGURE 3.11

Scatter plot: Heart—degeneration, myocardium. Observations based on light microscopic examination of paraffin-embedded specimens stained with hematoxylin and eosin.

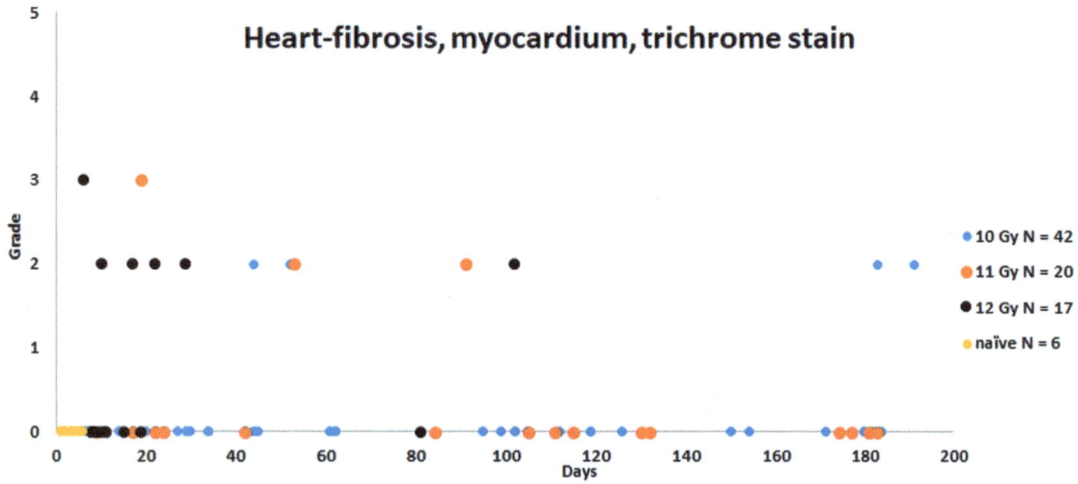

FIGURE 3.12

Scatter plot: Heart—fibrosis, myocardium. Observations based on light microscopic examination of paraffin-embedded specimens stained with Masson's trichrome stain.

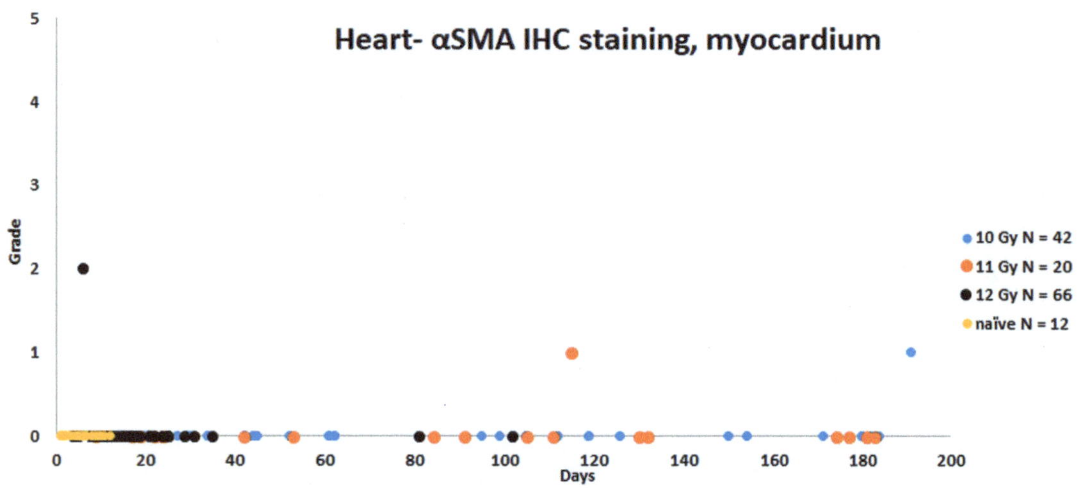

FIGURE 3.13

Scatter plot: Heart—alpha smooth muscle actin (α-SMA) staining. Observations based on light microscopic examination of paraffin-embedded specimens stained by immunohistochemical procedures for α-SMA with 3,3′-diaminobenzidine chromogen and hematoxylin counterstain.

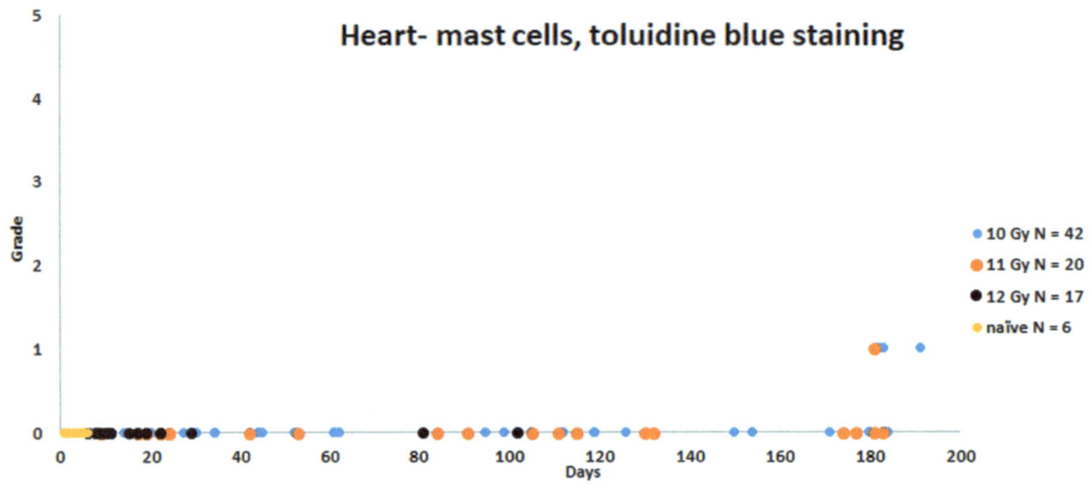

FIGURE 3.14

Scatter plot: Heart—mast cells, toluidine blue stain. Observations based on light microscopic examination of paraffin-embedded specimens stained with toluidine blue stain.

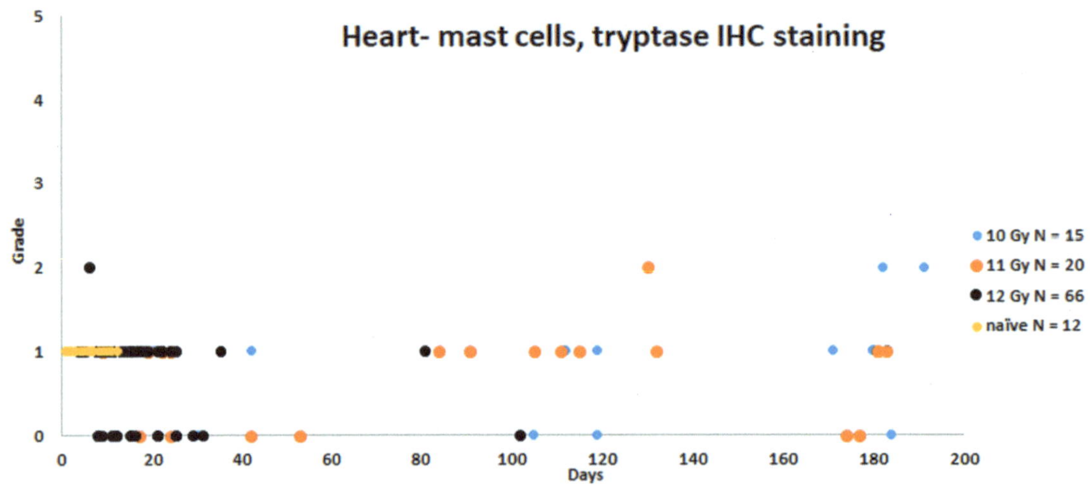

FIGURE 3.15

Scatter plot: Heart—degeneration, myocardium Scatter plot: Heart—mast cells, tryptase IHC staining. Observations based on light microscopic examination of paraffin-embedded specimens stained by immunohistochemical procedures for tryptase with 3,3′-diaminobenzidine chromogen and hematoxylin counterstain.

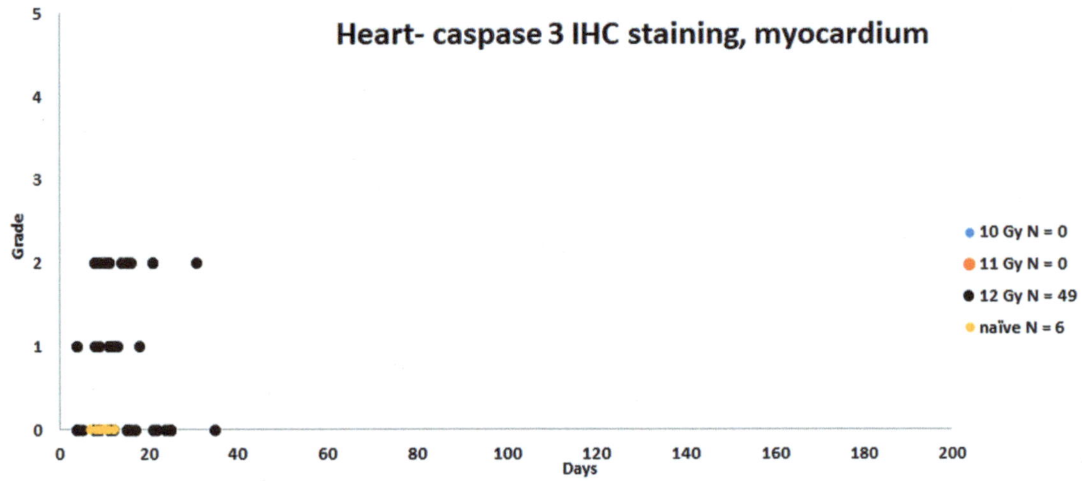

FIGURE 3.16

Scatter plot: Heart—caspase 3 staining. Observations based on light microscopic examination of paraffin-embedded specimens stained by immunohistochemical procedures for caspase 3 with 3,3′-diaminobenzidine chromogen and hematoxylin counterstain.

CHAPTER

Radiation-associated changes in the jejunum of rhesus macaques

4

Overview

The histologic evidence of intestinal mucosal injury was radiation dose-associated and largely related to effects on mitotically active cells of the mucosal crypts. Immediately after irradiation, there was a depletion of the epithelial elements of the mucosa which, in extreme cases, resulted in complete denudation of the mucosal surface. Effects on the jejunal mucosa were radiation dose associated, with the most pronounced changes in animals irradiated at 12 Gy. The epithelial depletion was associated with a distinct reduction in Ki-67$^+$ crypt cell populations, particularly in animals irradiated at 12 Gy. A number of irradiated animals had reduced populations of Bmi-1$^+$ crypt cells during the period from day 4 to day 17 postirradiation.

Repair of the radiation-associated epithelial depletion commenced within a few days, first consisting of a marked expansion of Ki-67$^+$ crypt cells. In situ hybridization (ISH) staining for Lgr5 revealed an associated increase in Lgr5 expression in crypt cell populations during the reparative phase. Even though the radiation-associated mucosal injury was largely repaired by day 45 postirradiation in most animals, histologic evidence of mucosal injury that persisted throughout the observation period in a subset of animals suggested that some animals did not completely recover from the radiation injury within 180 days.

The mucosal injury included a striking reduction in the population of Paneth cells, which produce molecules that are critical to the maintenance of the intestinal microbiome. Paneth cells were completely absent from the jejunal mucosa of a subset of animals necropsied up to 30 days following irradiation, and Paneth cell populations were reduced in a subset of animals necropsied throughout the observation period. Reduced expression of connective tissue growth factor (CTGF) in jejunal crypts, which was noted in animals necropsied through day 53 postirradiation, appeared to be related to radiation-associated depletion of Paneth cell populations. The effects on Paneth cells were radiation dose associated, with the most pronounced changes in animals irradiated at 12 Gy.

During the first 53 days following irradiation, there were pronounced reductions in the population of goblet cells that produce the protective mucin that is important to intestinal homeostasis. The goblet cell deficit was noted in a subset of animals necropsied throughout the observation period. The radiation-damaged small intestinal mucosa also had a substantial reduction in the CD13$^+$ brush border, which has important enzymatic activity. Decreased goblet cell populations and reduced CD13$^+$ brush borders were a feature of the immature status of regenerating mucosal epithelial cells. Reduced CD13$^+$ brush borders on mucosal epithelial cells persisted in a subset of animals necropsied throughout the observation period.

The jejunal mucosa of irradiated animals had a reduced population of activated (MxA$^+$) cells during the first 25 days postirradiation. However, there was a tendency for the population of MxA$^+$ cells to extend into the jejunal submucosa of irradiated animals.

Histopathology Atlas of Acute Radiation Syndrome and Delayed Effects in Rhesus Macaques. DOI: https://doi.org/10.1016/B978-0-323-91393-5.00004-9

Despite the histologic indications of severe radiation-associated mucosal injury, there was little histologic evidence of intestinal microbe penetration into the intestinal wall. Immunohistochemical (IHC) staining for lipopolysaccharide core antigen, a marker for enteric bacteria, revealed positive staining within the mucosa of only a small number of irradiated animals. This suggests the mucosal barrier function remained largely intact. However, the greater tendency for lipopolysaccharide (LPS) staining in mesenteric lymph nodes (see Chapter 6) suggests that multiple small breaks in intestinal mucosal integrity may not be revealed in the relatively sparse intestinal sampling.

Inflammatory cell infiltration was not a prominent feature of the radiation-associated intestinal lesion complex. IHC staining for myeloperoxidase (MPO) revealed a slight increase in the MPO^+ jejunal macrophage population immediately following irradiation and persisting in a low incidence through the observation period. IHC staining for CD38, a marker for pro-inflammatory macrophages, revealed no radiation-associated change in $CD38^+$ cell populations in the jejunum. By contrast, IHC staining revealed an increase in the population of pro-fibrotic $CD163^+$ macrophages in the intestinal mucosa, submucosa, and serosa of irradiated animals, particularly those irradiated at 12 Gy. There was a pronounced tendency for $CD163^+$ macrophages to extend into the jejunal submucosa, and the serosal surface of irradiated animals also had an increased population of $CD163^+$ macrophages. The combination of observations in MPO, CD38, and CD163 staining suggests the observed increase in $CD163^+$ macrophage populations may have involved phenotype switching in a preexisting macrophage population rather than the infiltration of a new macrophage population.

IHC staining for caspase 3 was performed only in naive animals and animals irradiated at 10 or 12 Gy. This staining revealed increased caspase 3 expression in mucosal epithelial cells in animals irradiated at 12 Gy and necropsied during the first 25 days postirradiation, and sporadically thereafter in animals irradiated at 10 Gy. Caspase 3 expression in the leukocyte population in the lamina propria was reduced in animals irradiated at 12 Gy and necropsied within 25 days of irradiation.

Regeneration of the mucosal injury commenced as early as day 7 postirradiation, but in many animals, the intestinal mucosa did not completely regain its normal histologic appearance during the observation period of approximately 180 days.

Fibrosis was multifocally distributed in the jejunum and was more prominent microscopically in the submucosa and/or serosa rather than the mucosa. The multifocal distribution of the fibrosis raised concerns that the relatively scant specimen collection may not have revealed the true incidence of radiation-associated intestinal fibrosis.

Variation in specimen fixation and processing resulted in variations in the demonstrable mRNA via ISH staining for Lgr5. Paired histologic sections of jejunum from each animal were stained with a positive control marker (PPIB) to ensure RNA integrity. Observations of Lgr5 staining were disregarded in jejunal specimens in which the PPIB ISH staining indicated a compromise in RNA integrity.

Staining and figures list

Routine histology stain-

Hematoxylin & eosin (H&E)—Figs. 4.1, 4.2, 4.3A, 4.3B, 4.4, 4.5, 4.6A, 4.6B, 4.7A, 4.7B, 4.8A, 4.8B, 4.8C, 4.8D, 4.9, 4.10, 4.11A, 4.11B, 4.11C, 4.12, 4.13, 4.14, 4.15, 4.16, 4.20A

Histochemical stains-

Lendrum—Figs. 4.27A, 4.27B, 4.27C, 4.27D
Masson's trichrome—Figs. 4.18, 4.19A, 4.19B, 4.20B, 4.20C, 4.20D, 4.20E
Periodic acid-Schiff/Alcian blue—Figs. 4.17A, 4.17B, 4.17C, 4.17D

IHC stains-

CD13 (aminopeptidase N)—Figs. 4.33A, 4.33B, 4.33C
CD163—Figs. 4.35, 4.36A, 4.36B, 4.37, 4.38, 4.39, 4.40
Collagen 1—Figs. 4.29, 4.30, 4.31, 4.32
Connective tissue growth factor—Figs. 4.28A, 4.28B, 4.28C
Ki-67—Figs. 4.21, 4.22, 4.23, 4.24A, 4.24B, 4.24C, 4.24D, 4.25A, 4.25B, 4.26
MxA—Figs. 4.6C, 4.6D, 4.34

ISH stains-

Lgr5—Figs. 4.25A, 4.25B, 4.26
(Figs. 4.21, 4.24C, 4.24D)

Graphic presentations (scatter plots) of individual histologic observations in the jejunum are presented in the following figures:

Fig. 4.66. Jejunum—Increased collagen 1 mucosal staining
Fig. 4.67. Jejunum—Increased collagen 1 submucosal staining
Fig. 4.68. Jejunum—Increased collagen 1 serosal staining

Detailed microscopic observations

Mucosal epithelial depletion

The most pronounced histologic changes were in animals irradiated at 12 Gy. Similar but less pronounced epithelial depletion was noted in animals which were irradiated at the 10 or 11 Gy levels. These changes were considered to be a major contributor to the shorter survival time of animals in this group. The jejunum of some animals had areas in which the surface epithelium was completely lost, which was recorded as denudation. Complete epithelial denudation was rarely noted in animals irradiated at 10 or 11 Gy.

Trichrome stain

Histologic evidence of fibrosis was noted in the mucosa, submucosa, and/or serosa of a small number of animals irradiated at 10, 11, or 12 Gy. The degree of mucosal fibrosis was notably sparse relative to the degree of radiation-associated mucosal injury. IHC staining for collagen 1 was a more sensitive method for the detection of fibrogenesis.

Periodic acid/Schiff (PAS)/Alcian blue stain

Animals necropsied within 45 days of irradiation had a pronounced reduction in the population of mucosal goblet cells, reflecting loss of preexisting mature mucosal epithelial cells and immaturity of the regenerative epithelial cells.

Leukocyte infiltration

Inflammatory cell infiltration was notably sparse relative to the degree of epithelial disruption.

Toluidine blue stain

A low-level mast cell population was present in the jejunal mucosa and submucosa of all naive control animals. There was no definitive radiation-related variation in the mast cell populations, but the absence of mast cells was most commonly noted in animals necropsied within 45 days of irradiation. It should be noted that the toluidine blue stain selectively stains the contents of mast cell granules, therefore, degranulated mast cells may not be revealed by toluidine blue staining.

Tryptase immunohistochemical stain

Tryptase staining revealed a slightly greater population of mast cells in the jejunum than were revealed by toluidine blue staining. This was expected, as the toluidine blue staining is known to reveal only the connective tissue mast cell population and does not include the mucosal mast cell population unless special fixatives are used at the time of tissue collection. The tryptase staining suggested the jejunal mast cell population was subject to radiation-associated effects at the 12 Gy radiation level. A substantial number of animals irradiated at 12 Gy and necropsied before day 35 postirradiation had no discernible mast cells in the jejunal mucosa/submucosa, but a few animals necropsied during that period had a slight increase in the jejunal mast cell population. It should be noted that the tryptase IHC stain targets components of mast cell granules, therefore, degranulated mast cells may not be revealed by tryptase IHC staining. As a result, the microscopic evidence of reduced mast cell populations may be due to mast cell degranulation.

Alpha smooth muscle actin immunohistochemical stain

Staining for α-SMA staining generally paralleled the histologic evidence of fibrosis as recorded under the trichrome stains.

MxA immunohistochemical stain

Mucosal staining for MxA, a marker for activated immunocytes, was noted in all the naive control animals. More pronounced MxA^+ cell populations were noted in a moderate number of irradiated animals. A smaller number of animals also had MxA staining of cells in the submucosa.

Lipopolysaccharide core antigen immunohistochemical stain

The naive control animals had no evidence of LPS^+ cells in either the lamina propria or the submucosa. The irradiated animals commonly had LPS^+ cells in the lamina propria but, despite the obvious radiation-associated damage to the mucosa, few animals had LPS^+ cells in the submucosa. Positively stained bacteria were sometimes present on the mucosal surface, thus serving as an internal control for the IHC staining procedure.

Ki-67 immunohistochemical stain

During the first few days following irradiation, there was a decrease in the number of positively stained crypt epithelial cells in animals irradiated at 12 Gy, and occasionally in animals irradiated at 10 or 11 Gy. This was followed by a robust regenerative response that included a substantial expansion of the $Ki\text{-}67^+$ crypt epithelial cells.

Lgr5 in situ hybridization stain

Chromogenic ISH staining for Lgr5 was performed only on naive control animals and animals irradiated at 10 or 12 Gy. An increased level of Lgr5 expression was present in crypt cells of animals necropsied within 35 days of irradiation and was noted sporadically in 10 Gy exposure animals that survived beyond 35 days postirradiation. The increased level of Lgr5 expression was more pronounced and more common in animals irradiated at 12 Gy as opposed to 10 Gy exposure.

The increased level of Lgr5 exposure appeared to be related to the degree of damage to jejunal epithelial cells, which was more pronounced in animals irradiated at 12 Gy.

Myeloperoxidase immunohistochemical stain

Low-level mucosal MPO^+ cell populations were present in most animals, including the naive controls. The absence of MPO^+ cells was noted only in irradiated animals and was most commonly observed in animals necropsied within 30 days of irradiation.

CD38 immunohistochemical stain

Staining for CD38 was performed only on naive control animals and animals irradiated at 10 or 12 Gy. There was a slight increase in the mucosal population of $CD38^+$ cells of a subset of animals necropsied within 45 days of irradiation at either 10 or 12 Gy.

CD163 immunohistochemical stain

Staining for CD163 was performed only on naive control animals and animals irradiated at 10 or 12 Gy. There was a substantial increase in the mucosal, submucosal, and serosal populations of $CD163^+$ cells in many animals exposed at 12 Gy and necropsied within 35 days of irradiation. This effect was not apparent in animals irradiated at 10 Gy.

CD3 immunohistochemical stain

Staining for $CD3^+$ lymphocytes primarily involved mucosal cell populations, with occasional $CD3^+$ cells in the submucosa or within lymphoid follicles when those structures were fortuitously included in the jejunal sections. The subjective evaluation revealed no alteration in the population of $CD3^+$ cells except that noted within mucosal lymphoid follicles.

CD13 immunohistochemical stain

Positive staining for CD13 was topographically subcategorized as staining of the superficial brush border of intestinal epithelial cells or staining of individual cells, consistent with leukocytes or macrophages, in the mucosa. Reduced CD13 staining of the epithelial cell brush border was observed in many animals during the first 40 days following irradiation, and sporadically thereafter. Reduction or absence of the epithelial cell brush border reflected the immature status, or complete absence, of epithelial cells. Positive CD13 staining was rarely noted in individual leukocytes in the mucosa.

Bmi-1 immunohistochemical stain

Positive staining for Bmi-1 was noted in a subpopulation of epithelial cells located near the deep aspect of jejunal crypts but typically were not located in the absolute bottom of crypts. Reduction in this population of Bmi-1^{+} cells was noted in a small number of animals during the first 35 days following irradiation, and sporadically thereafter. An increased number of Bmi-1^{+} cells was noted in a subset of animals up to 40 days postirradiation, and rarely thereafter.

IL22 immunohistochemical stain

A population of IL22^{+} leukocytes was commonly present in the mucosa of the jejunal specimens. There was no definitive indication of a radiation-associated effect on the population of IL22^{+} leukocytes in the jejunal mucosa, but it was noted that the complete absence of IL22^{+} cell populations was most commonly observed in animals irradiated at 12 Gy.

IL22R immunohistochemical stain

IL22R staining was commonly manifested in jejunal epithelial cells. There was no definitive indication of a radiation-associated effect on IL22R expression, other than the absence of IL22 expression when epithelial cells were depleted.

Collagen 1 immunohistochemical stain

Increased collagen staining was observed only sporadically in the jejunal mucosa. An apparent increase in mucosal collagen content noted in animals necropsied with 30 days of irradiation was due to loss of epithelial cells and collapse of the mucosa, resulting in a relative preponderance of the normal mucosal collagen. It should be noted that increased collagen staining in the various sublocations of the jejunum, but particularly the serosa, tended to be multifocally distributed. Based on the study protocol, only two specimens from the mid-jejunum were examined microscopically. Examination of additional jejunal specimens could reveal additional examples of increased collagen staining.

Transforming growth factor β immunohistochemical stain

Staining for transforming growth factor β (TGF-β) was noted in the deep aspects of mucosal crypts of the jejunum, where it appeared to be co-located with Paneth cells. Neutrophils stained positively for TGF-β thus served as an internal control for the technical quality of the stains. Absent or low-level TGF-β staining in jejunal crypts was commonly associated with depletion and/or denudation of the mucosal epithelium, particularly in animals that were necropsied within 30 days of radiation exposure.

Connective tissue growth factor immunohistochemical stain

Expression of CTGF was noted primarily in the jejunal crypt epithelial cells, with occasional staining in leukocyte/macrophage populations of the lamina propria. Absent or low-level CTGF staining in jejunal crypts was commonly associated with depletion and/or denudation of the mucosal epithelium, particularly in animals that were necropsied within 30 days of radiation exposure.

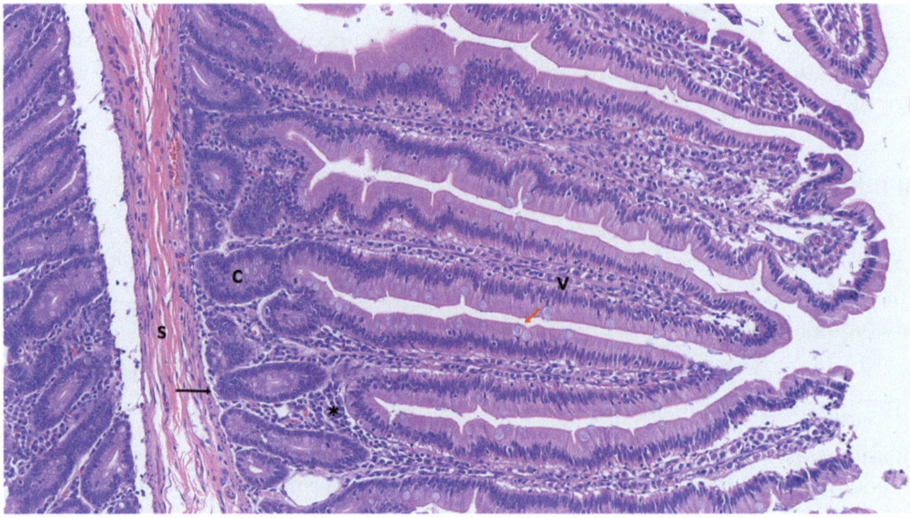

FIGURE 4.1

The jejunal mucosa of a naive male rhesus macaque has closely approximated finger-like villi (v) with basilar crypts (s) containing replicating epithelial cells. Bulbous amphophillic cells (red arrow) are goblet cells filled with mucin that is extruded to form a protective layer on the intestinal surface. The mucosal lamina propria normal contains a moderate lymphocytic population (*) that is a major component of mucosal immunity. A thin band of smooth muscle (black arrow) constitutes the muscularis mucosa, which separates the mucosa from the submucosa (s). Hematoxylin and eosin stain, 14.2 × objective magnification.

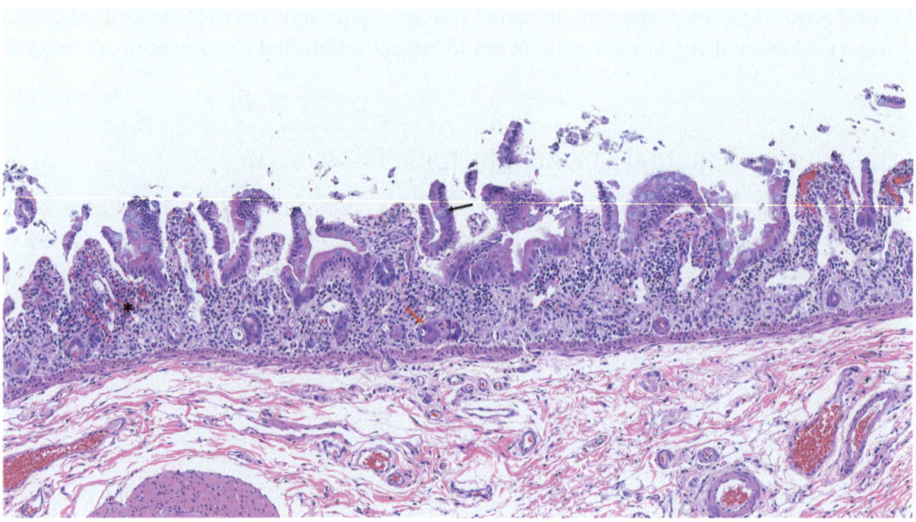

FIGURE 4.2

The jejunum of a male rhesus macaque necropsied 6 days after irradiation at 12 Gy shows loss of the normal villus structure, with detached rafts (black arrow) of degenerated epithelial cells and multifocal aggregates (*) of extravasated erythrocytes. Poorly defined remnants (red arrow) of crypts are present. Hematoxylin and eosin stain, 10 × objective magnification.

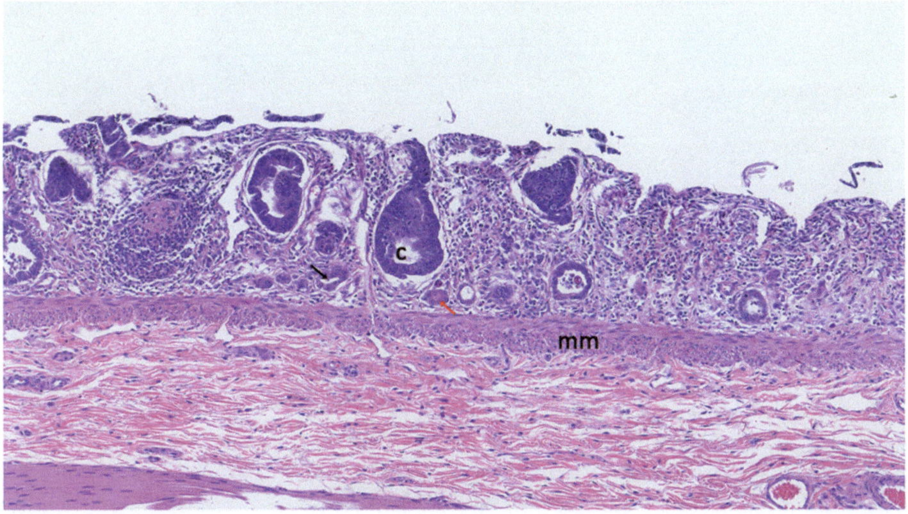

FIGURE 4.3A

The jejunum of a male rhesus macaque necropsied 7 days after irradiation at 10 Gy shows loss of the normal villus structure, with expansion of multiple crypt remnants (c) filled with mitotically active cells. Note the proliferative cells tend to be located some distance up the crypt, rather than at the absolute base of the crypt (red arrow). Many crypts (black arrow) remain at a relatively inactive status. The muscularis mucosa (mm) remains intact. Hematoxylin and eosin stain, 10 × objective magnification.

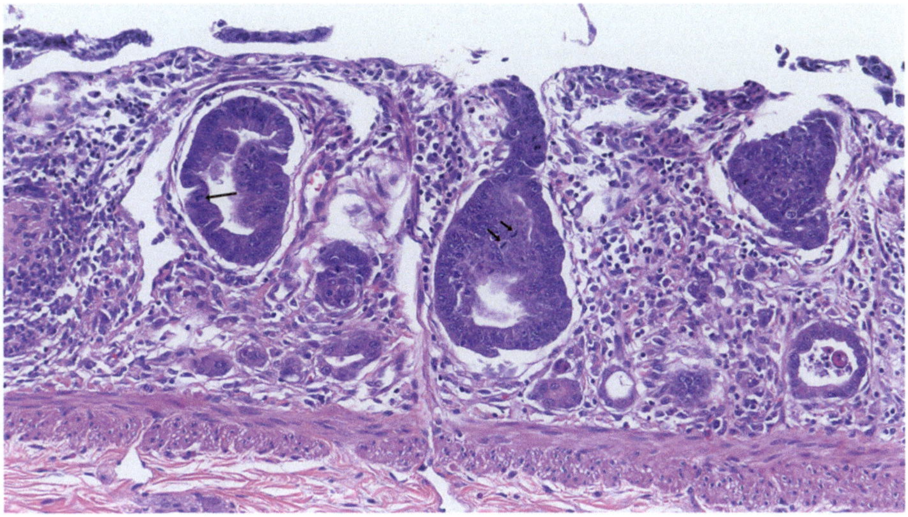

FIGURE 4.3B

Higher magnification of the previous specimen shows numerous mitotic figures (arrows) within the replicating cell population of regenerating crypts. Hematoxylin and eosin stain, 20 × objective magnification.

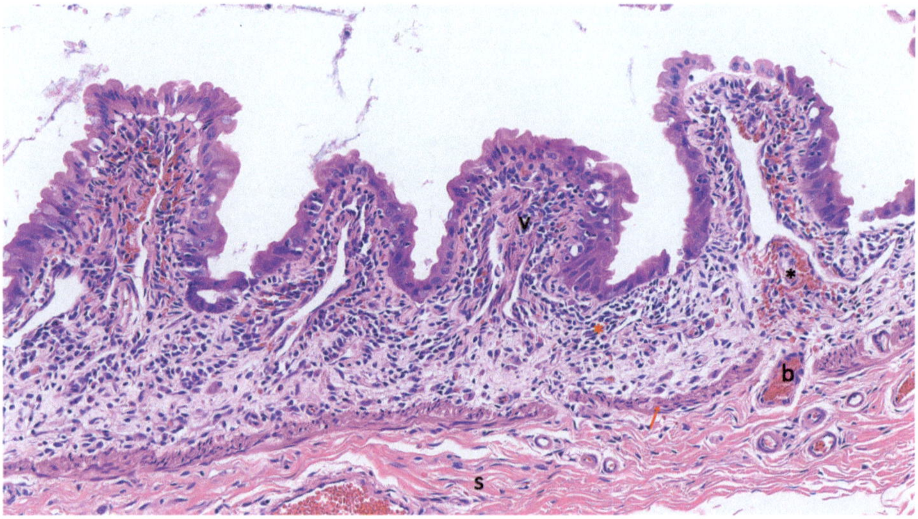

FIGURE 4.4

The jejunum of a male rhesus macaque necropsied 8 days after irradiation at 12 Gy shows short, blunted villi (v) are covered by immature epithelial cells. Note the extravasated blood (black *) and leukocytic population (red *) in the lamina propria. A discontinuity (in the muscularis mucosa (red arrow) is associated with penetration of a small blood vessel (b), but there is no evidence of leukocytic emigration through the opening into the underlying submucosa (s). Hematoxylin and eosin stain, 20 × objective magnification.

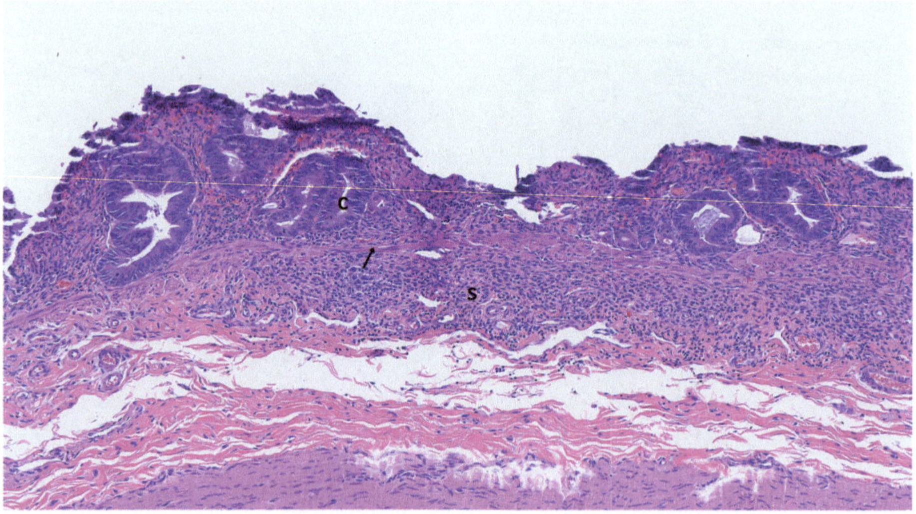

FIGURE 4.5

The jejunum of a male rhesus macaque necropsied 9 days after irradiation at 11 Gy shows extensive proliferation in crypts (c). Note the substantial leukocytic infiltration in the submucosa (s), though the overlying muscularis mucosa (arrow) appears to be intact. Hematoxylin and eosin stain, 10 × objective magnification.

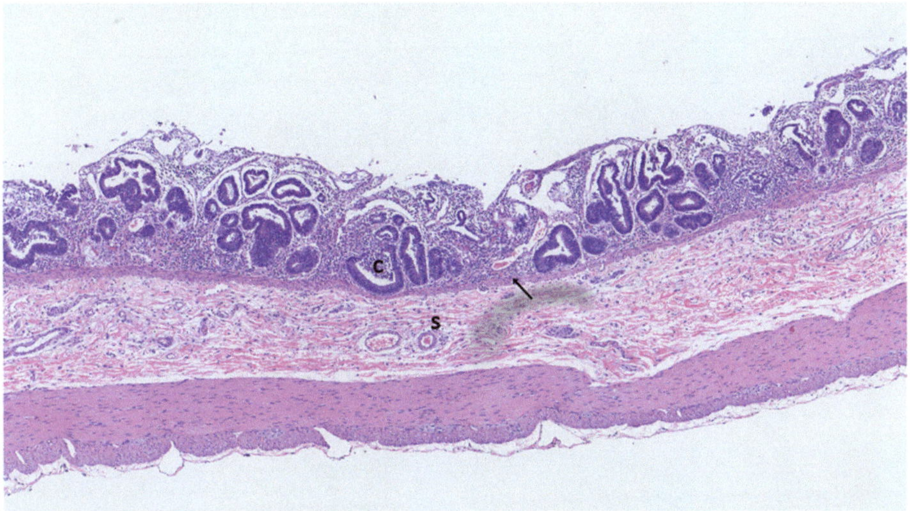

FIGURE 4.6A

The jejunum of a male rhesus macaque necropsied 11 days after irradiation at 10 Gy shows extensive cellular proliferation in virtually all crypts (c). The submucosa (s) is thickened by the presence of edema fluid. Note the intact muscularis mucosa (arrow) separating the mucosa from the submucosa. Hematoxylin and eosin stain, 5 × objective magnification.

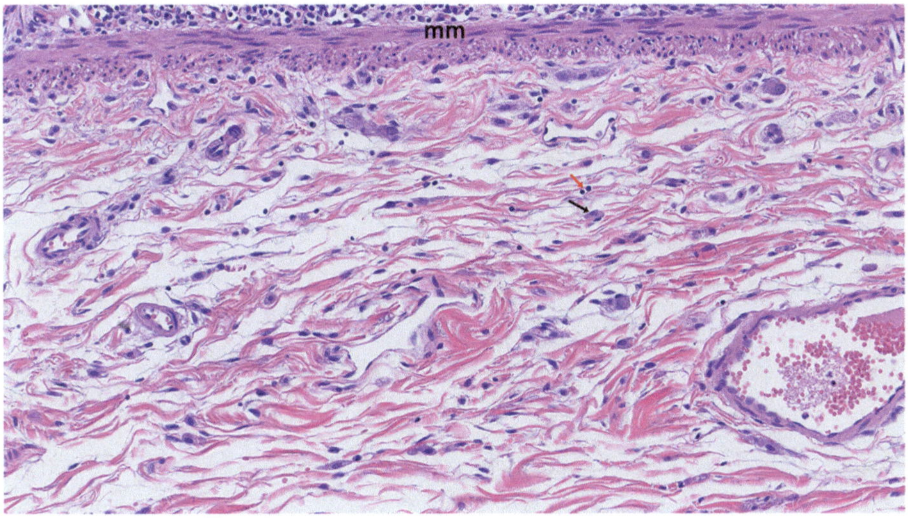

FIGURE 4.6B

Higher magnification of the previous specimen shows pink-stained collagen fibers separated by cleft-like spaces that have homogeneous, light pink edema fluid protein. A mixed cellular population in the submucosa consists of large mononuclear cells (black arrow) that are consistent with macrophages/dendritic cells and smaller, densely stained lymphocytes (red) arrow. The muscularis mucosa (MM) is intact. Hematoxylin and eosin stain, 20 × objective magnification.

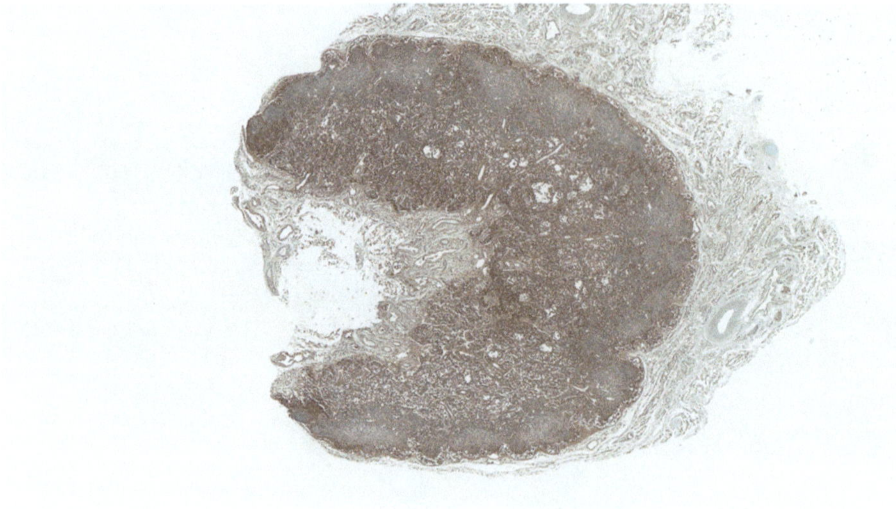

FIGURE 4.6C

An immunohistochemical stain for MxA performed on the mesenteric lymph node of the same animal shown in the previous images shows a large number of activated immunocytes throughout the lymph node. Stain performed with 3,3'-diaminobenzidine chromogen and hematoxylin counterstain, 1.27 × objective magnification.

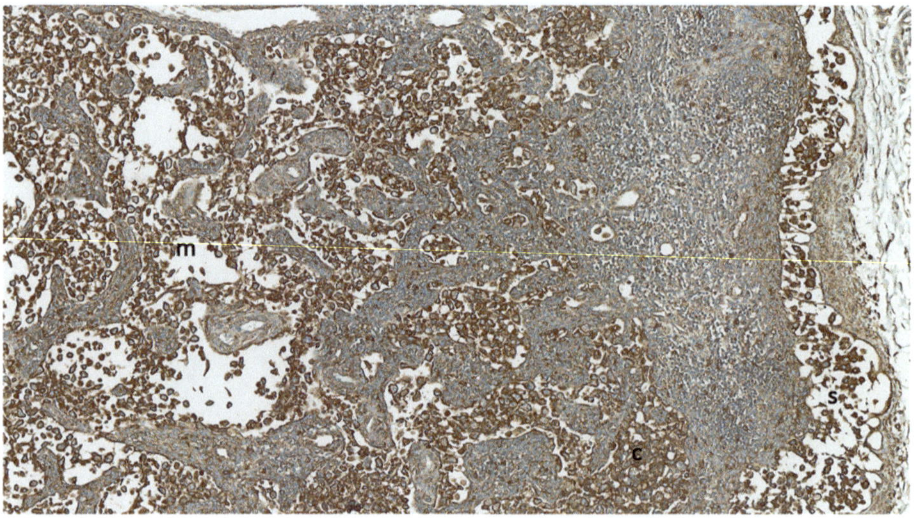

FIGURE 4.6D

Higher magnification of the previous specimen shows the MxA^+ activated cells consist primarily of large mononuclear cells, consistent with macrophages/dendritic cells. Note the MxA^+ cells are contained primarily in subcapsular, paracortical (c), and medullary (m) sinuses. Stain performed with 3,3'-diaminobenzidine chromogen and hematoxylin counterstain, 10 × objective magnification.

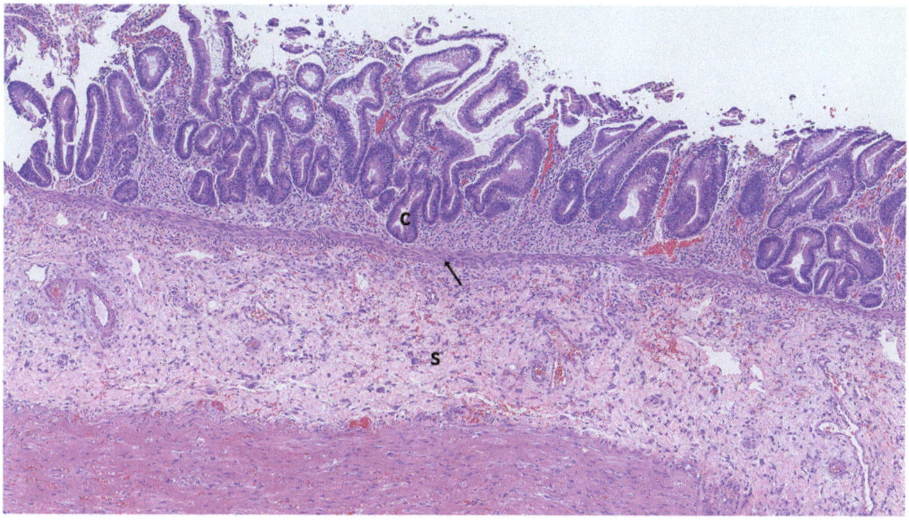

FIGURE 4.7A

The jejunum of a male rhesus macaque necropsied 20 days after irradiation at 10 Gy shows actively proliferating epithelial cells within crypts (c) and extending up toward the mucosal surface. The submucosa (s) is markedly thickened. Note the intact mucularis mucosa (arrow), which forms a barrier between the mucosa and submucosa. Hematoxylin and eosin stain, 5 × objective magnification.

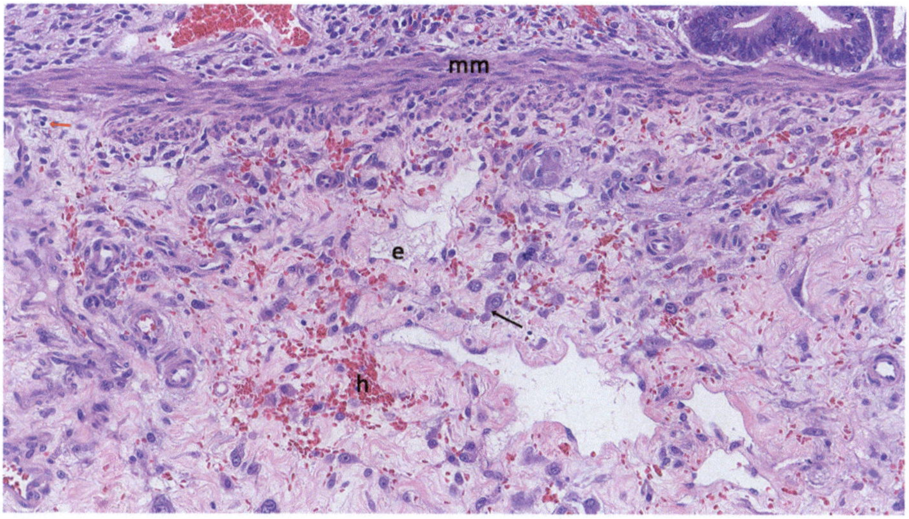

FIGURE 4.7B

Higher magnification of the previous specimen shows the thickened submucosa beneath an intact muscularis mucosa (mm). The submucosa has widespread extravasation of erythrocytes (hemorrhage-h) as well as pronounced edema indicated by the lightly stained edema fluid proteins (e). The submucosa has a pronounced population of large mononuclear cells (arrow) which have features of macrophages/dendritic cells intermixed with reactive stromal cells. Occasional infiltrating lymphocytes (red arrow) are noted. Hematoxylin and eosin stain, 20 × objective magnification.

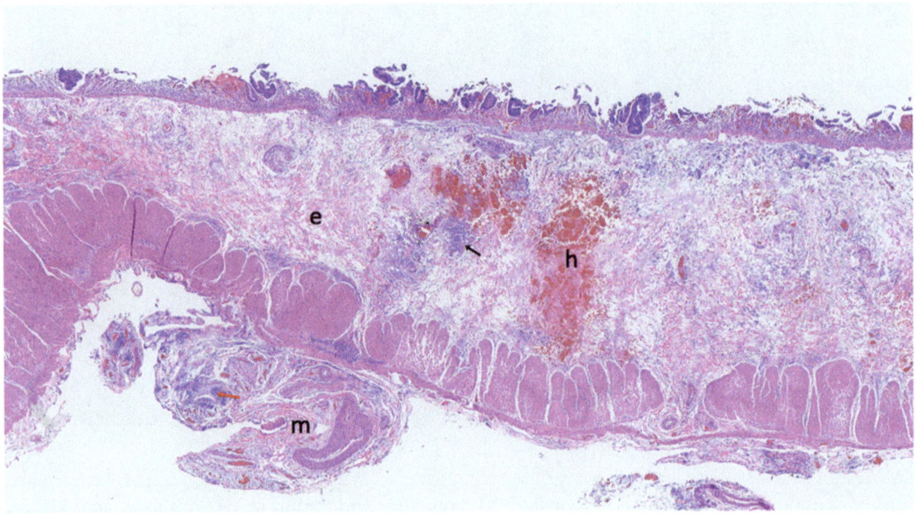

FIGURE 4.8A

The jejunal submucosa of a male rhesus macaque necropsied 22 days after irradiation at 12 Gy has pronounced hemorrhage (h), edema (e), and bacterial colonization (black arrow). Note the presence of bacterial colonies (red arrow) in the mesenteric attachment (m). Hematoxylin and eosin stain, 2.5 × objective magnification.

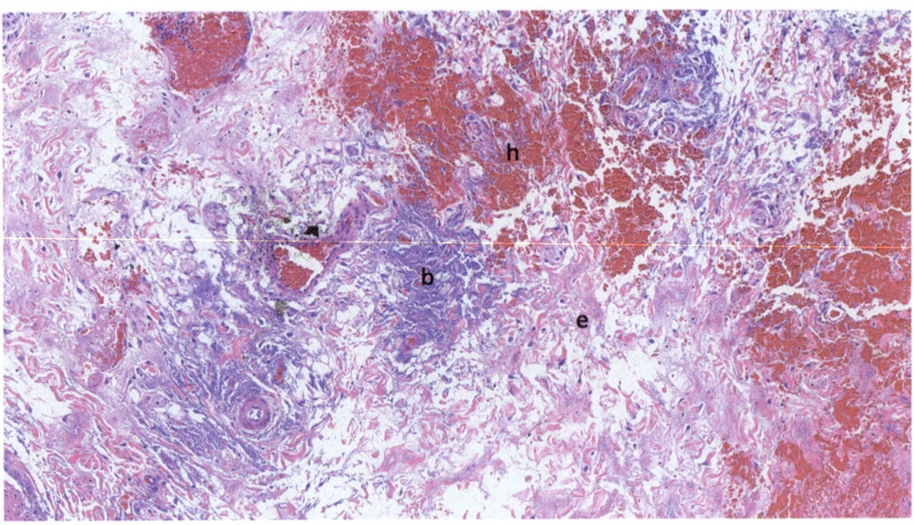

FIGURE 4.8B

Higher magnification of the previous specimen shows hemorrhage (h), edema (e), and bacterial colonies (b) in the submucosa. Note the absence of leukocytic infiltration associated with the bacterial colonies, which probably reflects radiation-associated impairment of leukocytopoiesis. Hematoxylin and eosin stain, 10 × objective magnification.

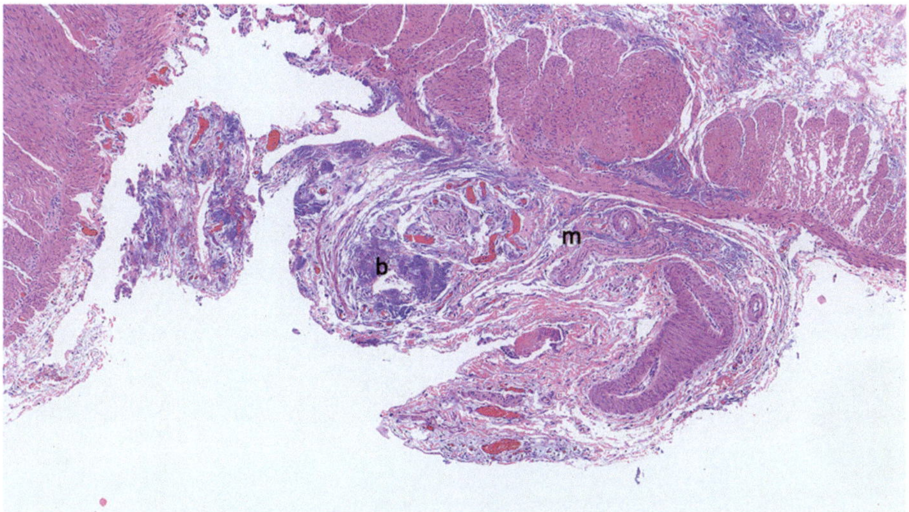

FIGURE 4.8C

An additional view of the previous specimen shows prominent bacterial colonies (b) in the mesenteric attachment (m). Hematoxylin and eosin stain, 5 × objective magnification.

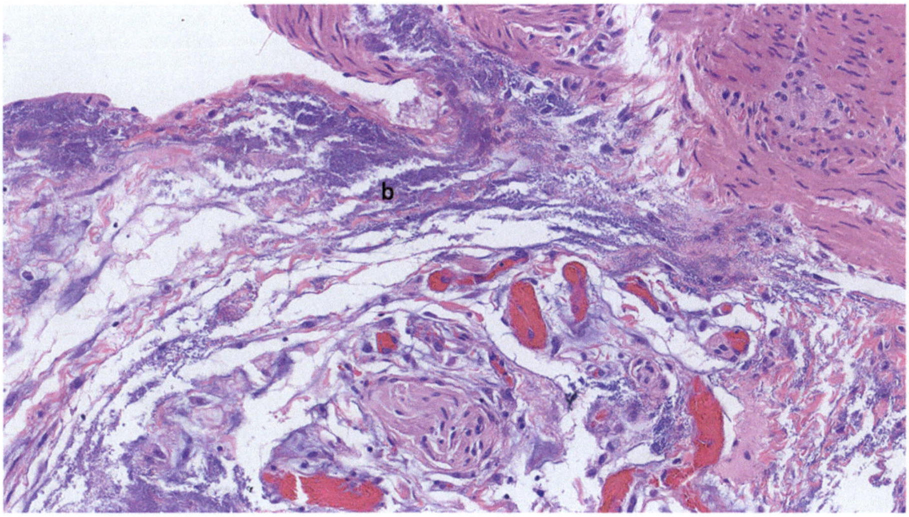

FIGURE 4.8D

Higher magnification of the previous specimen shows abundant particulate bacteria (b). Note the absence of leukocytic infiltration associated with bacterial colonization. Hematoxylin and eosin stain, 20 × objective magnification.

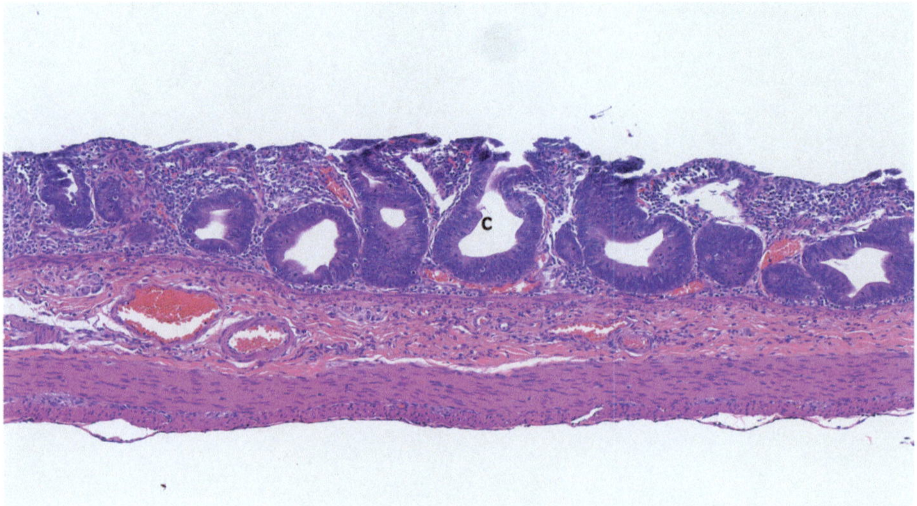

FIGURE 4.9

The jejunum of a male rhesus macaque necropsied 24 days after irradiation at 11 Gy shows all crypts (c) are lined by a thick layer of regenerating epithelial cells. Note the absence of villus structure. Hematoxylin and eosin stain, 10× objective magnification.

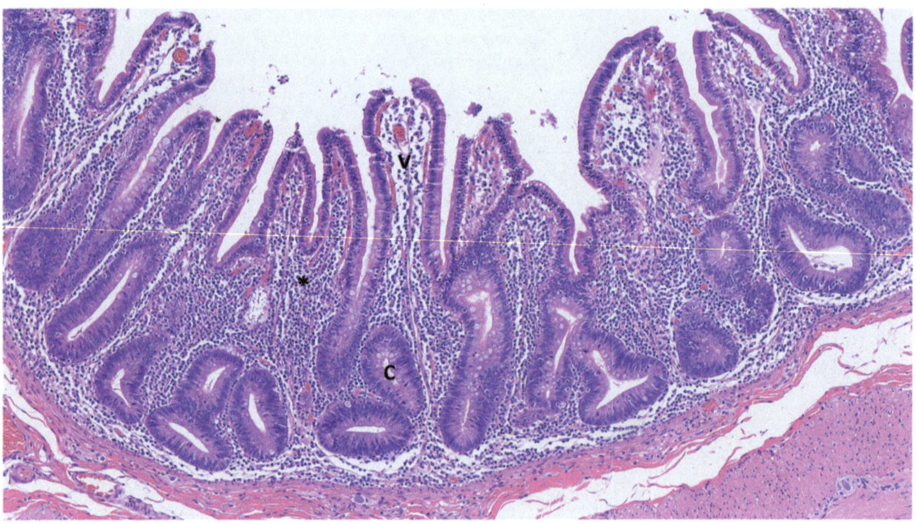

FIGURE 4.10

The jejunum of a male rhesus macaque necropsied 35 days after irradiation at 11 Gy shows villus structure (v) and a pronounced leukocytic population (*) in the mucosa. Villi are notably shorter than villi of naive male rhesus macaques. Crypt epithelial cells (c) remain highly proliferative, with numerous mitotic figures. Hematoxylin and eosin stain, 10× objective magnification.

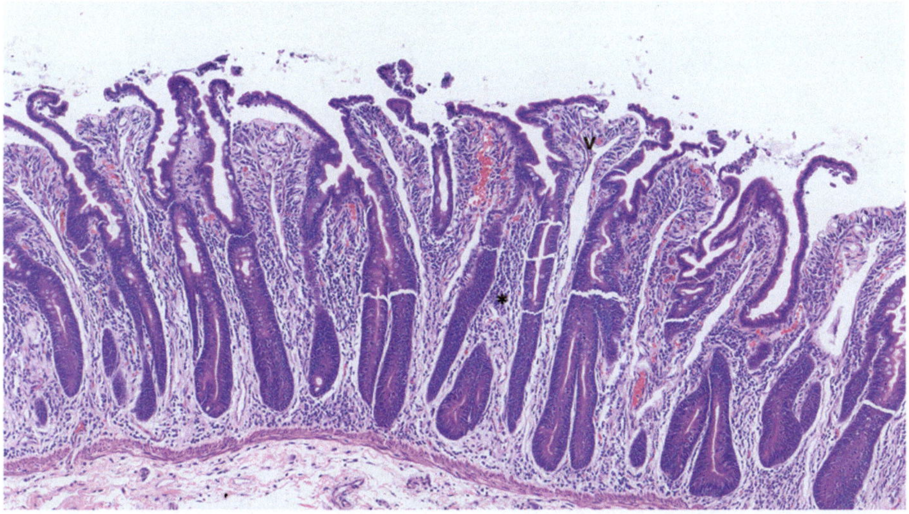

FIGURE 4.11A

The jejunum of a male rhesus macaque necropsied 42 days after irradiation at 11 Gy shows short, blunt villi (v) and a pronounced leukocytic population in the lamina propria (*). Hematoxylin and eosin stain, 10 × objective magnification.

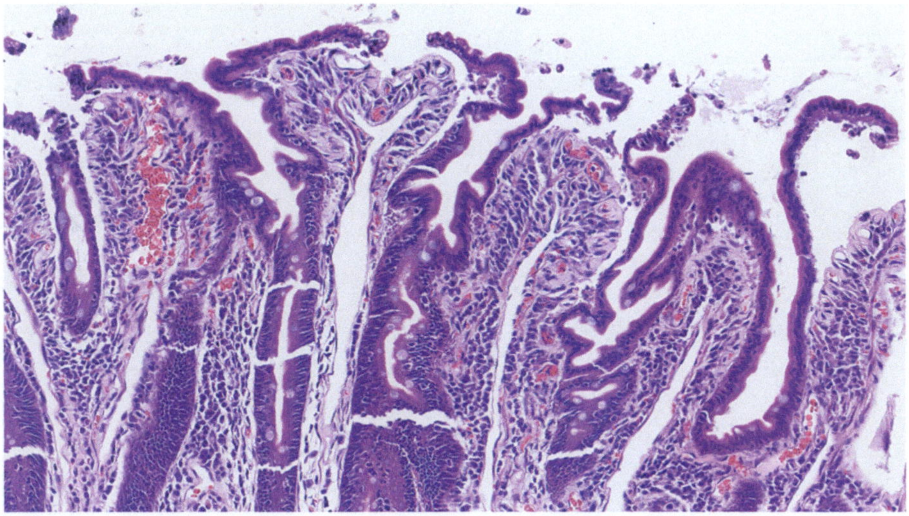

FIGURE 4.11B

Higher magnification of the previous specimen shows the short, blunted villi are covered by epithelial cells. Note the shape and staining characteristics of the epithelial cells at the tips of villi. The cuboidal shape and amphophillic staining of the cytoplasm are features of immature regenerative epithelial cells. The amphophillic staining characteristic of the epithelial cell cytoplasm reveals the presence of nucleic acids in addition to protein content, typical of replicating cells. Compare to the shape and staining characteristics of a naive male rhesus macaque shown in the following image. Hematoxylin and eosin stain, 20 × objective magnification.

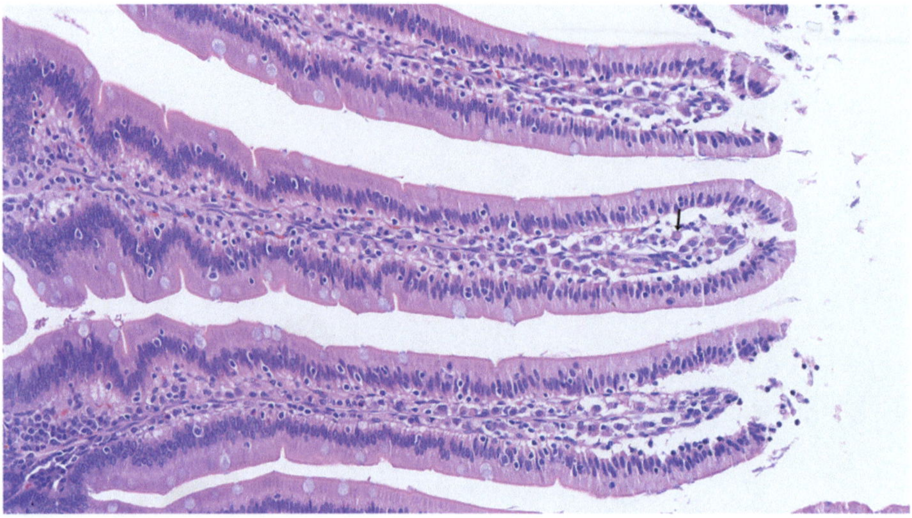

FIGURE 4.11C

For comparison, the jejunum of a naive male rhesus macaque has long, finger-like villi covered by high columnar epithelial cells. The epithelial cell cytoplasm is stained light pink, indicating predominant protein content, with blue-staining nucleic acids restricted to nuclei. Note the population of large mononuclear cells, consistent with macrophages or dendritic cells, within the lamina propria of villi (arrow). These cells emigrate to the mesenteric lymph nodes following immunological sampling of the intestinal environment. Hematoxylin and eosin stain, 20 × objective magnification.

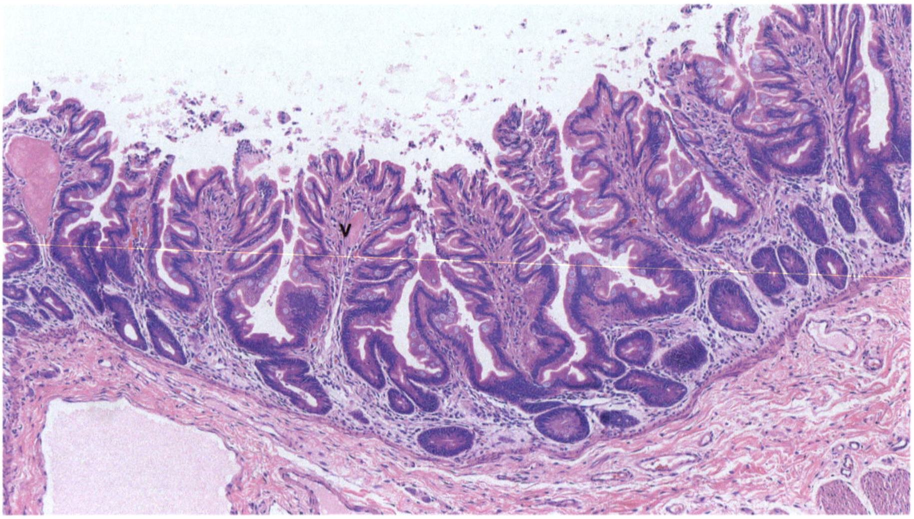

FIGURE 4.12

The jejunum of a male rhesus macaque necropsied 53 days after irradiation at 11Gy shows villi (v) completely covered with epithelial cells that have normal individual cellular morphology. The villi have the "leaf-shaped" contour that is sometimes encountered as has a nonspecific histologic presentation of uncertain genesis and significance. There is evidence the leaf-shaped morphology is more commonly observed in nonhuman primates receiving parenteral nutrition, thus may be an indication of reduced intestinal functioning. Hematoxylin and eosin stain, 10 × objective magnification.

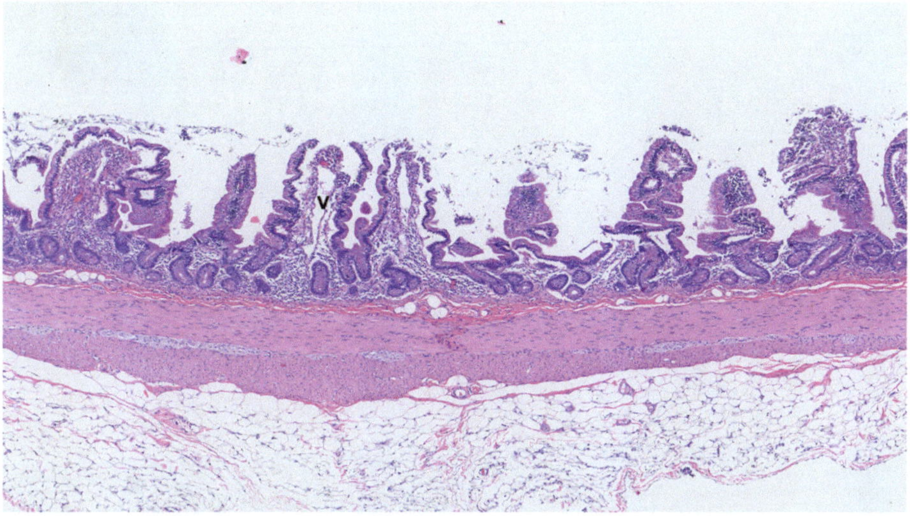

FIGURE 4.13

The jejunum of a male rhesus macaque necropsied 93 days after irradiation at 10 Gy shows villi (v) that are somewhat shortened and blunted but covered by epithelial cells that have normal histomorphology. The villi are not as closely approximated as the villi of the naive male rhesus macaque shown previously. Hematoxylin and eosin stain, 5 × objective magnification.

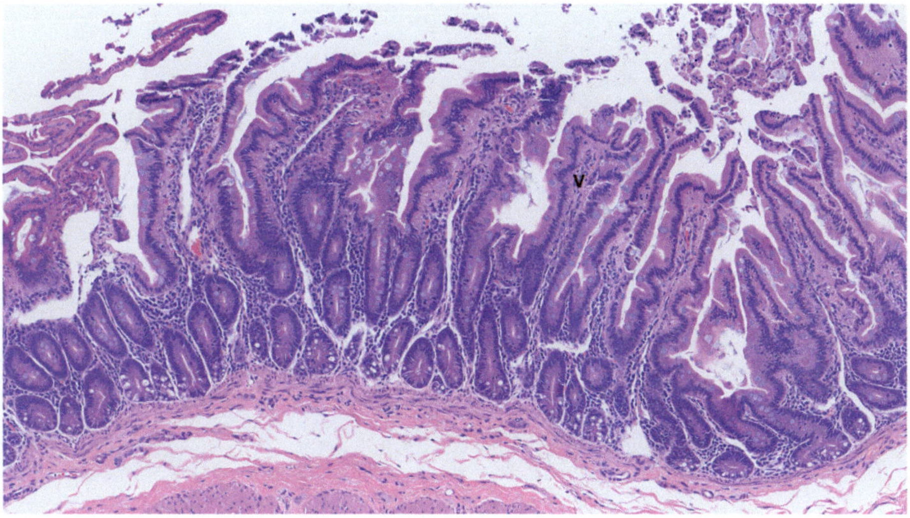

FIGURE 4.14

The jejunum of a male rhesus macaque necropsied 105 days after irradiation at 10 Gy shows villi (v) that are shorter than villi of the naive male rhesus macaque presented above but are otherwise histologically normal. Hematoxylin and eosin stain, 10 × objective magnification.

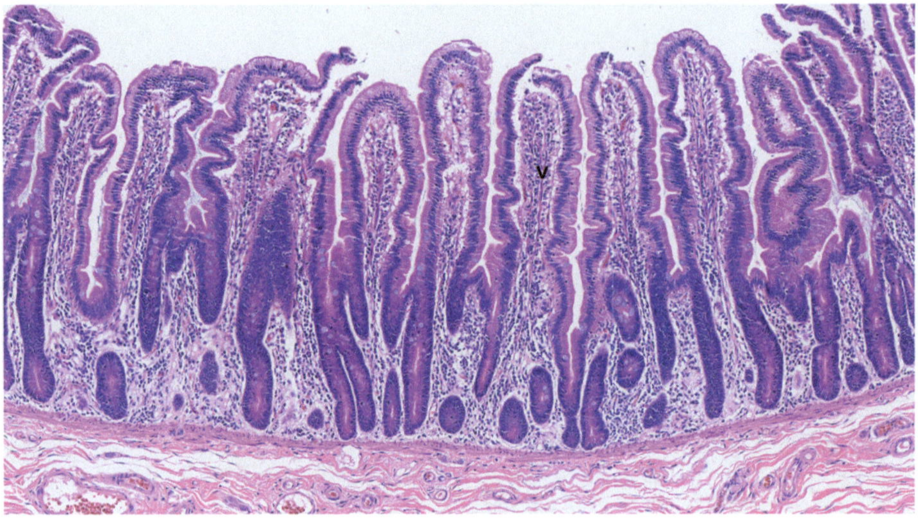

FIGURE 4.15

The jejunum of a male rhesus macaque necropsied 191 days after irradiation at 10 Gy has villi (v) that are somewhat shorter than villi of the naive male rhesus macaque presented above but are otherwise histologically normal. Hematoxylin and eosin stain, 10 × objective magnification.

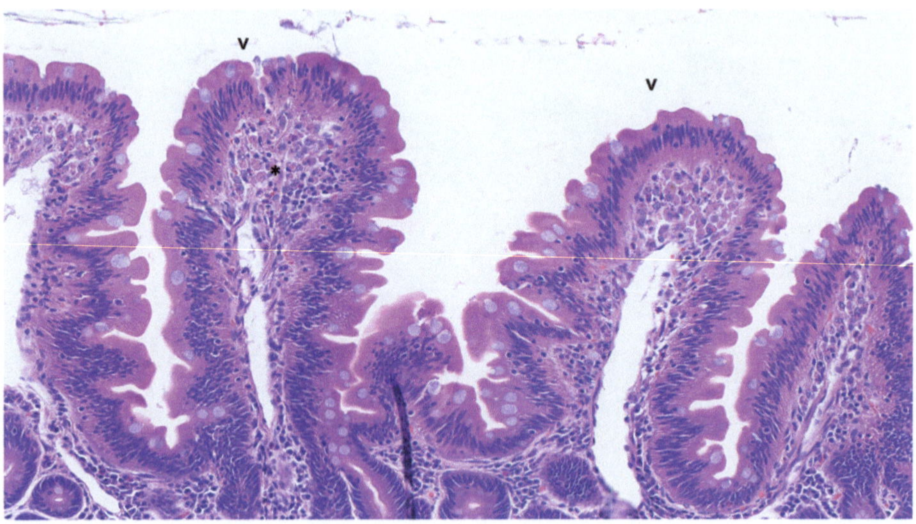

FIGURE 4.16

The jejunum of a male rhesus macaque necropsied 181 days after irradiation at 11 Gy has somewhat shortened villi (v) covered by microscopically normal epithelium. Note the population of large mononuclear cells (*) in the lamina propria of villus tips. The persistence of histomorphologic abnormalities throughout the 180-day observation period in some animal suggests the small intestine has not completely recovered from the radiation-associated injury. Hematoxylin and eosin stain, 20 × objective magnification.

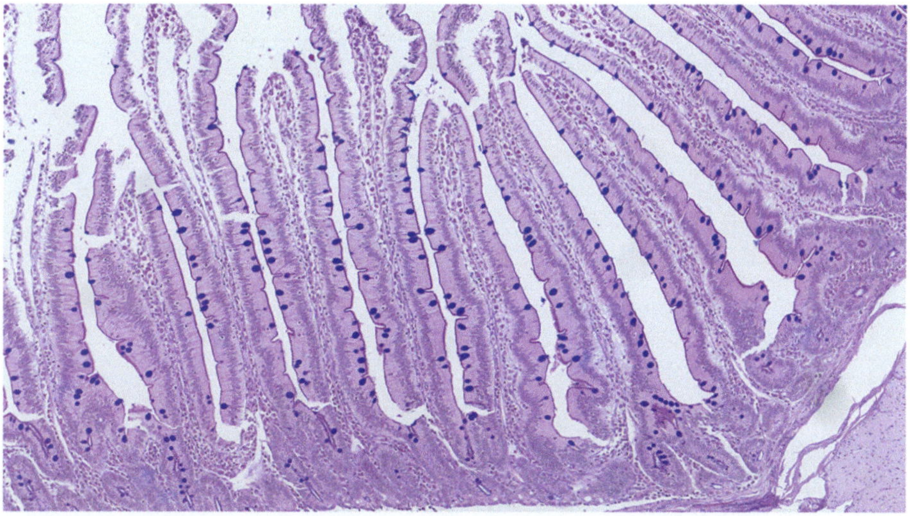

FIGURE 4.17A

Periodic acid-Schiff/Alcian blue stain performed on Jejunum of a naive male rhesus macaque shows numerous blue-stained goblet cells in the superficial epithelium. 10 × objective magnification.

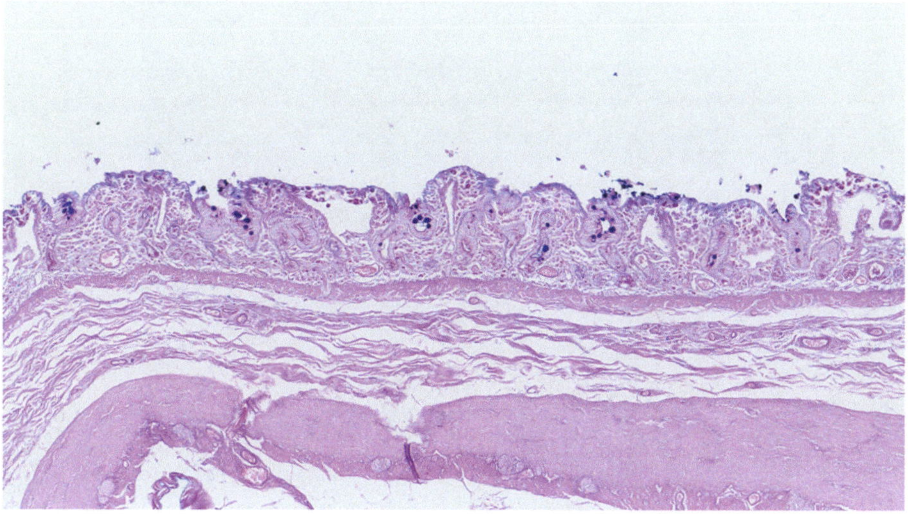

FIGURE 4.17B

Periodic acid-Schiff/Alcian blue stain performed on the Jejunum of a male rhesus macaque necropsied 6 days after irradiation at 10 Gy shows a marked reduction in blue-stained goblet cells in the superficial epithelium. 10 × objective magnification.

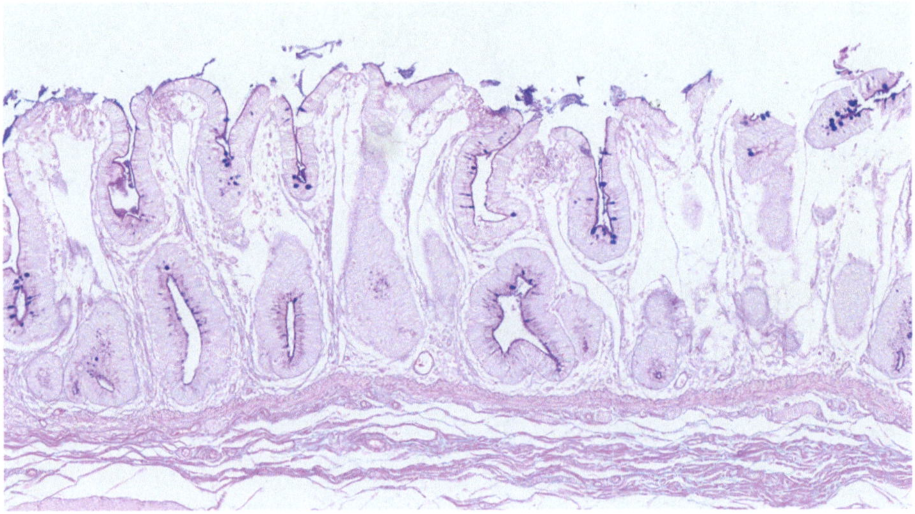

FIGURE 4.17C

Periodic acid-Schiff/Alcian blue stain performed on the Jejunum of a male rhesus macaque necropsied 29 days after irradiation at 10 Gy shows persistent reduction in the population of blue-stained goblet cells in the superficial epithelium. 10 × objective magnification.

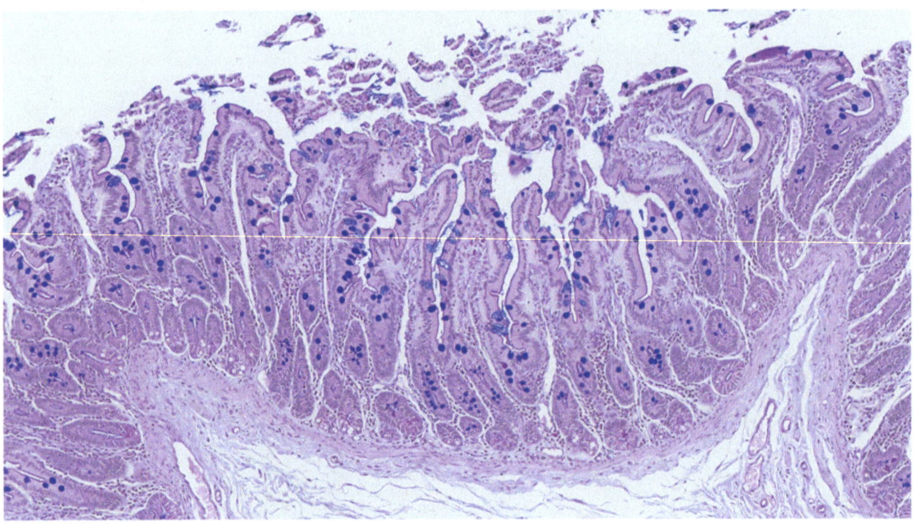

FIGURE 4.17D

Periodic acid-Schiff/Alcian blue stain performed on the Jejunum of a male rhesus macaque necropsied 105 days after irradiation at 10 Gy shows a normal population of blue-stained goblet cells in the superficial epithelium. 10 × objective magnification.

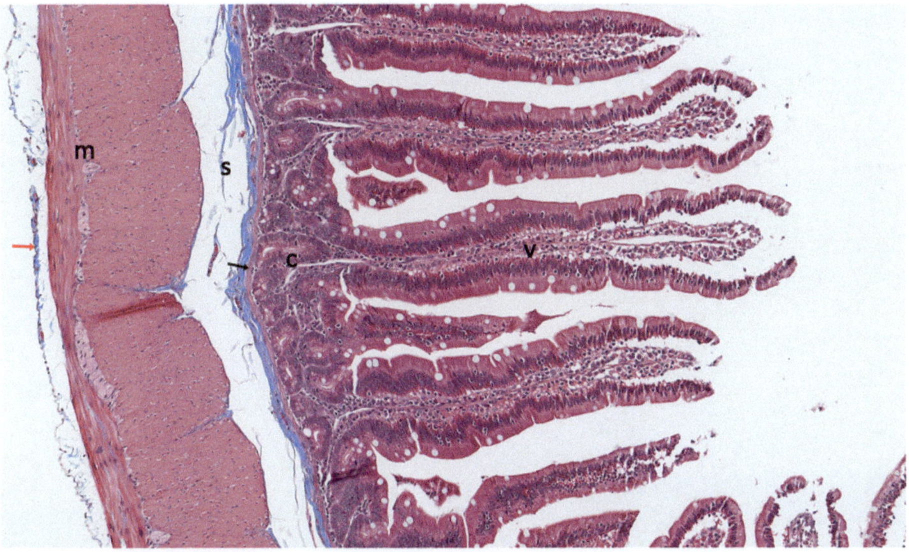

FIGURE 4.18

A Masson's trichrome stain performed on jejunum from a naive male rhesus macaque shows little blue-stained fibrous connective tissue in villi (v), crypts (c), muscularis mucosa (black arrow), or external muscular layer (m). The submucosa (s) and serosa (red arrow) contain a small amount of fibrous connective tissue. 10 × objective magnification.

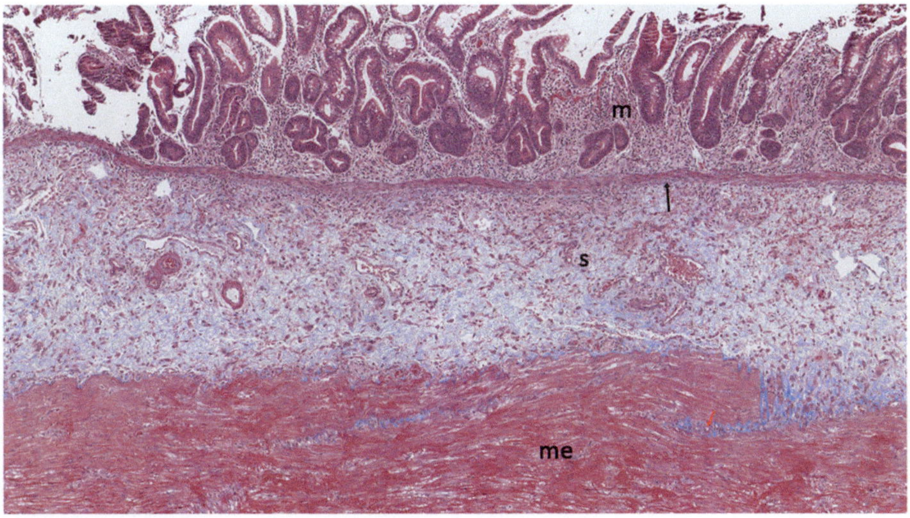

FIGURE 4.19A

Masson's trichrome stain on the jejunum of a male rhesus macaque necropsied 20 days after irradiation at 10 Gy shows a moderate amount of loosely woven, blue-stained fibrous connective tissue in the thickened submucosa (s). There is no evidence of fibrosis in the mucosa (m), or muscularis mucosa (black arrow). Fibrous connective tissue (red arrow) in the external muscular layer (me) is associated with vascular structures. 5 × objective magnification.

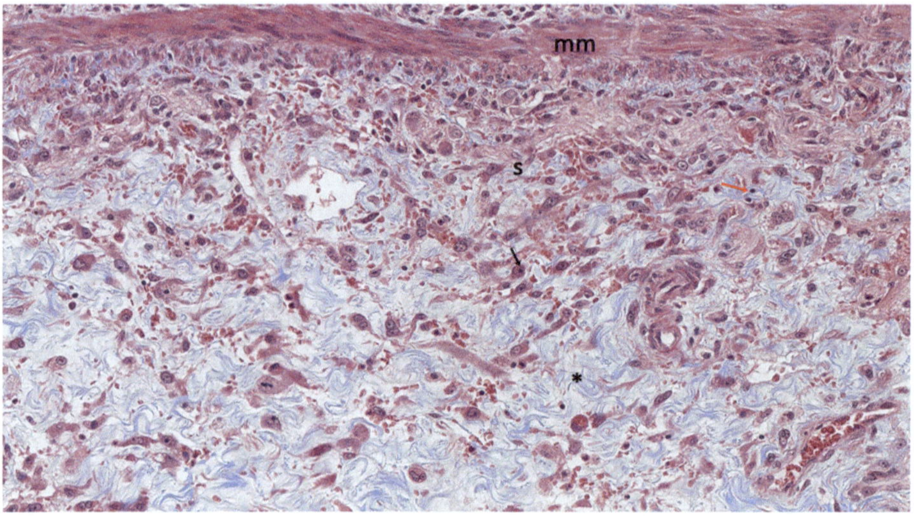

FIGURE 4.19B

Higher magnification of the previous specimen shows loosely woven immature fibrous connective tissue (*) in the jejunal submucosa. The submucosa has a moderate population of large, individualized mononuclear cells (arrow), consistent with macrophages or reactive stromal cells. Leukocytic infiltration (red arrow) is notably sparse. Note the intact muscularis mucosa (mm) separating the submucosa (s) from the overlying mucosa. Masson's trichrome stain, 20 × objective magnification.

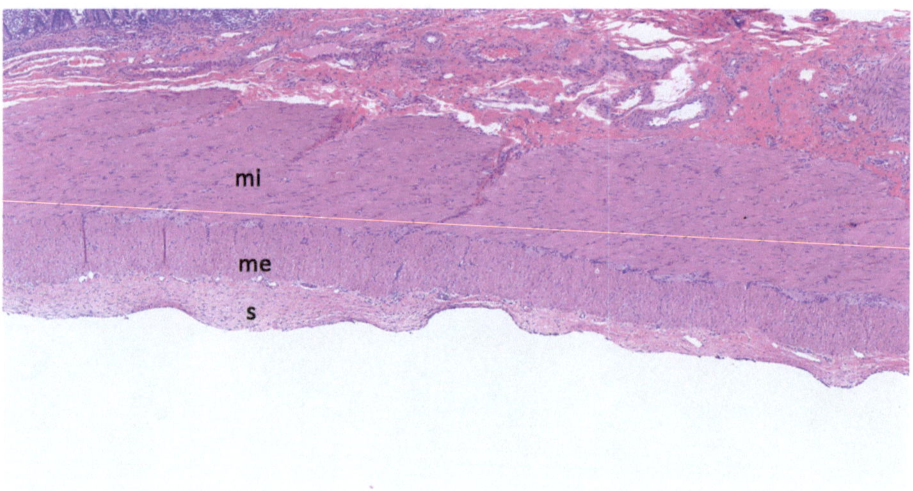

FIGURE 4.20A

The jejunal serosa of a male rhesus macaque necropsied 105 days after irradiation at 11 Gy has a plaque-like area of dense fibrous connective tissue (s). The external (me) and internal (mi) layers of the muscular wall have no discernible evidence of fibrosis. Hematoxylin and eosin stain, 3.68 × objective magnification.

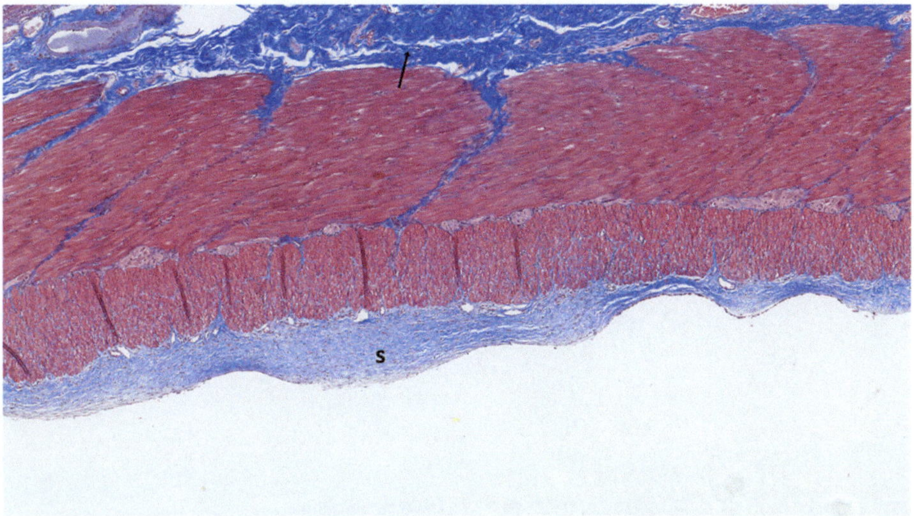

FIGURE 4.20B

Masson's trichrome stain performed on the specimen shown in the previous figure shows blue-stained fibrous connective tissue on the serosal surface (s) of the jejunum. Note the reduced intensity of blue staining in the newly formed fibrous connective tissue on the serosal surface versus the intense blue staining of the mature fibrous connective tissue in the submucosa (arrow). 5 × objective magnification.

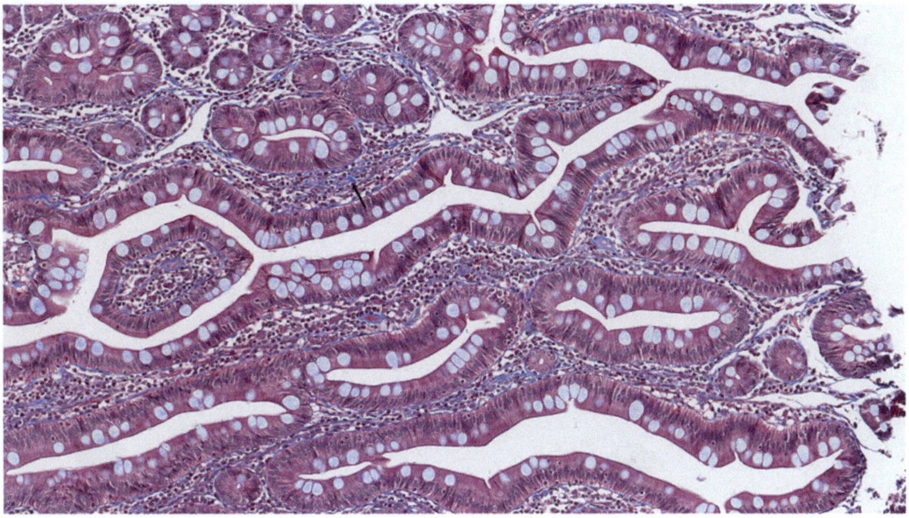

FIGURE 4.20C

Higher magnification of the mucosa of the previous specimen shows blue-stained fibrous connective tissue (arrow) in the mucosal lamina propria. Masson's trichrome stain, 20 × objective magnification.

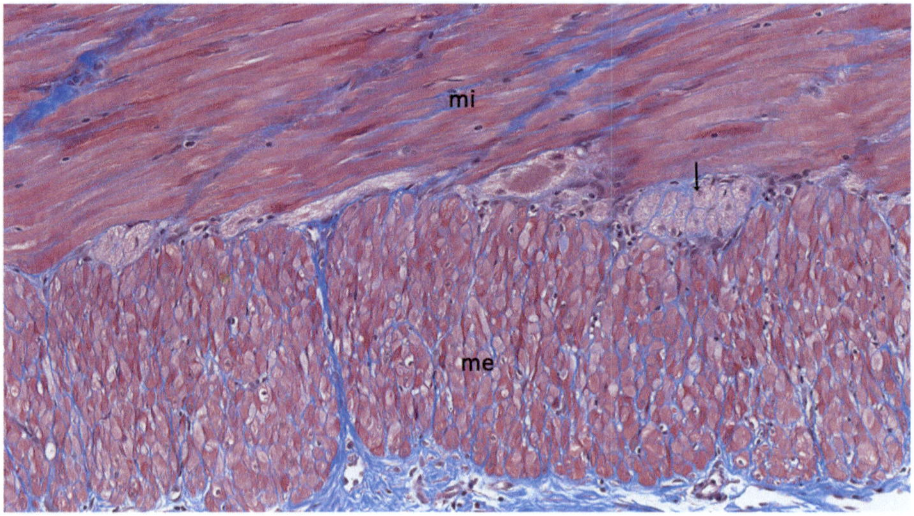

FIGURE 4.20D

Higher magnification of the previous trichrome-stained specimen shows thin filaments of fibrous connective surrounding myofibers of the external (me) and internal (mi) layers of the muscular wall, as well as the myenteric (Auerbach's) nerve plexus (arrow). Compare this image to the following image from the jejunum of a naive male rhesus macaque. 20 × objective magnification.

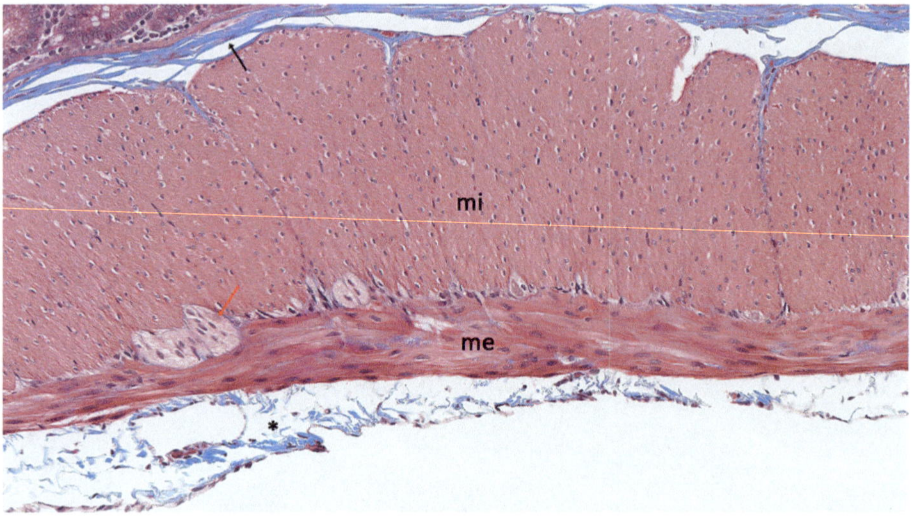

FIGURE 4.20E

Masson's trichrome stain performed on the jejunum of a naive male rhesus macaque shows little blue-stained fibrous connective tissue within the external (me) or internal (mi) layers of the muscular wall. A small amount of fibrous connective tissue is present in the serosa (*) and submucosa (black arrow). Fibrous connective tissue is not present in the myenteric (Auerbach's) nerve plexus (red arrow). 20 × objective magnification.

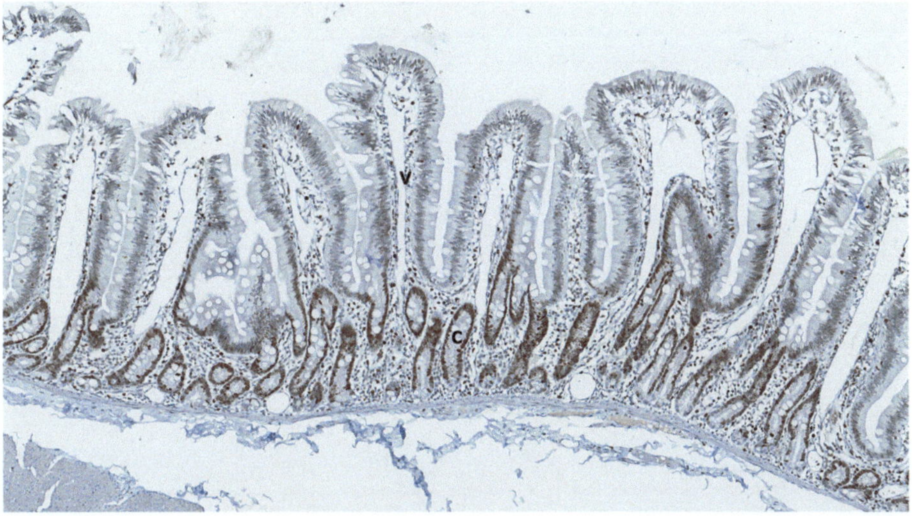

FIGURE 4.21

Immunohistochemical staining for Ki-67 performed on the jejunum of a naive male rhesus macaque shows positive staining of crypt cells (c), with no staining of epithelial cells on the sides or tips of villi (v). Stain performed with 3,3'-diaminobenzidine chromogen and hematoxylin counterstain, 10 × objective magnification.

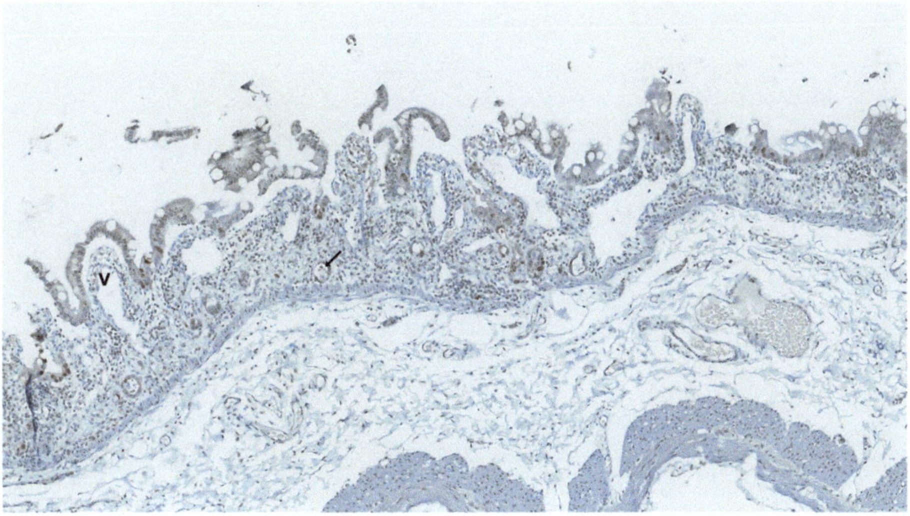

FIGURE 4.22

Immunohistochemical staining for Ki-67 performed on the jejunum of a male rhesus macaque necropsied 6 days after irradiation at 12 Gy reveals little evidence of cellular replication. Villi (v) are severely disrupted, and crypts (arrow) are barely discernible. Stain performed with 3,3'-diaminobenzidine chromogen and hematoxylin counterstain, 10 × objective magnification.

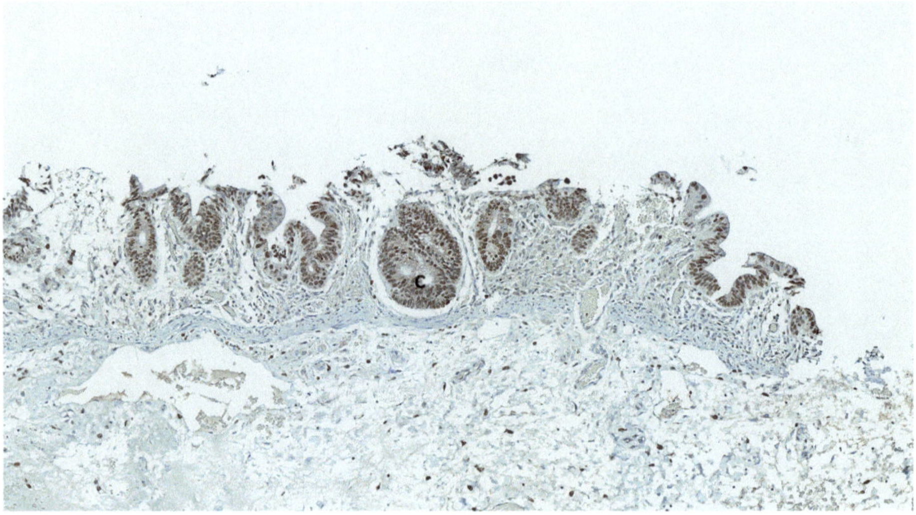

FIGURE 4.23

Immunohistochemical staining for Ki-67 performed on the jejunum of a male rhesus macaque necropsied 9 days after irradiation at 12 Gy reveals pronounced cellular proliferation within crypt remnants (c). Stain performed with 3,3'-diaminobenzidine chromogen and hematoxylin counterstain, 10 × objective magnification.

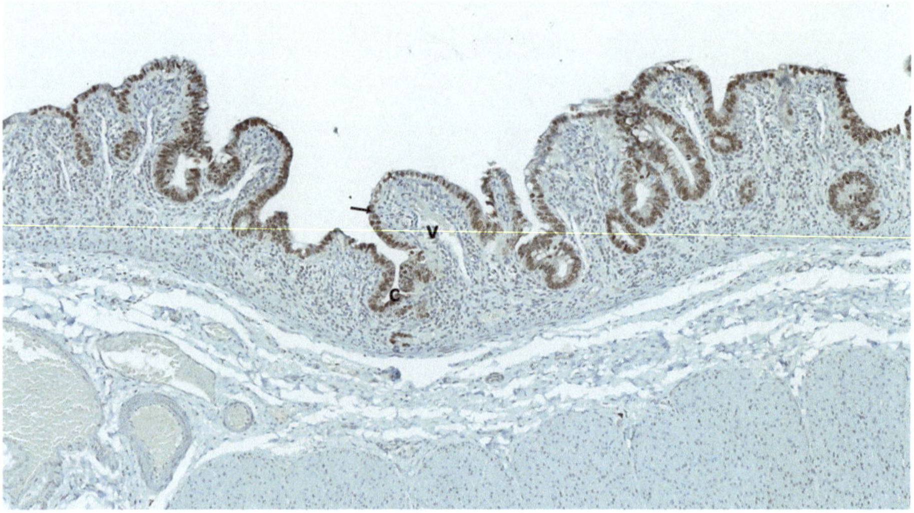

FIGURE 4.24A

Immunohistochemical staining for Ki-67 performed on the jejunum of a male rhesus macaque necropsied 10 days after irradiation at 12 Gy reveals extensive cellular replication within crypts (c). Ki-67^{+} cells (arrow) are present on the sides and tips of reforming villi (v). Stain performed with 3,3'-diaminobenzidine chromogen and hematoxylin counterstain, 10 × objective magnification.

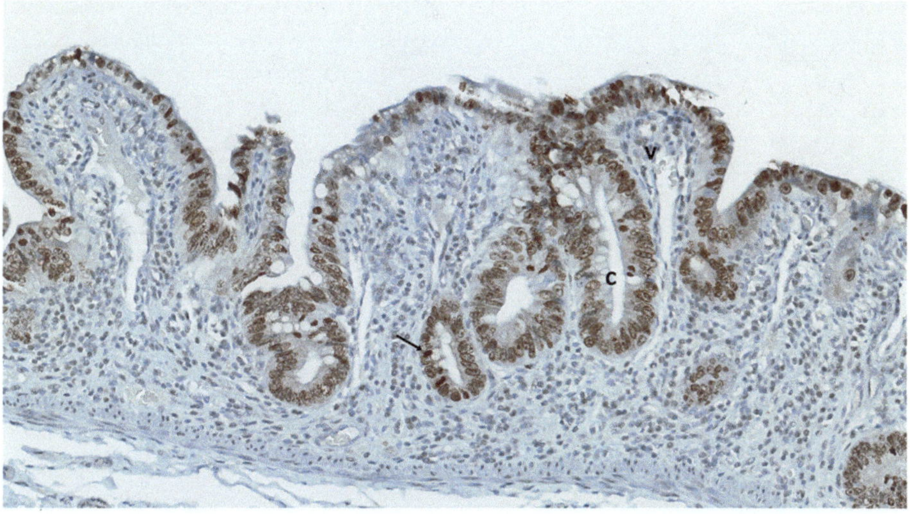

FIGURE 4.24B

Higher magnification of the previous specimen shows Ki-67$^+$ cells within crypts (c) and covering reforming villi (v). Mitotic figures (arrow) are restricted to the crypts. Immunohistochemical stain for Ki-67 with 3,3'-diaminobenzidine chromogen and hematoxylin counterstain, 20 × objective magnification.

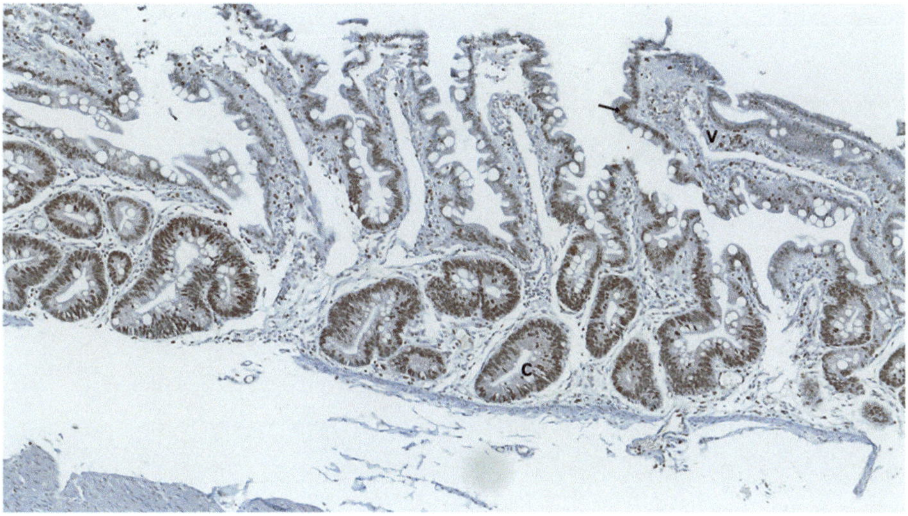

FIGURE 4.24C

Immunohistochemical staining for Ki-67 performed on the jejunum of a male rhesus macaque necropsied 29 days after irradiation at 12 Gy shows labeled cells are most prominent in crypts (c), though clusters of translocated labeled cells (arrow) are present on the sides and tips of villi (v). Stain performed with 3,3'-diaminobenzidine chromogen and hematoxylin counterstain, 10 × objective magnification.

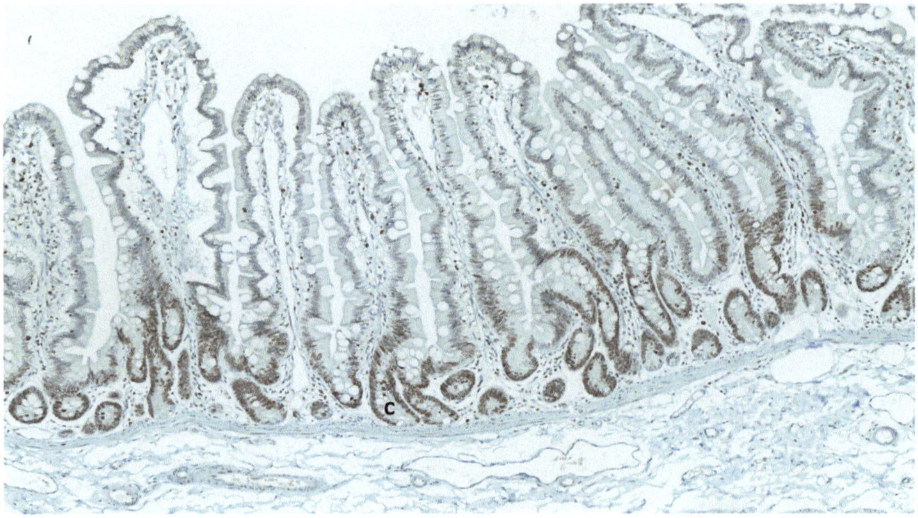

FIGURE 4.24D

Immunohistochemical staining for Ki-67 performed on the jejunum of a male rhesus macaque necropsied 111 days after irradiation at 11 Gy shows labeled cells are restricted to the crypts (c). Stain performed with 3,3’-diaminobenzidine chromogen and hematoxylin counterstain, 10 × objective magnification.

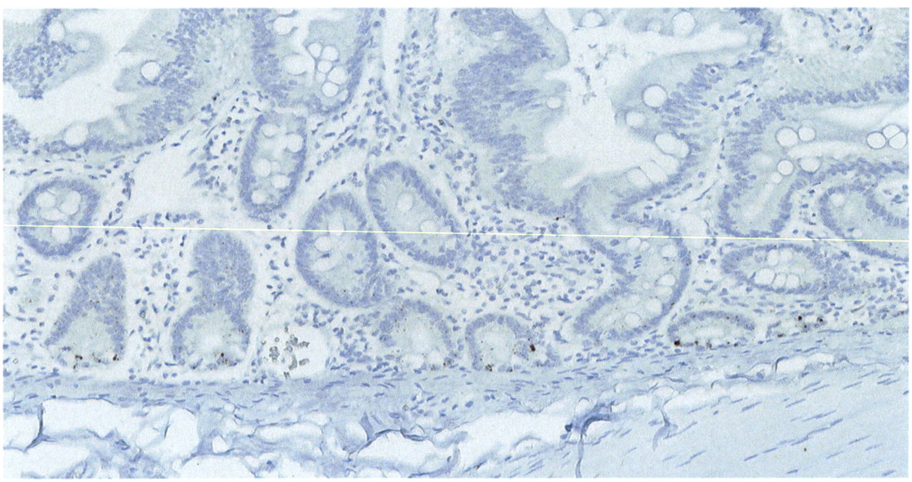

FIGURE 4.25A

In situ hybridization staining for Lgr5 performed on jejunum of naive control male rhesus macaque. Note the minimal amount of black-stained Lgr^+ cells located in the deep aspect of intestinal crypts, indicating sparse expression of Lgr5. Stain performed with 3,3’-diaminobenzidine chromogen and hematoxylin counterstain. 20 × objective magnification.

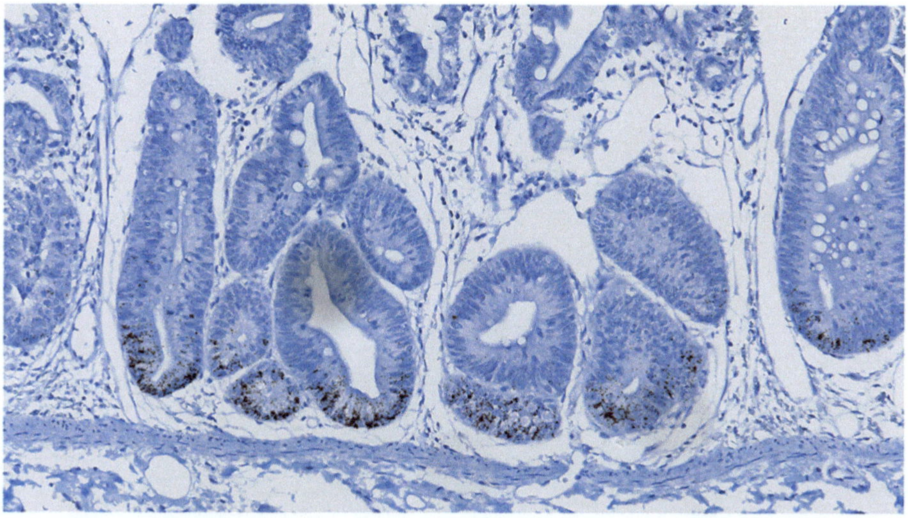

FIGURE 4.25B

In situ hybridization staining for Lgr5 performed on jejunum of a male rhesus macaque necropsied 20 days after irradiation at 10 Gy shows an increased number of $Lgr5^{+}$ cells in the intestinal crypts. Stain performed with 3,3'-diaminobenzidine chromogen and hematoxylin counterstain. 20 × objective magnification.

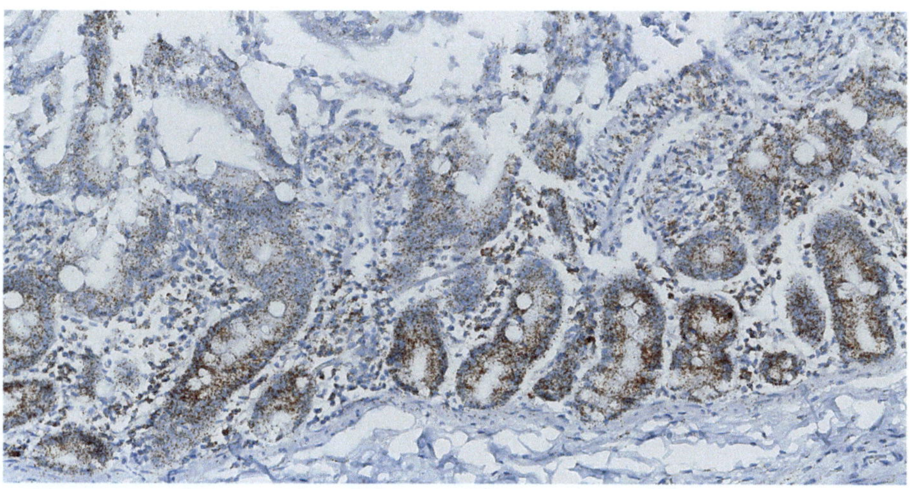

FIGURE 4.26

In situ hybridization staining for positive control marker (PPIB) (RNA integrity control) performed on jejunum of a naive control male rhesus macaque shows extensive positive staining, thus confirming the integrity of RNA in the specimen. RNA integrity of each intestinal specimen was confirmed by parallel in situ staining for PPIB. Stain performed with 3,3'-diaminobenzidine chromogen and hematoxylin counterstain. 20 × objective magnification.

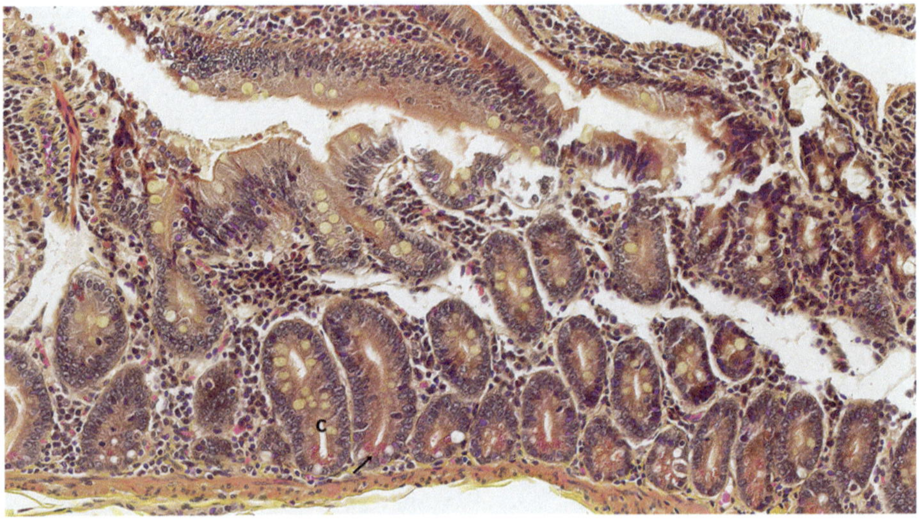

FIGURE 4.27A

A Lendrum stain performed on the small intestine of a nonhuman primate reveals bright red cytoplasmic granules in Paneth cells (arrow) located in the depth of crypts (c). Erythrocytes within capillaries are also stained by the Lendrum stain, 20 × objective magnification.

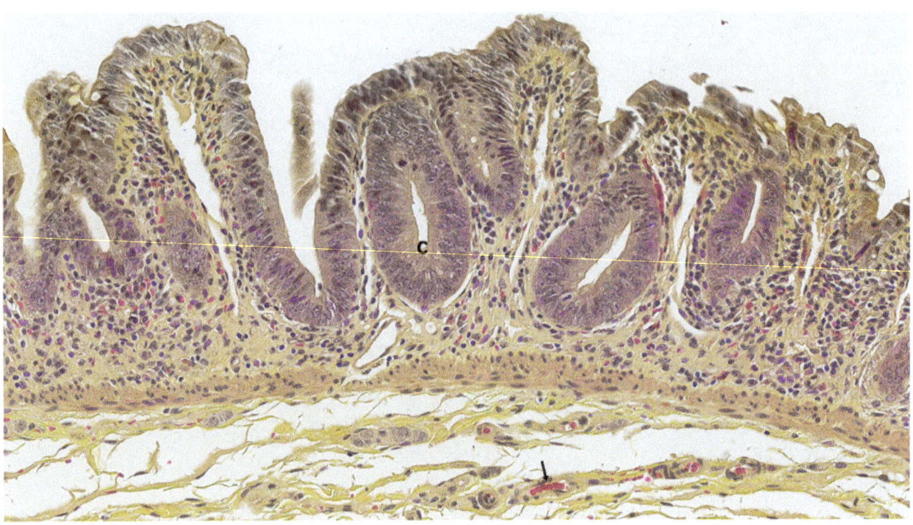

FIGURE 4.27B

A Lendrum stain performed on the jejunum of a male rhesus macaque necropsied 10 days after irradiation at 12 Gy reveals no remaining Paneth cells in crypts (c). Note erythrocytes in blood vessels (arrow) are positively stained. 20 × objective magnification.

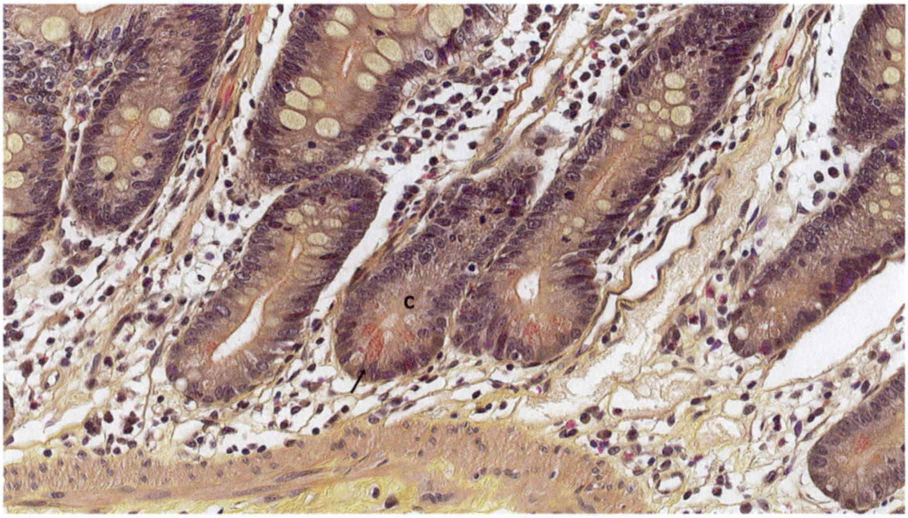

FIGURE 4.27C

A Lendrum stain performed on the jejunum of a male rhesus macaque necropsied 91 days after irradiation at 11 Gy reveals a few Paneth cells (arrow) in crypts (c). Note erythrocytes in blood vessels (arrow) are positively stained. 30.3 × objective magnification.

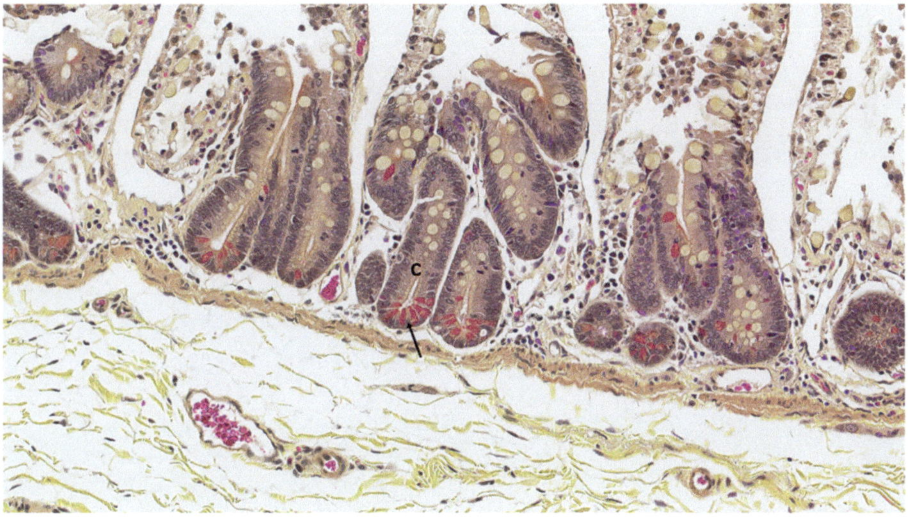

FIGURE 4.27D

A Lendrum stain performed on the jejunum of a male rhesus macaque necropsied 105 days after irradiation at 10 Gy reveals numerous Paneth cells (arrow) in crypts (c). Note erythrocytes in blood vessels (arrow) are positively stained. 20 × objective magnification.

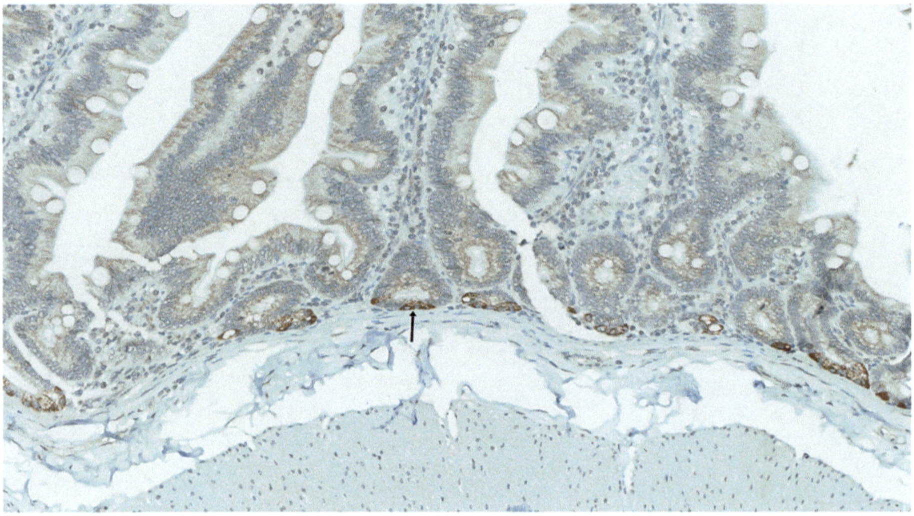

FIGURE 4.28A

Immunohistochemical staining for connective tissue growth factor performed on the jejunum of a naive male rhesus macaque reveals a population of positively stained cells (arrow) in the depths of intestinal crypts, consistent with Paneth cells. Stain performed with 3,3'-diaminobenzidine chromogen and hematoxylin counterstain, 20 × objective magnification.

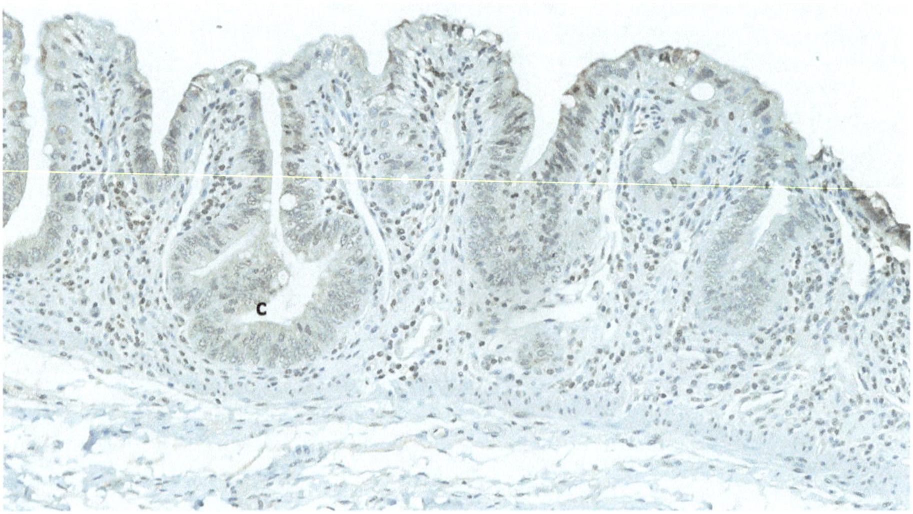

FIGURE 4.28B

Immunohistochemical staining for connective tissue growth factor (CTGF) on a male rhesus macaque necropsied 10 days after irradiation at 11 Gy (same specimen used for Lendrum staining, above) reveals no $CTGF^+$ cells in crypts. (c) Stain performed with 3,3'-diaminobenzidine chromogen and hematoxylin counterstain, 20 × objective magnification.

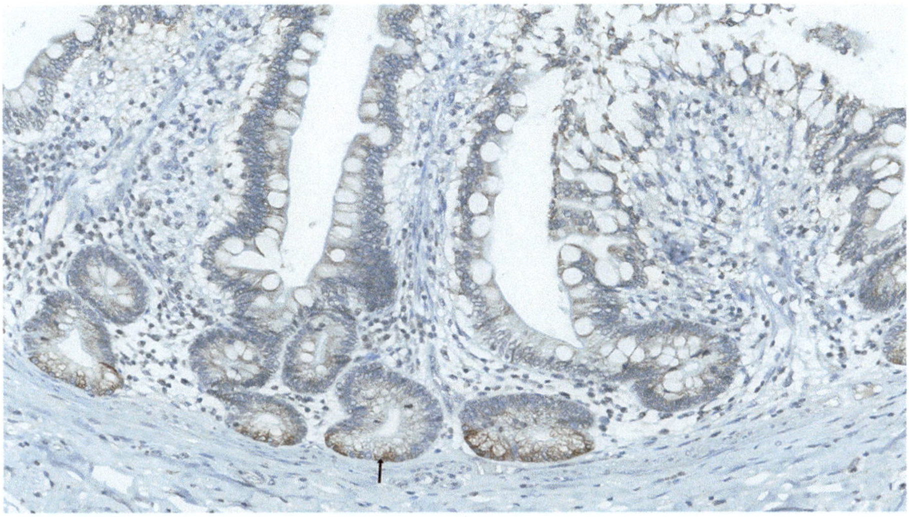

FIGURE 4.28C

Immunohistochemical staining for connective tissue growth factor (CTGF) on a male rhesus macaque necropsied 91 days after irradiation at 11 Gy (same specimen used for Lendrum staining, above) reveals a population of $CTGF^+$ cells (arrow) in crypts. Stain performed with 3,3'-diaminobenzidine chromogen and hematoxylin counterstain, 20 × objective magnification.

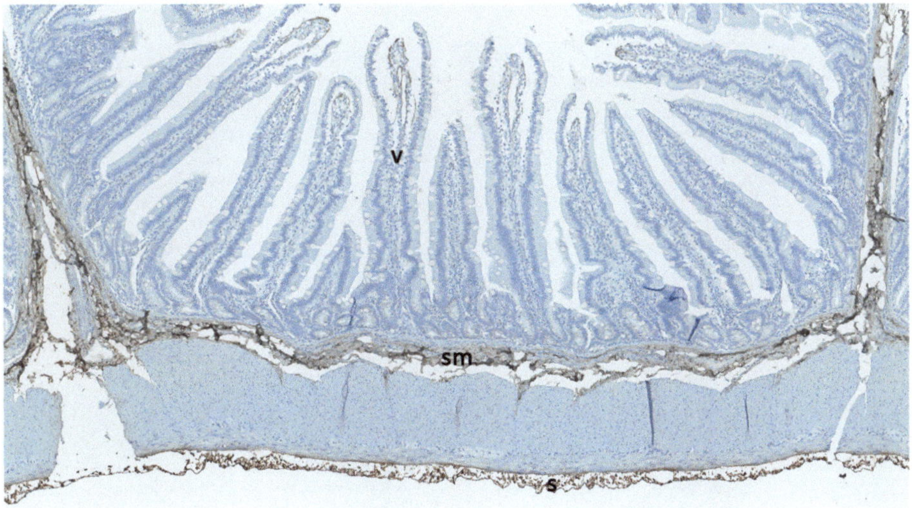

FIGURE 4.29

Immunohistochemical staining for collagen-1 on the jejunum of a naïve male rhesus macaque shows collagen staining in the serosa (s) and submucosa (sm), with faintly discernible staining in the tips of villi (v). Stain performed with 3,3'-diaminobenzidine chromogen and hematoxylin counterstain, 6.36 × objective magnification.

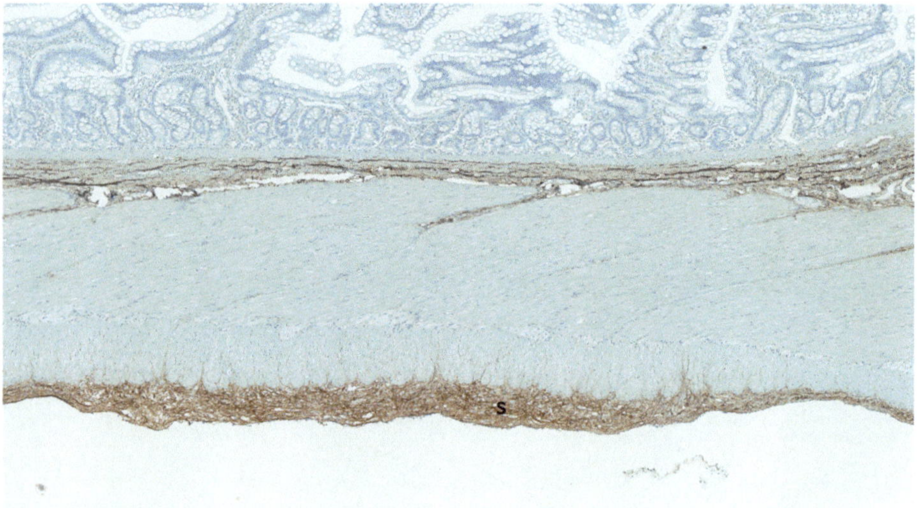

FIGURE 4.30

Immunohistochemical staining for collagen-1 on a male rhesus macaque necropsied 183 days after irradiation at 11 Gy shows a thick band of collagenous tissue on the serosal surface (s) of the jejunum. Stain performed with 3,3'-diaminobenzidine chromogen and hematoxylin counterstain, 5 × objective magnification.

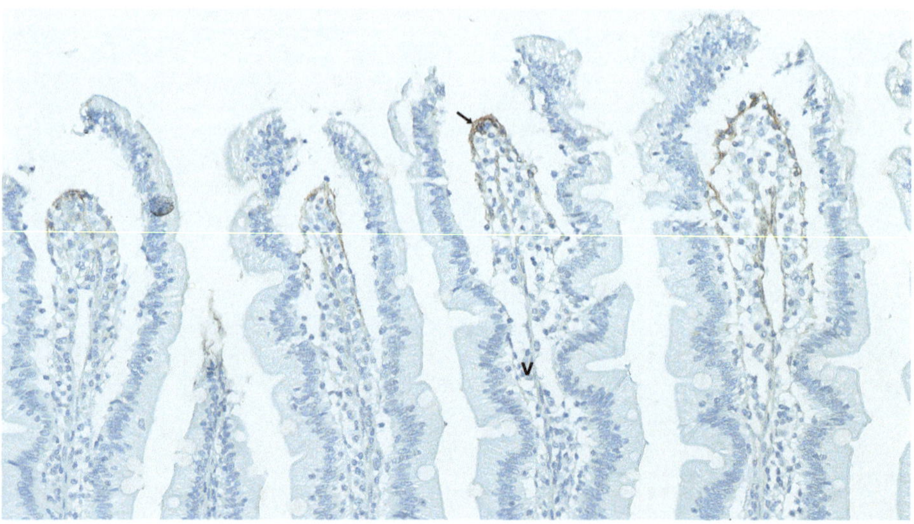

FIGURE 4.31

Immunohistochemical staining for collagen-1 performed on jejunum of a naïve male rhesus macaque shows a minimal amount of collagen staining (arrow) in the lamina propria at the tips of villi (v). Stain performed with 3,3'-diaminobenzidine chromogen and hematoxylin counterstain, 20 × objective magnification.

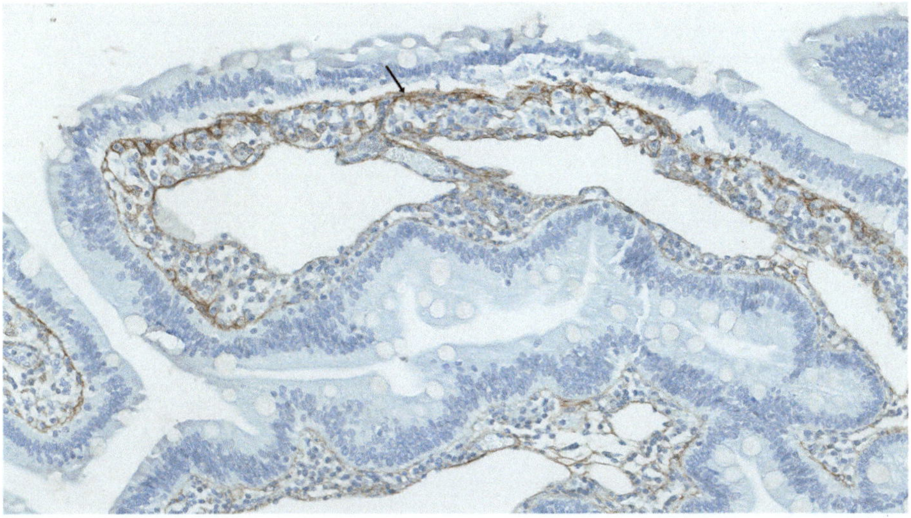

FIGURE 4.32

Immunohistochemical staining for collagen-1 on a male rhesus macaque necropsied 107 days after irradiation at 11 Gy reveals an increased amount of collagen (arrow) in the lamina propria of villi. Stain performed with 3,3'-diaminobenzidine chromogen and hematoxylin counterstain, 20 × objective magnification.

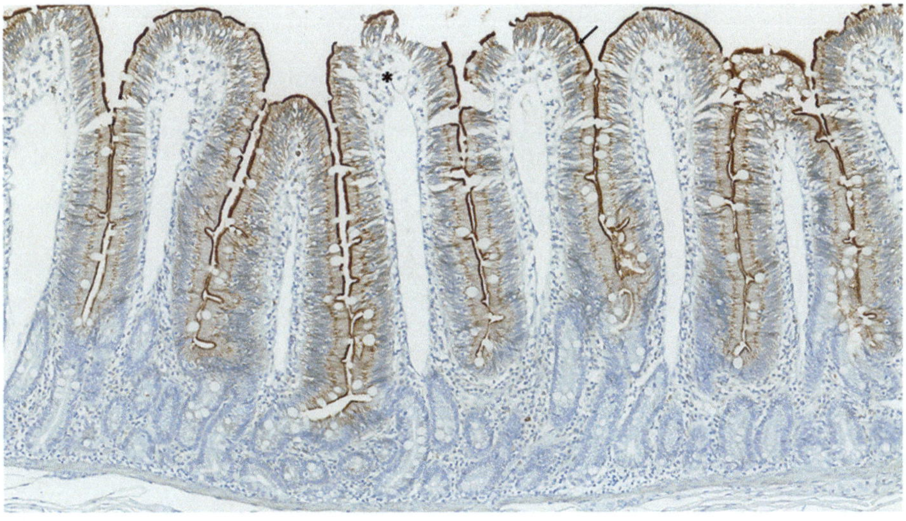

FIGURE 4.33A

Immunohistochemical staining for CD13 (aminopeptidaase N) performed on jejunum of a naive male rhesus macaque shows distinct staining of the brush border (arrow) on the luminal surface of jejunal epithelial cells. Note absence of staining of cell populations in lamina propria of villus tipes (*). Stain performed with 3,3'-diaminobenzidine chromogen and hematoxylin counterstain, 12.4 × objective magnification.

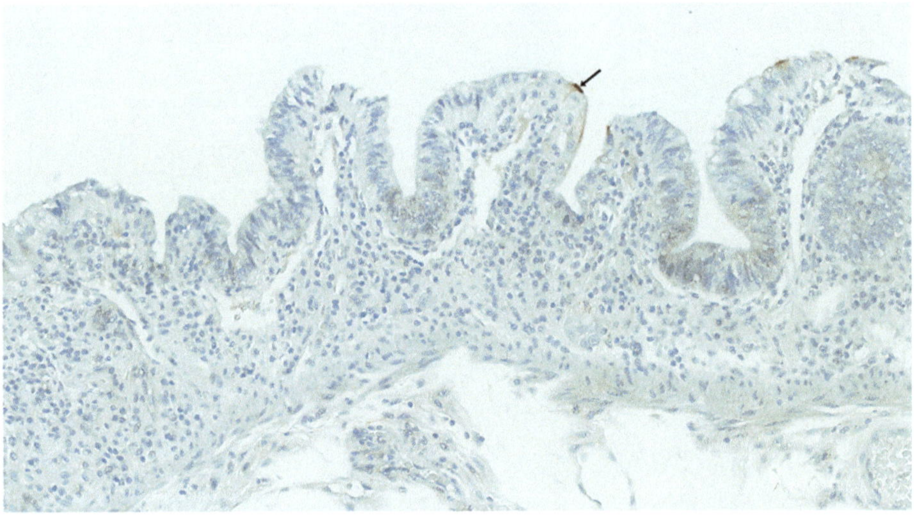

FIGURE 4.33B

Immunohistochemical staining for CD13 performed on jejunum of a male rhesus macaque necropsied 10 days after irradiation at 12 Gy has only scattered remnants of $CD13^+$ brush border elements (arrow). Stain performed with 3,3'-diaminobenzidine chromogen and hematoxylin counterstain, 20 × objective magnification.

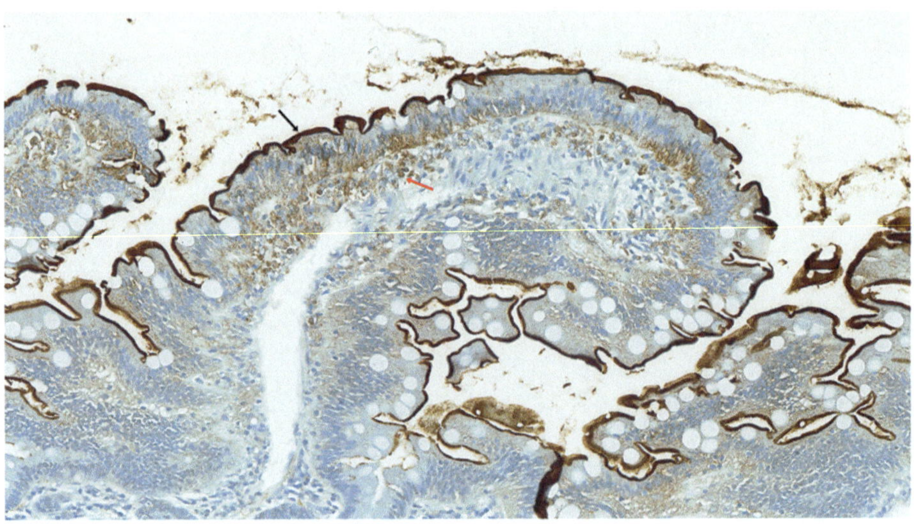

FIGURE 4.33C

Immunohistochemical staining for CD13 performed on jejunum of a male rhesus macaque necropsied 182 days after irradiation at 10 Gy has continuous $CD13^+$ brush border (black arrow) on the luminal surface of villus epithelial cells. Note the positive staining of macrophages/dendritic cells (red arrow) in the lamina propria. Stain performed with 3,3'-diaminobenzidine chromogen and hematoxylin counterstain, 20 × objective magnification.

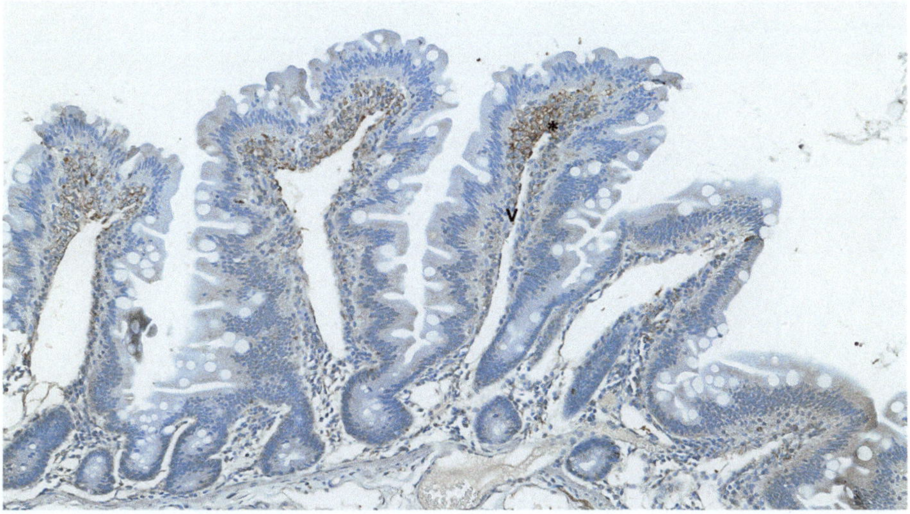

FIGURE 4.34

Immunohistochemical staining for MxA performed on the same CD13-stained jejunal specimen reveals a substantial population of activated immunocytes (*) in the lamina propria at the tips of villi (v). Stain performed with 3,3'-diaminobenzidine chromogen and hematoxylin counterstain, 15.7 × objective magnification.

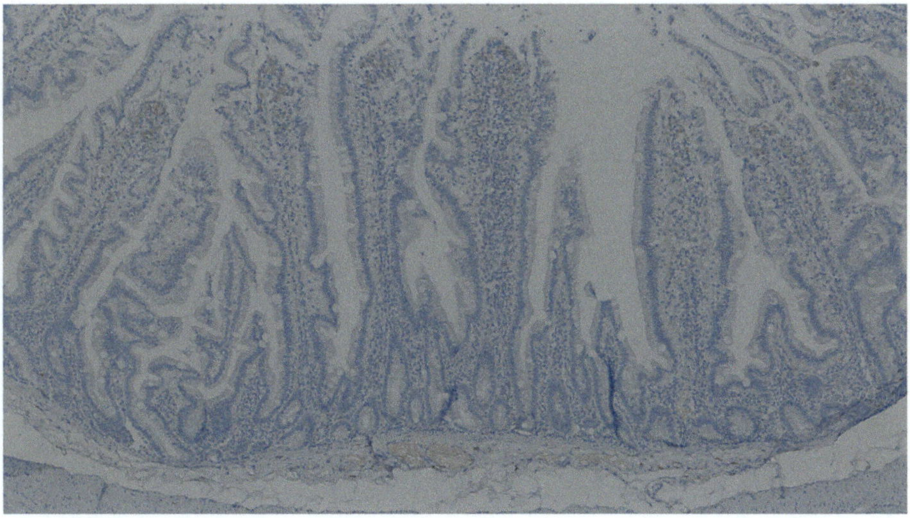

FIGURE 4.35

Immunohistochemical staining for CD163 performed on jejunum of a naive male rhesus macaque. Note the absence of positively stained cells. Stain performed with 3,3'-diaminobenzidine chromogen and hematoxylin counterstain. 10 × objective magnification.

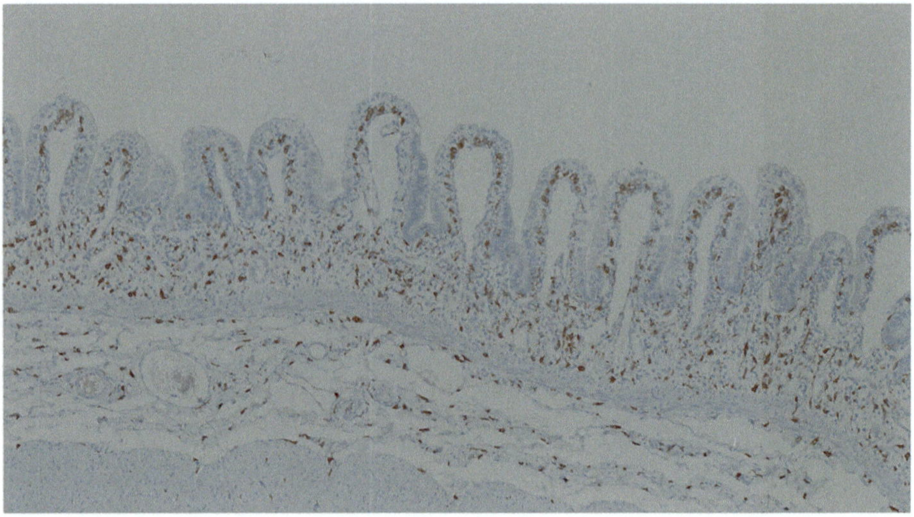

FIGURE 4.36A

Immunohistochemical staining for CD163 performed on jejunum of a male rhesus macaque necropsied 8 days after irradiation at 12 Gy has a marked increase in the population of $CD163^+$ cells in the mucosa and submucosa. Stain performed with 3,3'-diaminobenzidine chromogen and hematoxylin counterstain. 10 × objective magnification.

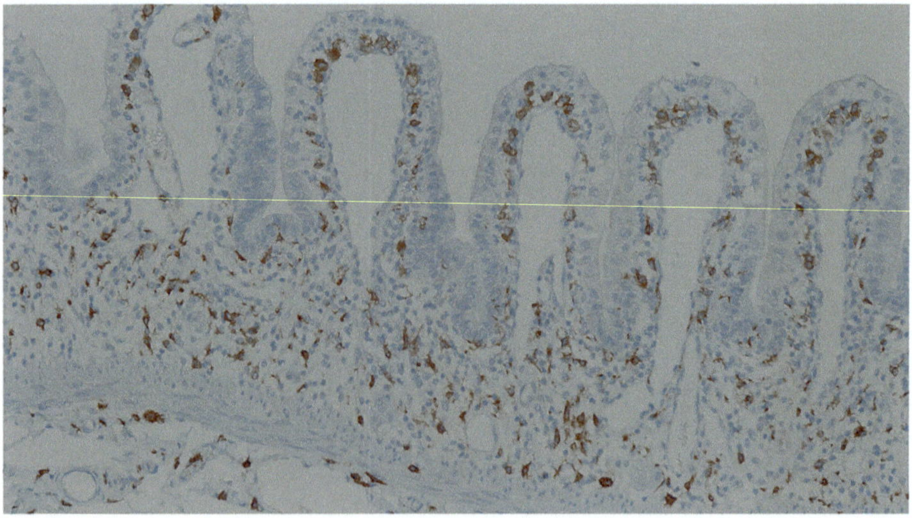

FIGURE 4.36B

Higher magnification of specimen shown in Fig. 4.36A showing localization of $CD163^+$ cells in the mucosal lamina propria and submucosa.

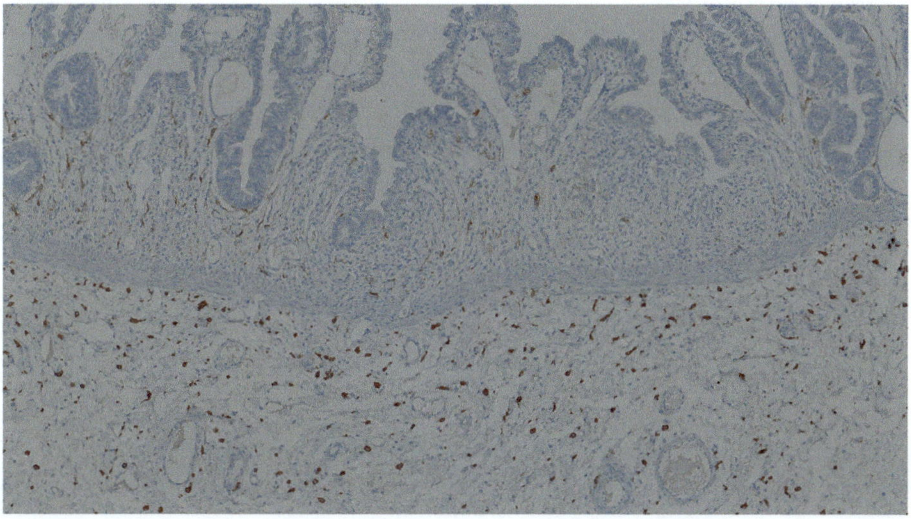

FIGURE 4.37

Immunohistochemical staining for CD163 performed on jejunum of a male rhesus macaque necropsied 12 days after irradiation at 12 Gy has a marked increase in the population of CD163$^+$ cells in the submucosa, with fewer positively stained cells in the newly regenerated mucosa. Stain performed with 3,3'-diaminobenzidine chromogen and hematoxylin counterstain. 10 × objective magnification.

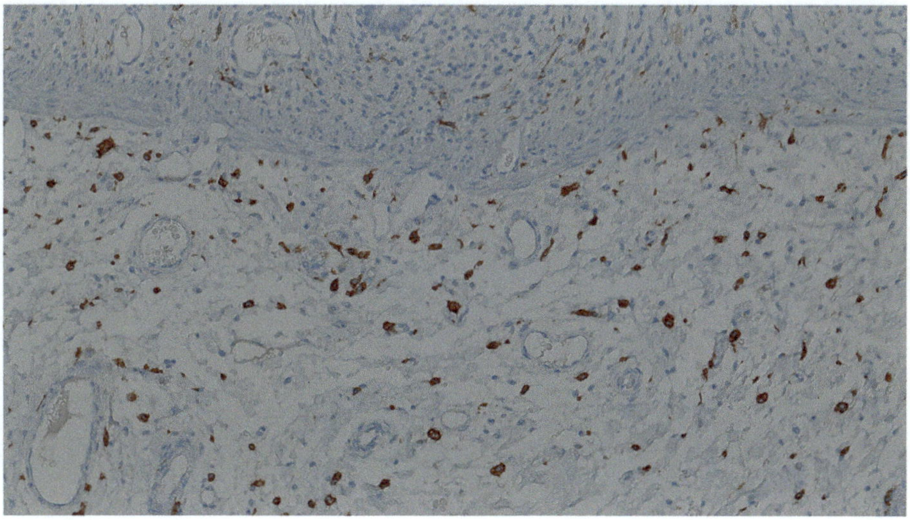

FIGURE 4.38

Immunohistochemical staining for CD163 performed on jejunum of a male rhesus macaque necropsied 12 days after irradiation at 12 Gy. The positively stained cell population in the submucosa consists primarily of large mononuclear cells, consistent with macrophages. Stain performed with 3,3'-diaminobenzidine chromogen and hematoxylin counterstain. 20 × objective magnification.

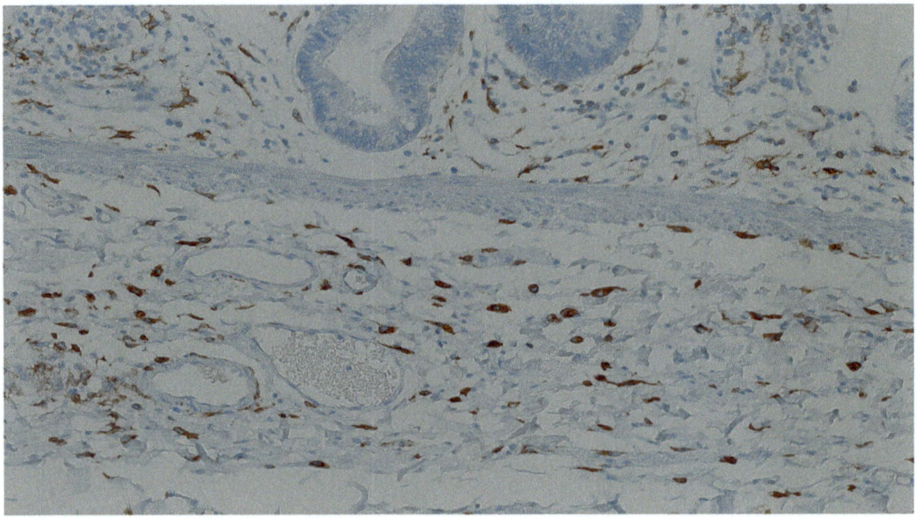

FIGURE 4.39

Immunohistochemical staining for CD163 performed on jejunum of a male rhesus macaque necropsied 21 days after irradiation at 12 Gy has a marked increase in the population of CD163$^+$ cells in the mucosa and submucosa. Stain performed with 3,3'-diaminobenzidine chromogen and hematoxylin counterstain. 20 × objective magnification.

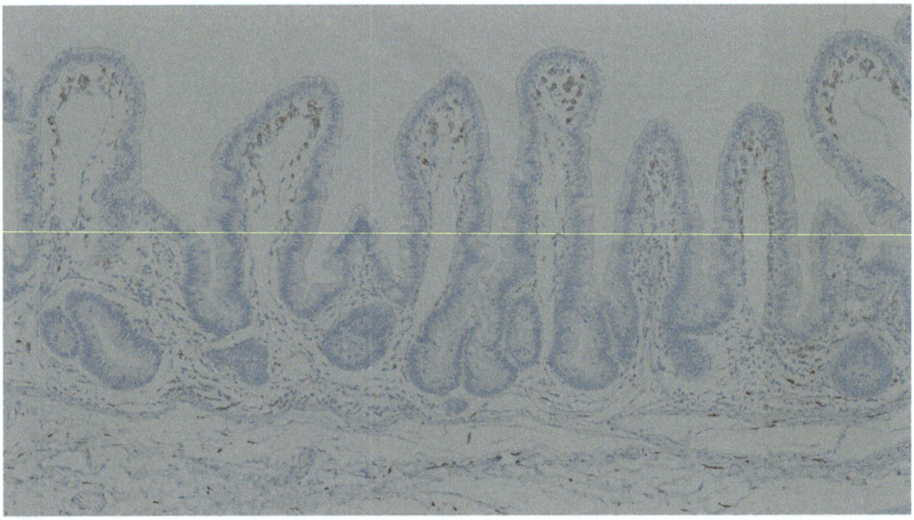

FIGURE 4.40

Immunohistochemical staining for CD163 performed on jejunum of a male rhesus macaque necropsied 31 days after irradiation at 12 Gy. The population of CD163$^+$ cells is markedly reduced in comparison to previouis post-irradiation intervals. Stain performed with 3,3'-diaminobenzidine chromogen and hematoxylin counterstain. 20 × objective magnification.

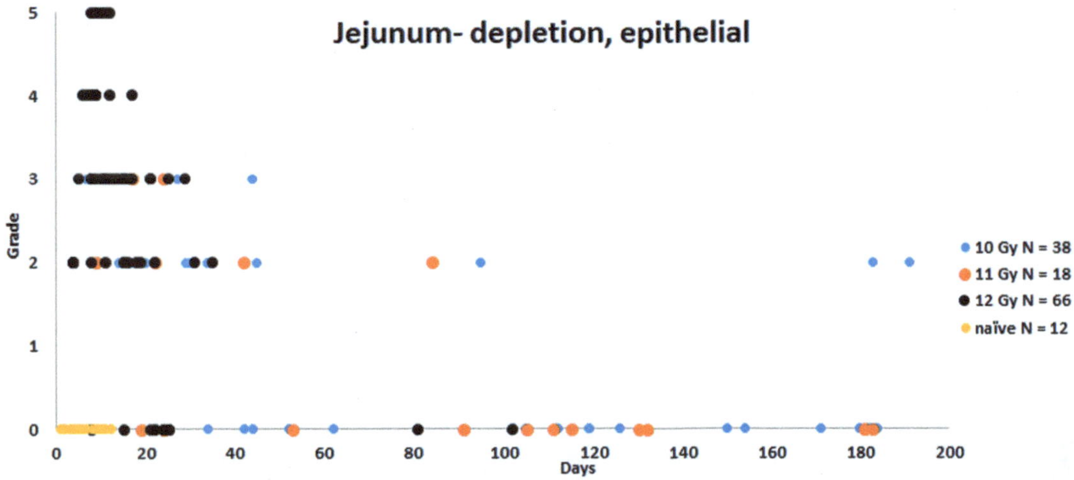

FIGURE 4.41

Scatter plot: Jejunum—depletion, epithelial. Observations based on light microscopic examination of paraffin-embedded specimens stained with hematoxylin and eosin.

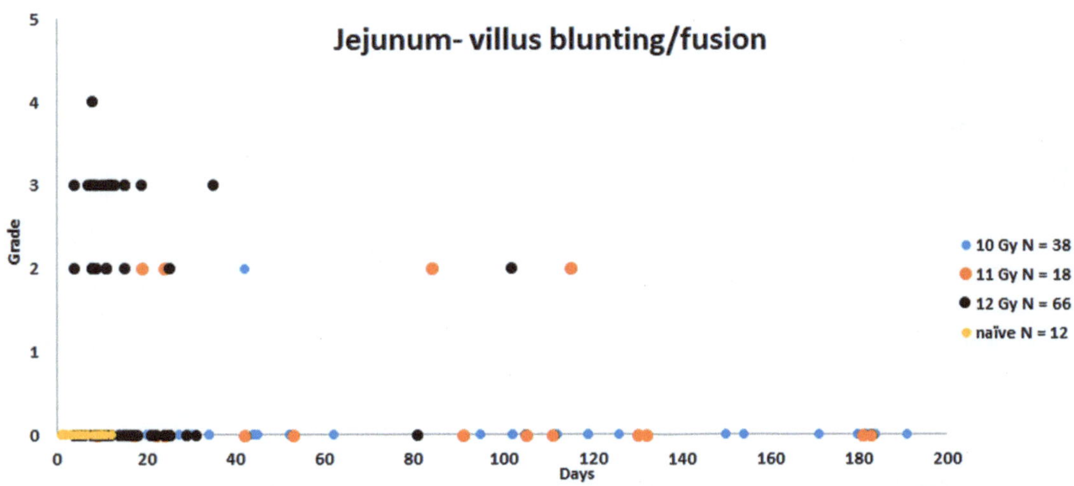

FIGURE 4.42

Scatter plot: Jejunum—villus blunting/fusion. Observations based on light microscopic examination of paraffin-embedded specimens stained with hematoxylin and eosin.

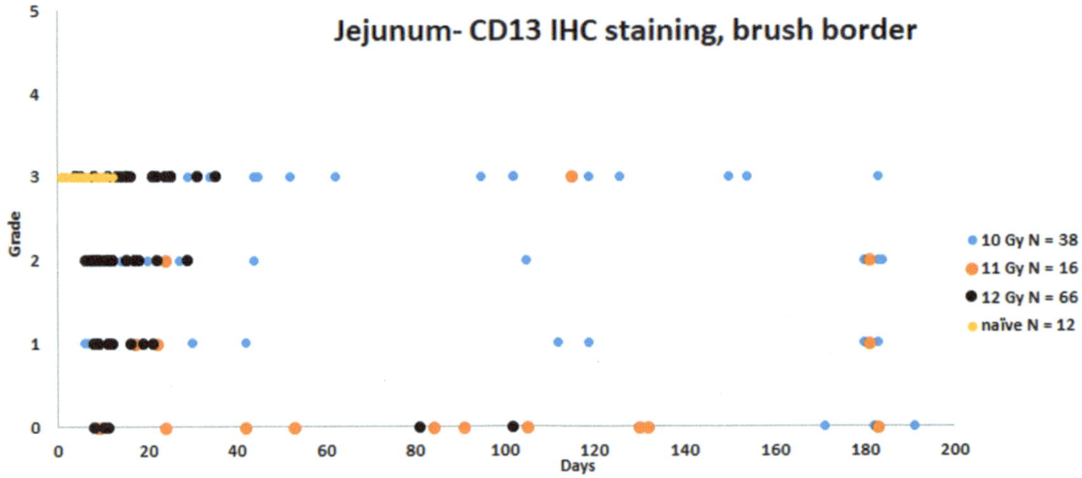

FIGURE 4.43

Scatter plot: Jejunum—CD13 staining, brush border. Observations based on light microscopic examination of paraffin-embedded specimens stained by immunohistochemical procedures for CD13 with 3,3'-diaminobenzidine chromogen and hematoxylin counterstain.

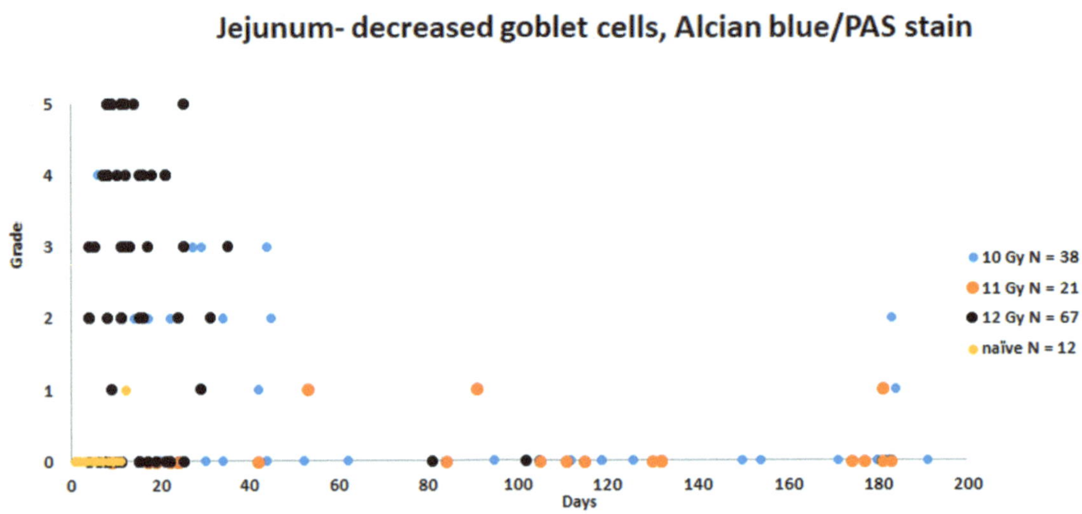

FIGURE 4.44

Scatter plot: Jejunum—decreased goblet cell population. Observations based on light microscopic examination of paraffin-embedded specimens stained with alcian blue/periodic acid-Schiff stain.

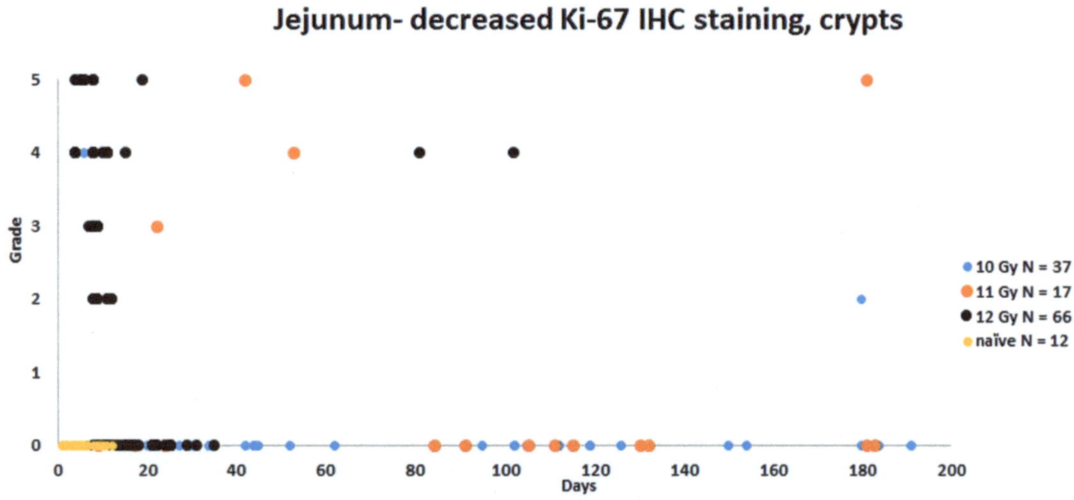

FIGURE 4.45

Scatter plot: Jejunum—decreased Ki67 staining of crypt epithelial cells. Observations based on light microscopic examination of paraffin-embedded specimens stained by immunohistochemical procedures for Ki67 with 3,3'-diaminobenzidine chromogen and hematoxylin counterstain.

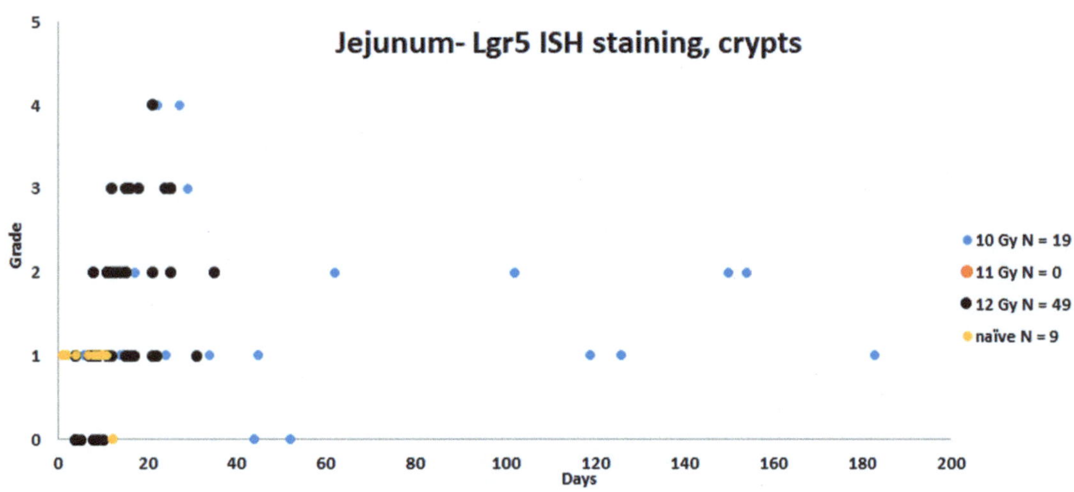

FIGURE 4.46

Scatter plot: Jejunum—Lgr5 staining of crypt epithelial cells. Observations based on light microscopic examination of paraffin-embedded specimens stained by in-situ hybridization procedures for Lgr5 with 3,3'-diaminobenzidine chromogen and hematoxylin counterstain.

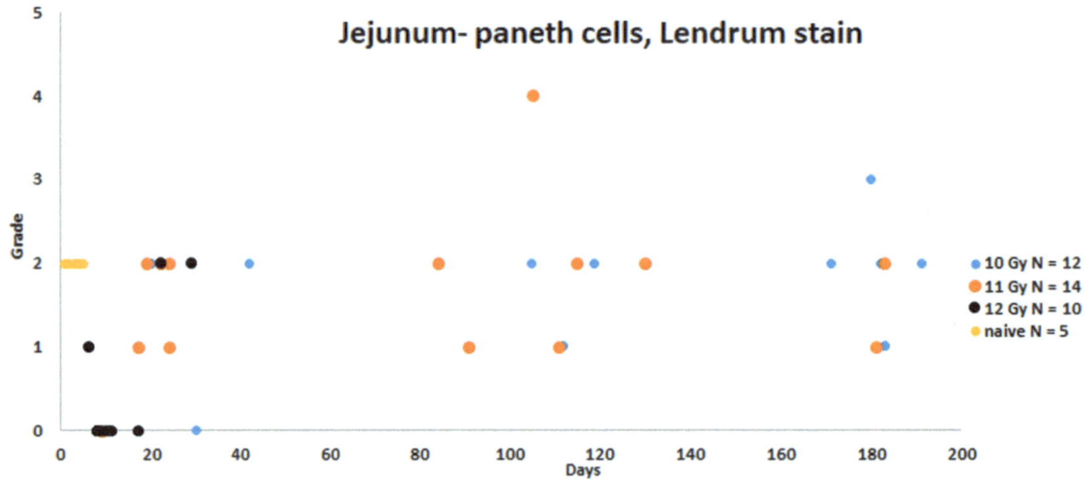

FIGURE 4.47

Scatter plot: Jejunum—Crypt Paneth cell populations. Observations based on light microscopic examination of paraffin-embedded specimens stained with Lendrum stain.

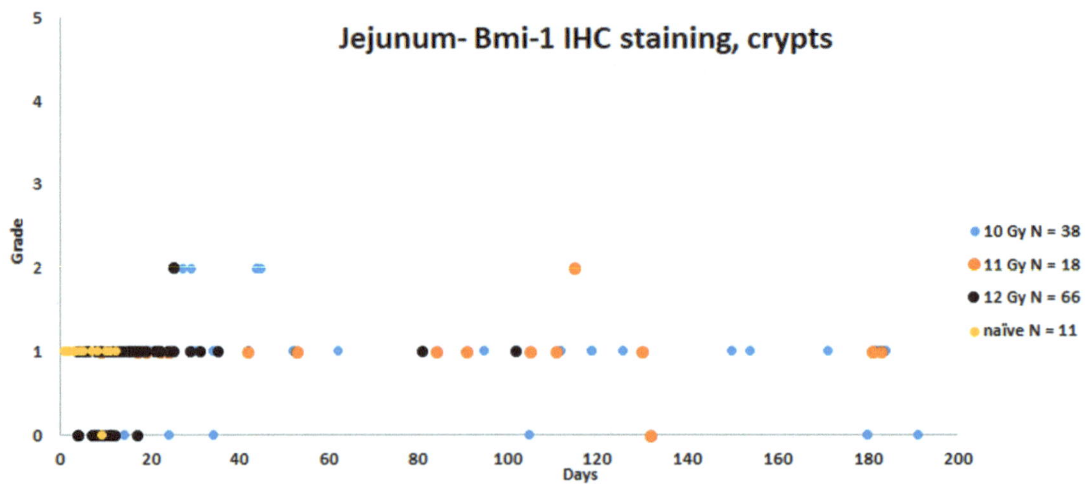

FIGURE 4.48

Scatter plot: Jejunum—Bmi-1$^+$ crypt cell populations. Observations based on light microscopic examination of paraffin-embedded specimens stained by immunohistochemical procedures for Bmi-1 with 3,3'-diaminobenzidine chromogen and hematoxylin counterstain.

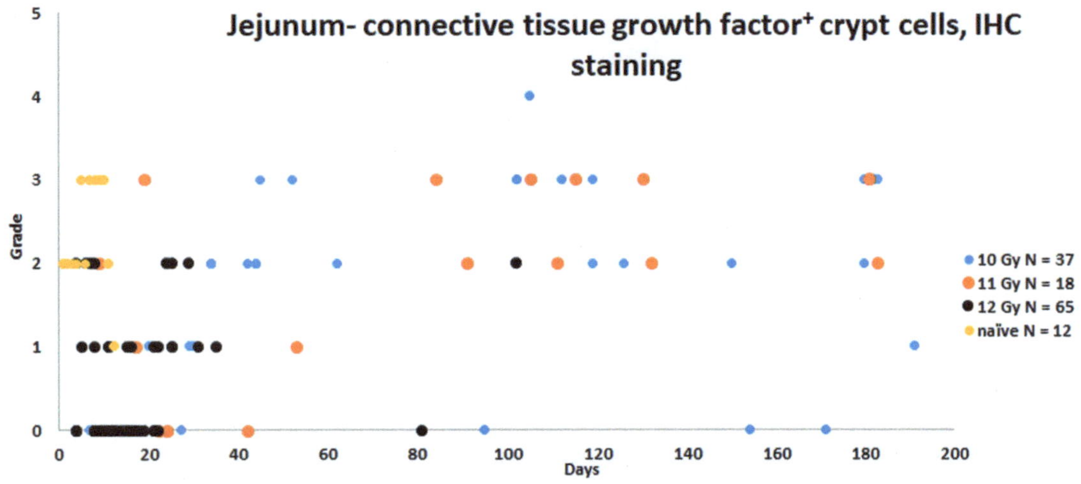

FIGURE 4.49

Scatter plot: Jejunum—connective tissue growth factor (CTGF)$^+$ crypt cell populations. Observations based on light microscopic examination of paraffin-embedded specimens stained by immunohistochemical procedures for CTGF with 3,3'-diaminobenzidine chromogen and hematoxylin counterstain.

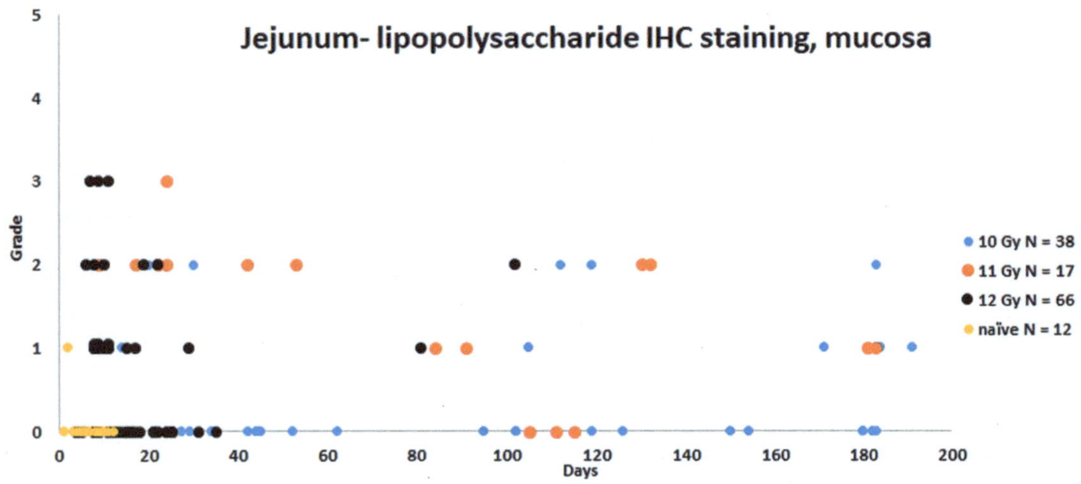

FIGURE 4.50

Scatter plot: Jejunum—lipopolysaccharide core antigen (LPS)$^+$ mucosal staining. Observations based on light microscopic examination of paraffin-embedded specimens stained by immunohistochemical procedures for LPS with 3,3'-diaminobenzidine chromogen and hematoxylin counterstain.

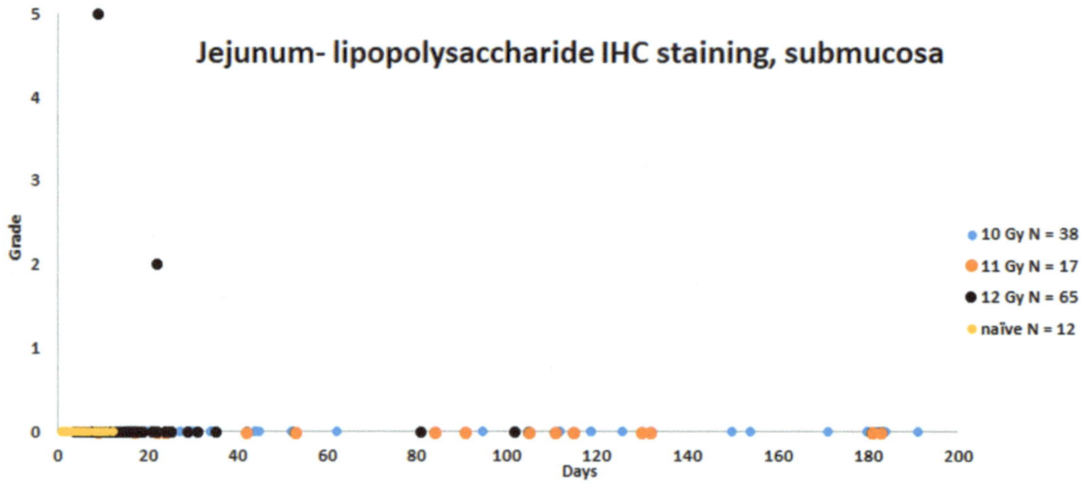

FIGURE 4.51

Scatter plot: Jejunum—lipopolysaccharide core antigen (LPS)$^{+}$ submucosal staining. Observations based on light microscopic examination of paraffin-embedded specimens stained by immunohistochemical procedures for LPS with 3,3'-diaminobenzidine chromogen and hematoxylin counterstain.

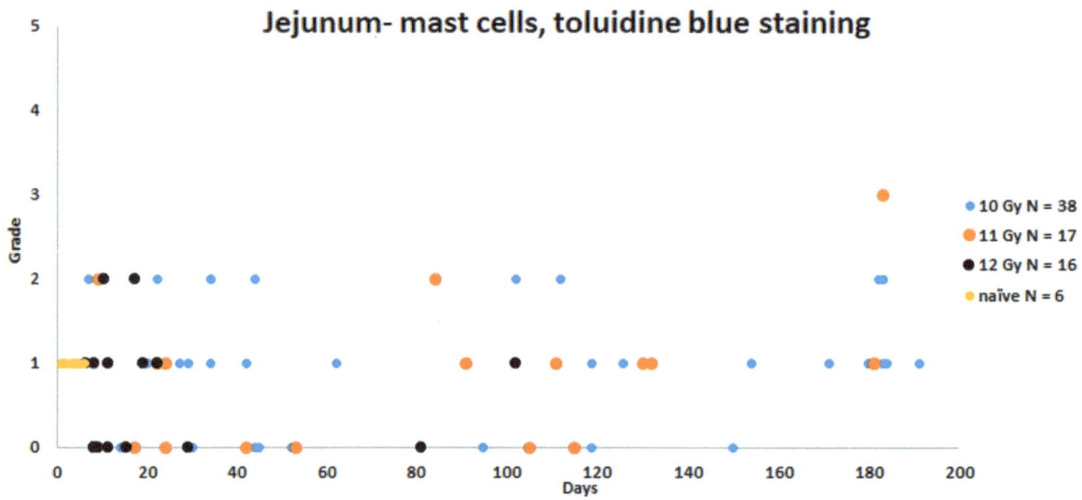

FIGURE 4.52

Scatter plot: Jejunum—mucosal and submucosal mast cell populations. Observations based on light microscopic examination of paraffin-embedded specimens stained with toluidine blue stain.

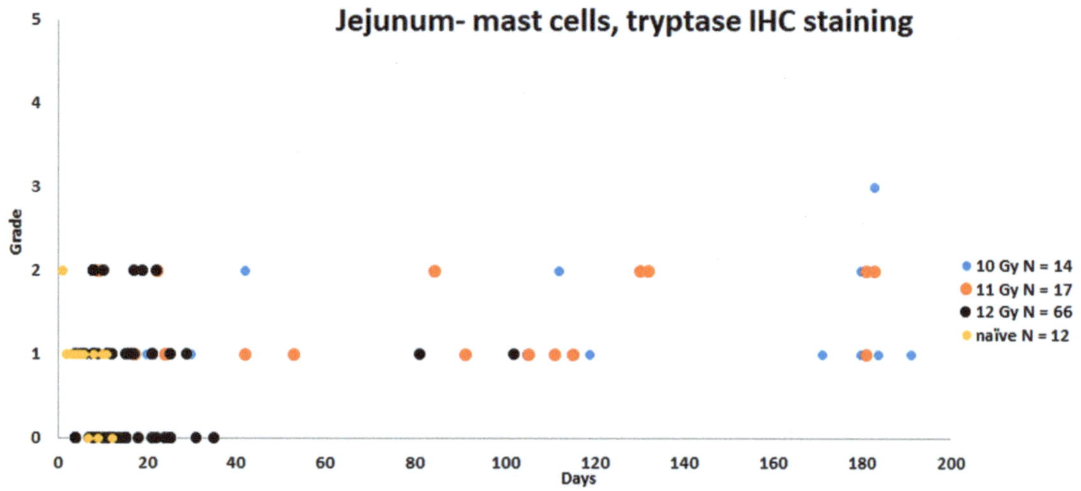

FIGURE 4.53

Scatter plot: Jejunum—mucosal and submucosal mast cell populations. Observations based on light microscopic examination of paraffin-embedded specimens stained by immunohistochemical procedures for tryptase with 3,3'-diaminobenzidine chromogen and hematoxylin counterstain.

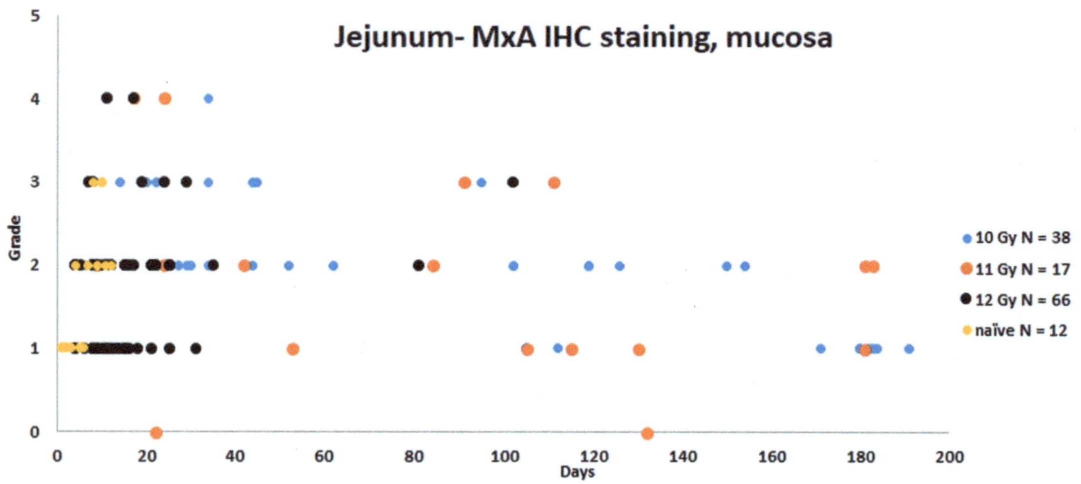

FIGURE 4.54

Scatter plot: Jejunum—mucosal populations of MxA^+ cells. Observations based on light microscopic examination of paraffin-embedded specimens stained by immunohistochemical procedures for MxA with 3,3'-diaminobenzidine chromogen and hematoxylin counterstain.

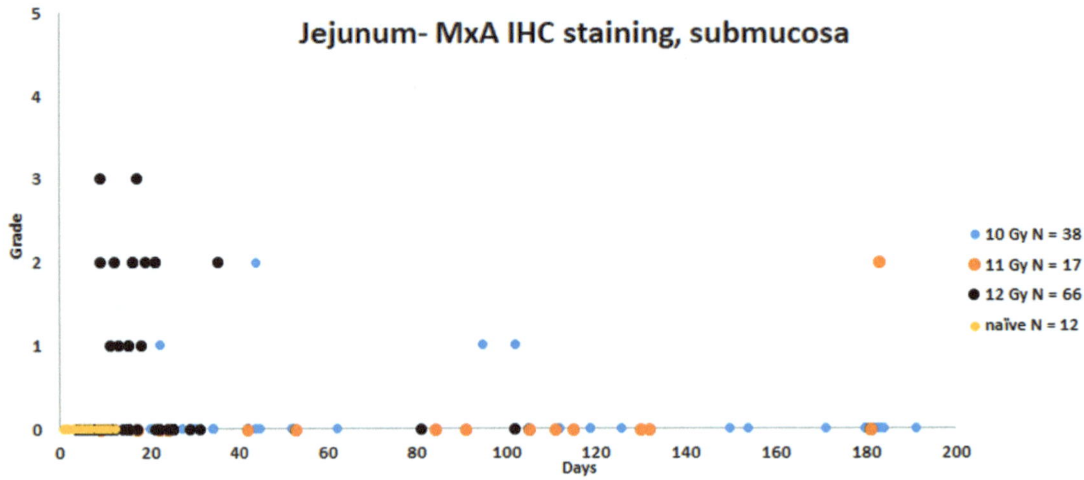

FIGURE 4.55

Scatter plot: Jejunum—submucosal populations of MxA^{+} cells. Observations based on light microscopic examination of paraffin-embedded specimens stained by immunohistochemical procedures for MxA with 3,3'-diaminobenzidine chromogen and hematoxylin counterstain.

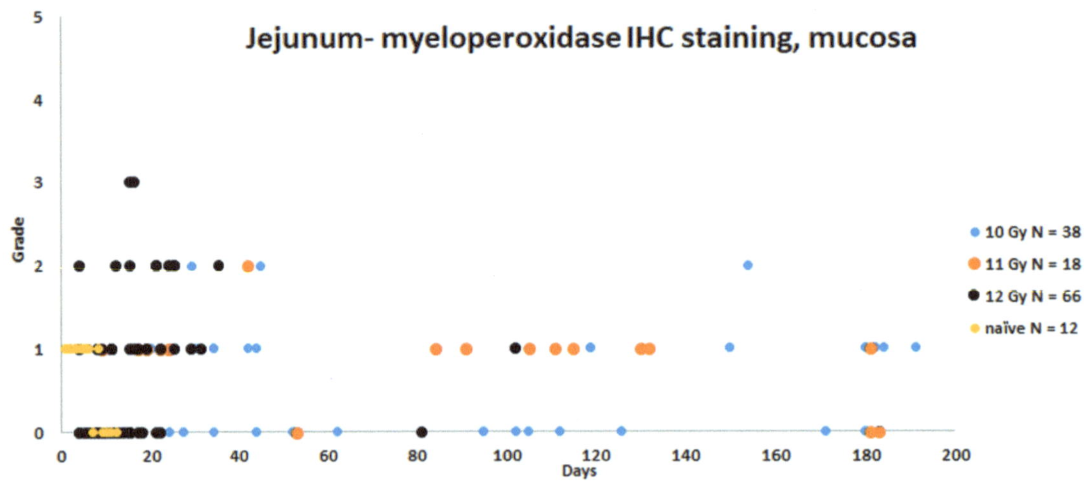

FIGURE 4.56

Scatter plot: Jejunum—mucosal populations of myeloperoxidase^{+} cells. Observations based on light microscopic examination of paraffin-embedded specimens stained by immunohistochemical procedures for myeloperoxidase with 3,3'-diaminobenzidine chromogen and hematoxylin counterstain.

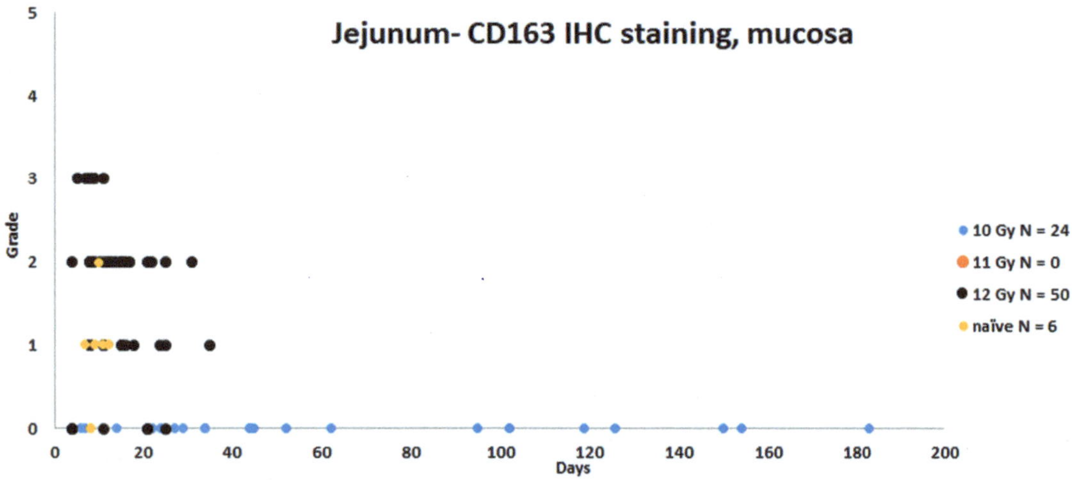

FIGURE 4.57

Scatter plot: Jejunum—mucosal populations of CD163$^+$ cells. Observations based on light microscopic examination of paraffin-embedded specimens stained by immunohistochemical procedures for CD163 with 3,3'-diaminobenzidine chromogen and hematoxylin counterstain.

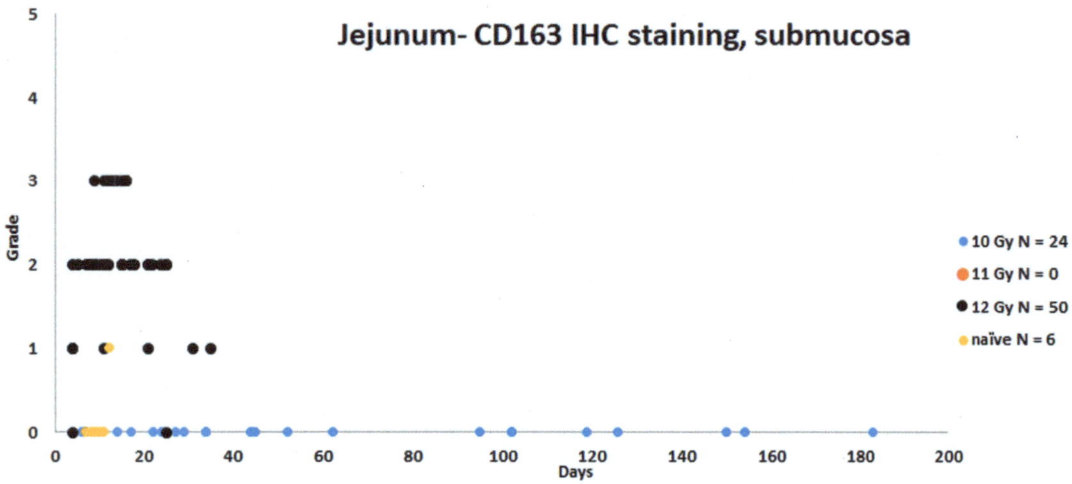

FIGURE 4.58

Scatter plot: Jejunum—submucosal populations of CD163$^+$ cells. Observations based on light microscopic examination of paraffin-embedded specimens stained by immunohistochemical procedures for CD163 with 3,3'-diaminobenzidine chromogen and hematoxylin counterstain.

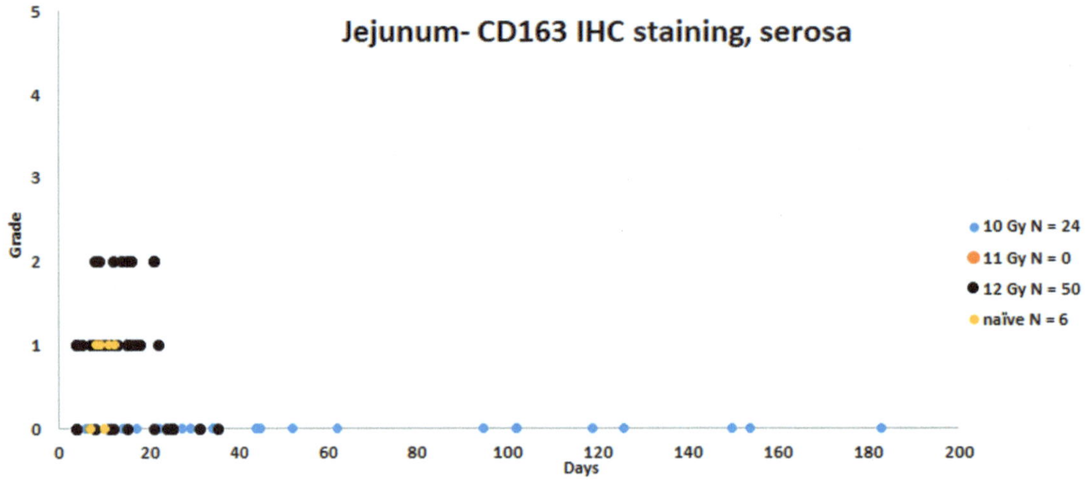

FIGURE 4.59

Scatter plot: Jejunum—serosal populations of CD163^{+} cells. Observations based on light microscopic examination of paraffin-embedded specimens stained by immunohistochemical procedures for CD163 with 3,3'-diaminobenzidine chromogen and hematoxylin counterstain.

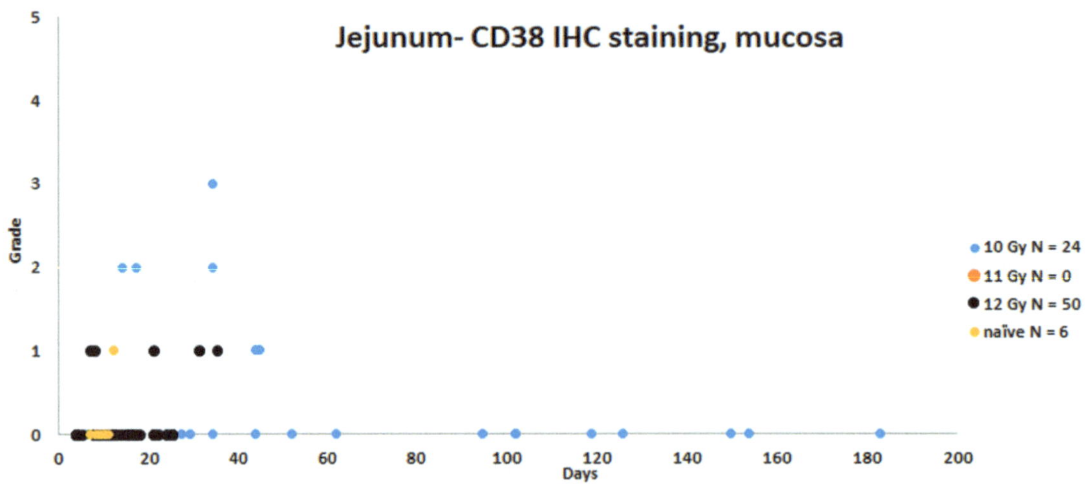

FIGURE 4.60

Scatter plot: Jejunum—mucosal populations of CD38^{+} cells. Observations based on light microscopic examination of paraffin-embedded specimens stained by immunohistochemical procedures for CD38 with 3,3'-diaminobenzidine chromogen and hematoxylin counterstain.

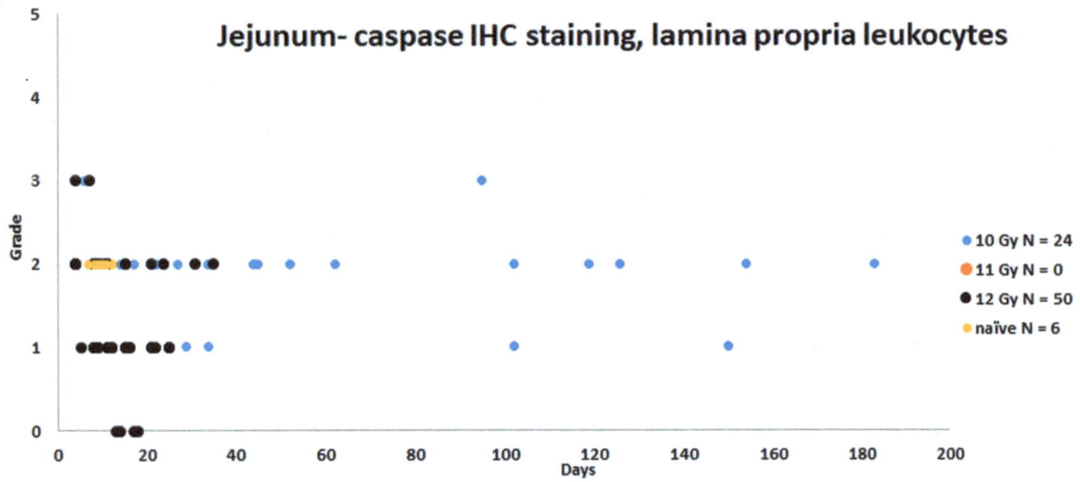

FIGURE 4.61

Scatter plot: Jejunum—mucosal (laminal propria) populations of caspase 3^+ leukocytes. Observations based on light microscopic examination of paraffin-embedded specimens stained by immunohistochemical procedures for caspase with 3,3'-diaminobenzidine chromogen and hematoxylin counterstain.

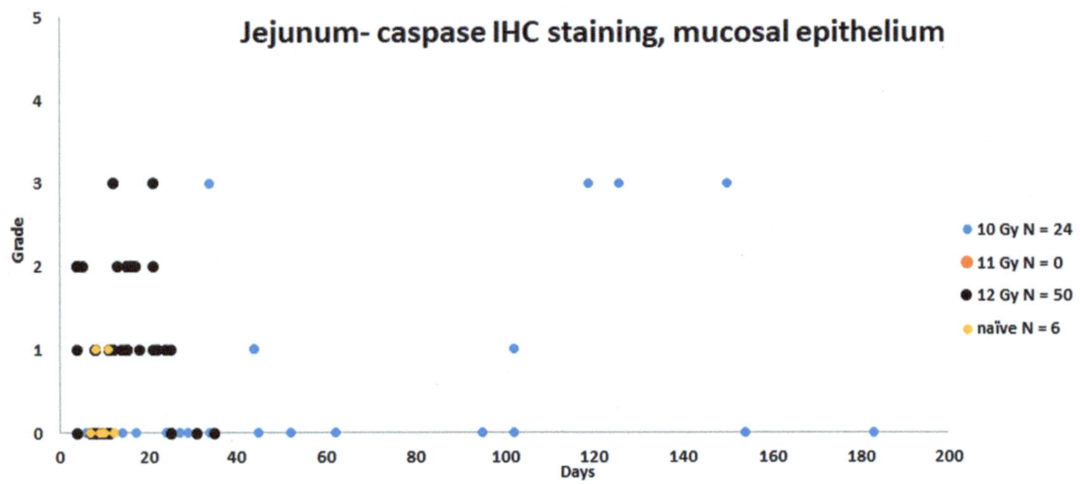

FIGURE 4.62

Scatter plot: Jejunum—mucosal populations of caspase 3^+ epithelial cells. Observations based on light microscopic examination of paraffin-embedded specimens stained by immunohistochemical procedures for caspase with 3,3'-diaminobenzidine chromogen and hematoxylin counterstain.

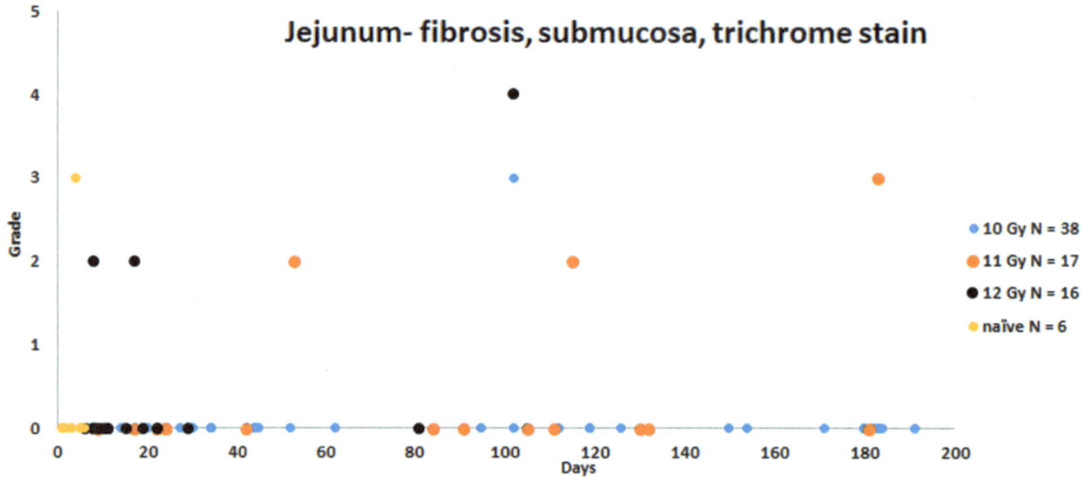

FIGURE 4.63

Scatter plot: Jejunum—submucosal fibrosis. Observations based on light microscopic examination of paraffin-embedded specimens stained with Masson's trichrome stain.

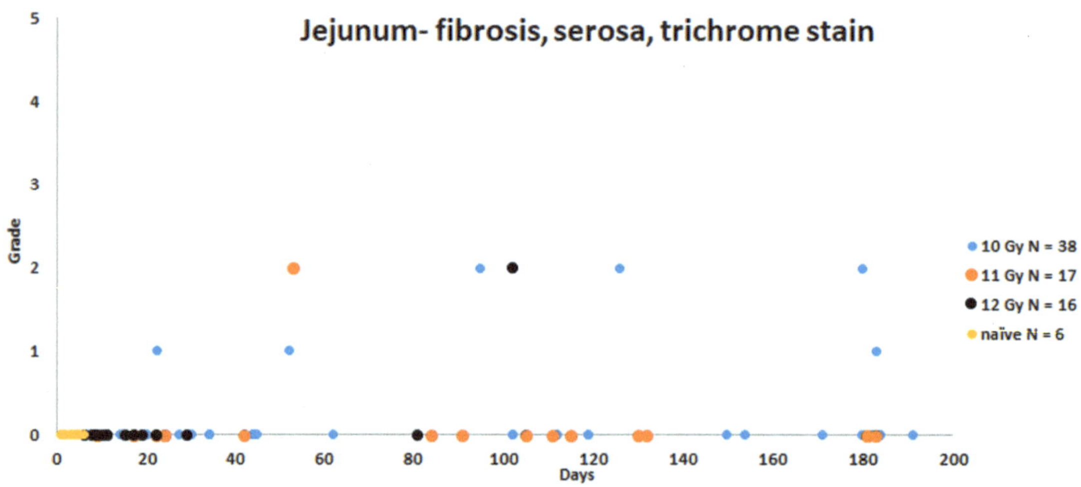

FIGURE 4.64

Scatter plot: Jejunum—serosal fibrosis. Observations based on light microscopic examination of paraffin-embedded specimens stained with Masson's trichrome stain.

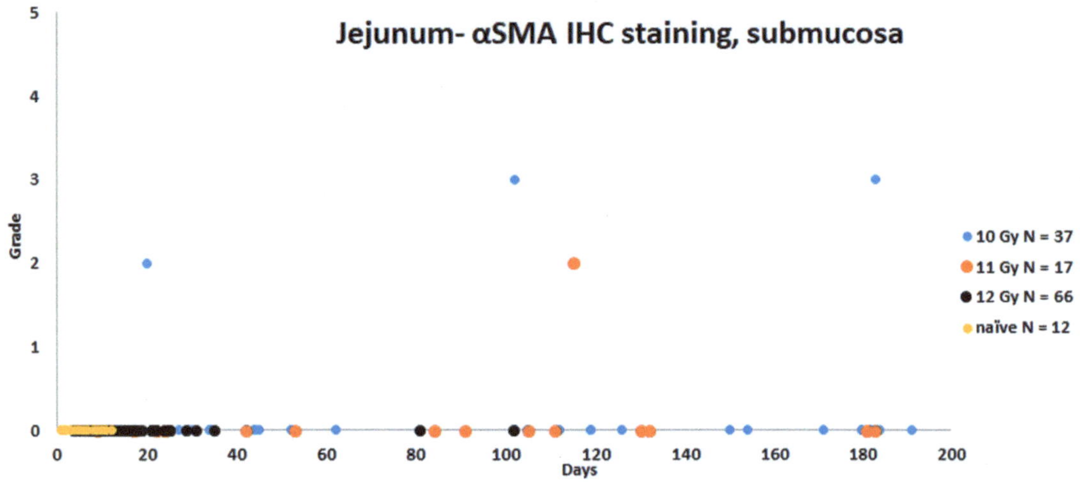

FIGURE 4.65

Scatter plot: Jejunum—alpha smooth muscle actin (αSMA) submucosal staining. Observations based on light microscopic examination of paraffin-embedded specimens stained by immunohistochemical procedures for αSMA with 3,3'-diaminobenzidine chromogen and hematoxylin counterstain.

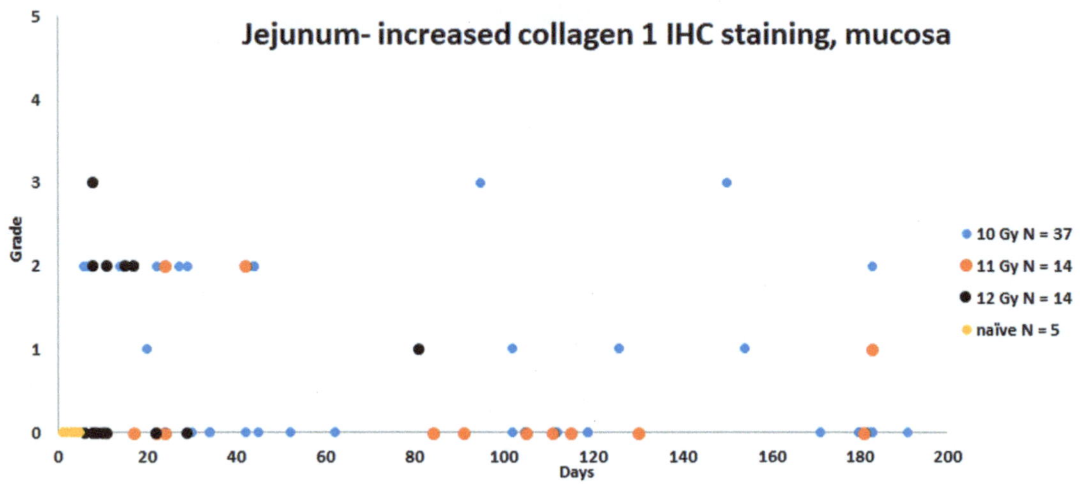

FIGURE 4.66

Scatter plot: Jejunum—increased collagen 1 mucosal staining. Observations based on light microscopic examination of paraffin-embedded specimens stained by immunohistochemical procedures for collagen with 3,3'-diaminobenzidine chromogen and hematoxylin counterstain.

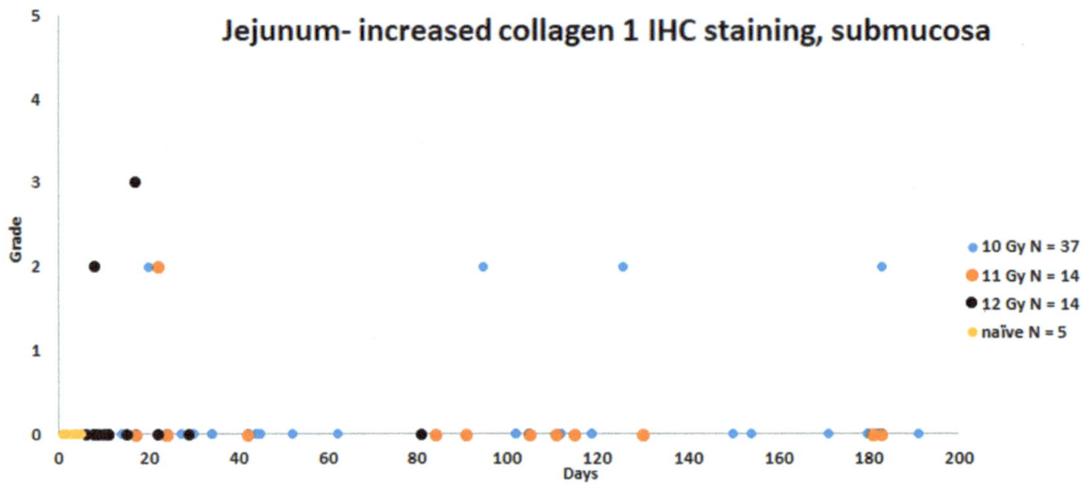

FIGURE 4.67

Scatter plot: Jejunum—increased collagen 1 submucosal staining. Observations based on light microscopic examination of paraffin-embedded specimens stained by immunohistochemical procedures for collagen with 3,3'-diaminobenzidine chromogen and hematoxylin counterstain.

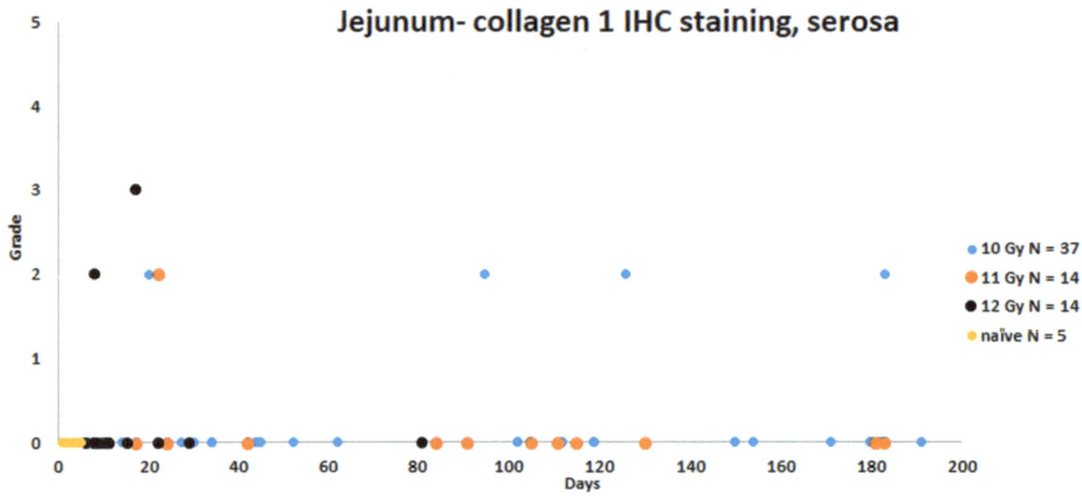

FIGURE 4.68

Scatter plot: Jejunum—increased collagen 1 serosal staining. Observations based on light microscopic examination of paraffin-embedded specimens stained by immunohistochemical procedures for collagen with 3,3'-diaminobenzidine chromogen and hematoxylin counterstain.

CHAPTER

Radiation-associated changes in the colon of rhesus macaques

5

Overview

Epithelial depletion of the colonic mucosa was commonly observed in animals irradiated at 12 Gy, and less commonly in animals irradiated at 10 or 11 Gy. These changes were considered to be a major contributor to the shorter survival time of animals in this group. Similar but less pronounced histologic changes were present in animals irradiated at 10 or 11 Gy.

Decreased populations of goblet cells, as revealed by PAS/Alcian blue staining, were a reflection of the immature status of regenerating mucosal epithelial cells.

The colonic mucosa of irradiated animals had an increased population of activated (MxA^+) immunocytes, though examination of routine hematoxylin & eosin (H&E) stains rarely indicated an increase in the overall immunocyte populations of the colonic mucosa. This suggests a radiation-associated change in activation status rather than an influx of immunocyte populations. Activated immunocyte populations in irradiated animals tended to extend into the colonic submucosa.

Immunohistochemical (IHC) staining for bacterial lipopolysaccharide (LPS) core antigen rarely revealed penetration of enteric bacteria into the colonic mucosa. The prominent populations of LPS^+ cells in mesenteric lymph nodes suggest that minor breaks in mucosal integrity may not be reflected in the sparse sampling of colon and jejunum.

IHC staining for Bmi-1 revealed Bmi-1^+ cells existing individually or in small clusters in the colonic lamina propria of animals with pronounced mucosal epithelial depletion. These Bmi-1^+ cells were considered to be remnants of colonic crypts that no longer had the typical crypt morphology but retained cellular regenerative capacity. An increased number of Bmi-1 + cells was noted in crypts of a minor subset of animals necropsied after postirradiation day 27, but an increase in BMI-1^+ cells was not a prominent feature of the colonic mucosal response.

As was seen in the jejunum, IHC staining for caspase 3 revealed radiation-associated caspase 3 expression in degenerating mucosal epithelial cells as well as a reduction in the normal level of apoptosis in mucosal populations of immunocytes, the latter a reflection of radiation-associated reductions in existing immunocyte populations.

IHC staining for myeloperoxidase (MPO) revealed an increased population of MPO^+ cells, largely macrophages, in the colonic mucosa of irradiated animals. IHC staining for CD13 also revealed an increased population of $CD13^+$ leukocytes/macrophages in the colonic mucosa of irradiated animals. These increased cell populations were not overtly apparent upon examination of routine H&E-stained histologic sections.

IHC staining for CD38, a marker for pro-inflammatory macrophages, rarely revealed $CD38^+$ cell populations in the colonic mucosa, submucosa, or serosa. By contrast, staining for CD163, a marker for profibrotic macrophages, revealed a distinct increase in populations of $CD163^+$

Histopathology Atlas of Acute Radiation Syndrome and Delayed Effects in Rhesus Macaques. DOI: https://doi.org/10.1016/B978-0-323-91393-5.00005-0

macrophages in the colonic mucosa, submucosa, and serosa. While it is possible the increased population of $CD163^+$ macrophages resulted from an influx of precursor cells from the bone marrow, the distribution and other microscopic features of the $CD163^+$ cell population suggest a radiation-associated shift of preexisting macrophages to a $CD163^+$ (pro-fibrotic) phenotype.

Staining and figure lists

Routine histology stain-

Hematoxylin & eosin—Figs. 5.1A, 5.1B, 5.2, 5.3A, 5.3B, 5.4A, 5.4B, 5.4C, 5.5, 5.6, 5.7, 5.8, 5.9, 5.10A, 5.10B, 5.10C, 5.11AA, 5.11B, 5.12A, 5.12B, 5.12C
Histochemical stains—5.6

Masson's trichrome—Figs. 5.14A, 5.14B, 5.14C, 5.14D, Fig. 5.14E.
Periodic acid-Schiff/Alcian blue—Figs. 5.13A, 5.13B

IHC stains—

CD163—Figs. 5.24A, 5.24B, 5.24C, 5.24D, 5.24E, 5.24F
Collagen 1—Figs. 5.20, 5.21A, 5.21B
Connective tissue growth factor—Figs. 5.22, 5.23

Ki-67—Figs. 5.15, 5.16, 5.17, 5.18, 5.19
Myeloperoxidase—Figs. 5.25A, 5.25B, 5.25CC

Graphic presentations (scatter plots) of individual histologic observations in the colon are presented in the following figures:
Fig. 5.26. Colon—depletion, epithelial, mucosa.
Fig. 5.27. Colon—reduced Ki-67 + crypt epithelial cells.
Fig. 5.28. Colon—decreased goblet cell population.
Fig. 5.29. Colon—caspase 3 staining, mucosal epithelial cells.
Fig. 5.30. Colon—Bmi-1 + crypt epithelial cells.
Fig. 5.31. Colon—lipopolysaccharide core antigen staining, mucosa.
Fig. 5.32. Colon—lipopolysaccharide core antigen staining, submucosa.
Fig. 5.33. Colon—infiltration, lymphocytic.
Fig. 5.34. Colon—cellular MxA staining, mucosa.
Fig. 5.35. Colon—cellular MxA staining, submucosa.
Fig. 5.36. Colon—cellular CD3 staining, mucosa.
Fig. 5.37. Colon—cellular myeloperoxidase staining, mucosa.
Fig. 5.38. Colon—mast cell population, mucosa, toluidine blue staining.
Fig. 5.39. Colon—mast cell population, mucosa, tryptase staining
Fig. 5.40. Colon—CD13 cellular staining, mucosa.
Fig. 5.41. Colon—caspase 3 staining, mucosal lamina propria.
Fig. 5.42. Colon—fibrosis, serosa.
Fig. 5.43. Colon—fibrosis, submucosa.
Fig. 5.44. Colon—increased collagen 1 staining, mucosa.

Fig. 5.45. Colon—increased collagen 1 staining, submucosa.
Fig. 5.46. Colon—increased collagen 1 staining, serosa.
Fig. 5.47. Colon—increased alpha-smooth muscle actin (αSMA) staining, submucosa.
Fig. 5.48. Colon—increased alpha-smooth muscle actin (αSMA) staining, serosa.
Fig. 5.49. Colon—connective tissue growth factor staining, mucosa.
Fig. 5.50. Colon—connective tissue growth factor staining, submucosa.
Fig. 5.51. Colon—connective tissue growth factor staining, serosa. Observations based on light microscopic examination of paraffin-embedded specimens stained.

Detailed colon observations

Leukocyte infiltration

Inflammatory cell (lymphocytes, neutrophils, and macrophages) infiltration in the colon was notably sparse relative to the degree of epithelial disruption. This was presumed to be partially a result of radiation effects on the hematopoietic system during the early postirradiation period when the colonic mucosa was most severely debilitated.

Trichrome stain

Fibrosis of the submucosa and/or serosa was noted sporadically in animals from all radiation dose levels. The fibrotic lesions tended to be focal or multifocal, suggesting that more extensive specimen sampling could reveal additional fibrotic lesions. The trichrome stain did not reveal mucosal fibrosis, though increased mucosal collagen 1 content was discernible with collagen 1 IHC staining.

PAS/Alcian blue stain

There was a marked reduction in goblet cells in animals necropsied during the first 35 days following irradiation, and rarely thereafter. The reduced number of goblet cells reflected the depleted or immature status of the damaged or regenerating mucosal epithelium.

Toluidine blue stain

Mast cells were essentially absent from the colon of a substantial number of irradiated animals, particularly the animals irradiated at 12 Gy. The absence of discernible mast cells in the toluidine blue-stained sections persisted in the colon of the two 12 Gy exposure animals that survived to 81 or 102 days postirradiation. The toluidine blue stain demonstrates the contents of mast cell granules; thus the stain does not distinguish between an actual reduction in mast cell numbers vs mast cell degranulation.

Tryptase immunohistochemical stain

As with the jejunum, tryptase staining revealed a slightly greater population of mast cells in the colon than were revealed by toluidine blue staining. This was expected, as the toluidine blue staining is known to reveal only the connective tissue mast cell population and does not include the mucosal mast cell population unless special fixatives are used at the time of tissue collection. As identified by tryptase ICH staining, mast cells were absent from the colon mucosa/submucosa of a substantial subset of animals irradiated at 12 Gy. It was interesting to note that tryptase IHC staining revealed a low-level mast cell population in the two 12 Gy exposure animals that survived to 81 and 102 days postirradiation compared to the absence of mast cells in those specimens when stained with toluidine blue. This suggests the persistent mast cells in those two animals were mucosal mast cells rather than connective tissue mast cells. Tryptase IHC staining is based on mast cell granule contents, thus does not distinguish between numerical depletion and mast cell degranulation.

Alpha smooth muscle actin immunohistochemical stain

There was very little evidence of radiation-associated increase in αSMA staining in the colon.

MxA immunohistochemical stain

Staining for MxA in the mucosa was increased in a substantial subset of animals necropsied within 45 days of irradiation. The increased population of MxA^+ cells was more commonly noted in animals irradiated at 12 Gy than 10 or 11 Gy. Increased populations of MxA^+ cells in the colonic submucosa were restricted to animals irradiated at 12 Gy.

Lipopolysaccharide core antigen immunohistochemical stain

Irradiated animals commonly had LPS^+ cells in the mucosa but, despite the obvious radiation-associated damage to the mucosa, only two animals had LPS^+ cells in the submucosa. LPS^+ bacteria were commonly present on the mucosal surface, thus serving as an internal control for the staining procedure.

Ki-67 immunohistochemical stain

The stain revealed reduced Ki-67 expression in a moderate number of animals necropsied within 22 days of irradiation, and sporadically thereafter. The postirradiation reduction in cellular proliferation was followed by a rebound proliferation as mucosal repair commenced.

Myeloperoxidase immunohistochemical stain

Staining for MPO was limited to mucosal staining, primarily indicating MPO-positive leukocyte or macrophage populations. There was an increase in the population of MPO^+ cells in animals necropsied up to 35 days after irradiation, and sporadically thereafter.

CD3 immunohistochemical stain

Staining for $CD3^+$ lymphocytes in the colon primarily involved mucosal cell populations, with occasional $CD3^+$ cells in the submucosa. There was no discernible radiation-associated effect on the diffuse $CD3^+$ cell population in the mucosa. This analysis excludes the depletion of all lymphocyte populations in mucosal lymphoid follicles, which were severely depleted by radiation exposure.

CD13 immunohistochemical stain

Staining for CD13 was restricted to individual cells, consistent with leukocytes or macrophages. There was a distinct increase in the population of CD13 + cells in the mucosa of animals necropsied within 35 days of irradiation, and sporadically thereafter.

Bmi-1 immunohistochemical stain

Staining of crypt cell populations was present at low level in three of six naive control animals and absent in the remaining three naive control animals. There was no definitive evidence of radiation-associated effects on the population of $Bmi\text{-}1^+$ cells in the colonic mucosa.

IL-22 immunohistochemical stain

IHC staining revealed $IL22^+$ leukocytes were uncommonly present in the colonic mucosa.

IL-22R immunohistochemical stain

Staining for IL-22R was commonly present in colonic epithelial cells. There was no discernible alteration in IL-22R staining relative to irradiation.

Collagen 1 immunohistochemical stain

Increased collagen staining was topographically subcategorized as mucosal, submucosal, or serosal. Increased mucosal collagen 1 staining was present in a subset of irradiated animals up to 45 days after irradiation, and sporadically thereafter. Increased staining for collagen 1 was noted only sporadically in the colonic submucosa and serosa. It should be noted that increased collagen staining in the various sublocations of the colon tended to be multifocally distributed, and only two specimens from the colon were examined microscopically. It is likely examination of additional colonic specimens would reveal additional examples of increased collagen staining.

Transforming growth factor β immunohistochemical stain

IHC staining for transforming growth factor β (TGF-β) was rarely noted in the mucosa of the colon.

Connective tissue growth factor immunohistochemical stain

An increased level of connective tissue growth factor (CTGF) staining was noted in the colonic mucosa of a substantial subset of animals necropsied within 42 days of irradiation, and sporadically thereafter. Staining for CTGF was rarely noted in the colonic submucosa or serosa.

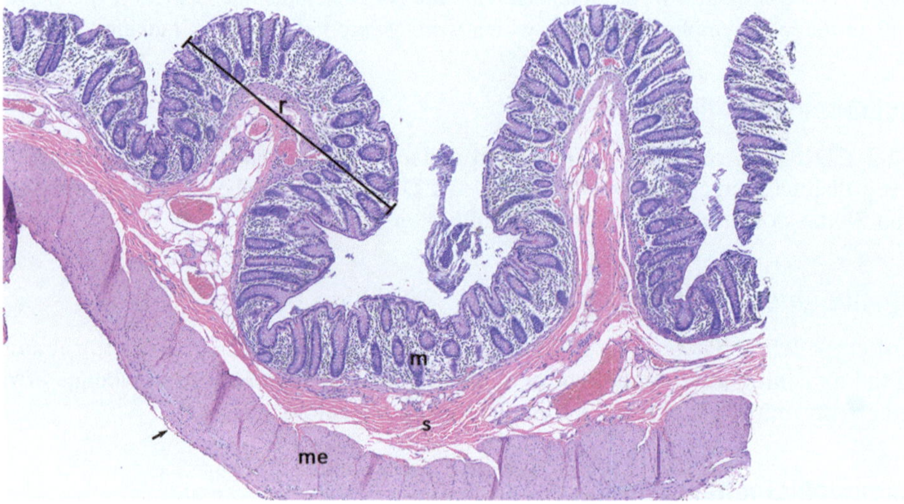

FIGURE 5.1A

The colon of a naive male rhesus macaque has mucosal folds (rugae-r) covered by mucosa (m). The submucosa (s) separates the mucosa from the external muscle layers (muscularis externa-me), which consist of a thick inner circular layer and an external longitudinal layer. The thin serosal layer (arrow) consists of mesothelial cells and a scant amount of fibrous connective tissue. Hematoxylin and eosin stain, 4.78 × objective magnification.

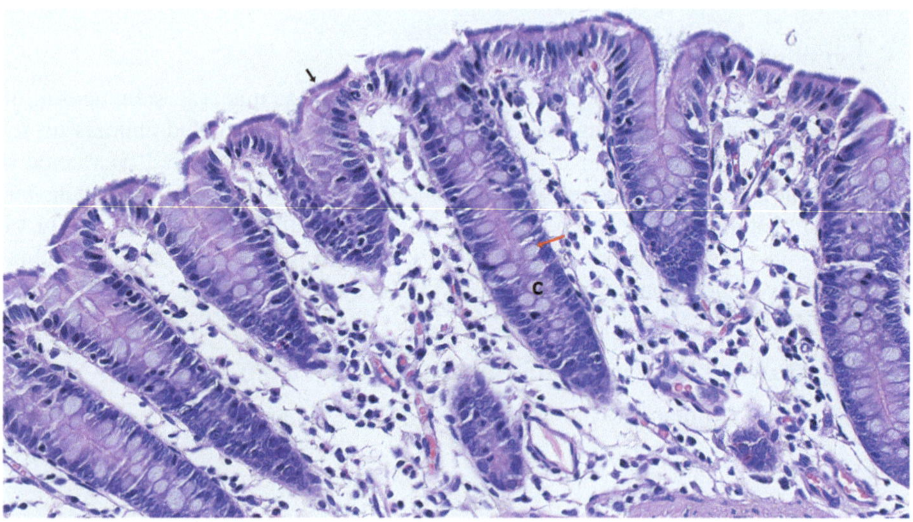

FIGURE 5.1B

Higher magnification of colonic mucosa from a naive male rhesus macaque shows crypts (c) with mucin-filled goblet cells (red arrow). The mucin is released to form a protective coating on the luminal surface of the colon. Note the thick particulate and filamentous layer (black arrow) composed of bacterial colonies on the luminal surface. Hematoxylin and eosin stain, 30.2 × objective magnification.

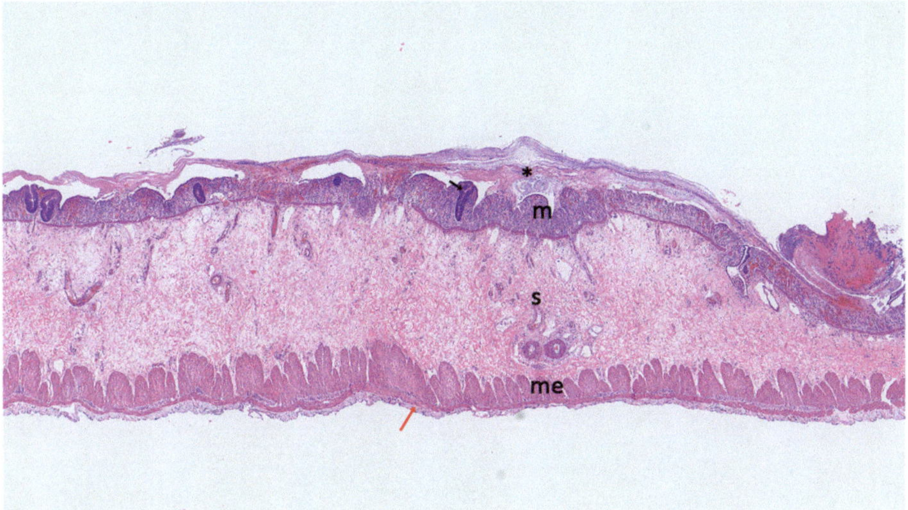

FIGURE 5.2

The colon of a male rhesus macaque necropsied 8 days after irradiation at 12 Gy has a thick layer (*) of extravasated blood, fibrin, and cellular debris on the luminal surface. The epithelial component of the mucosa (m) is reduced to widely spaced regenerative crypts (black arrow). The submucosa (s) has marked edematous thickening. The external muscular layer (muscularis externa-me) and serosa (red arrow) are largely unaffected. Hematoxylin and eosin stain, 2.5 × objective magnification.

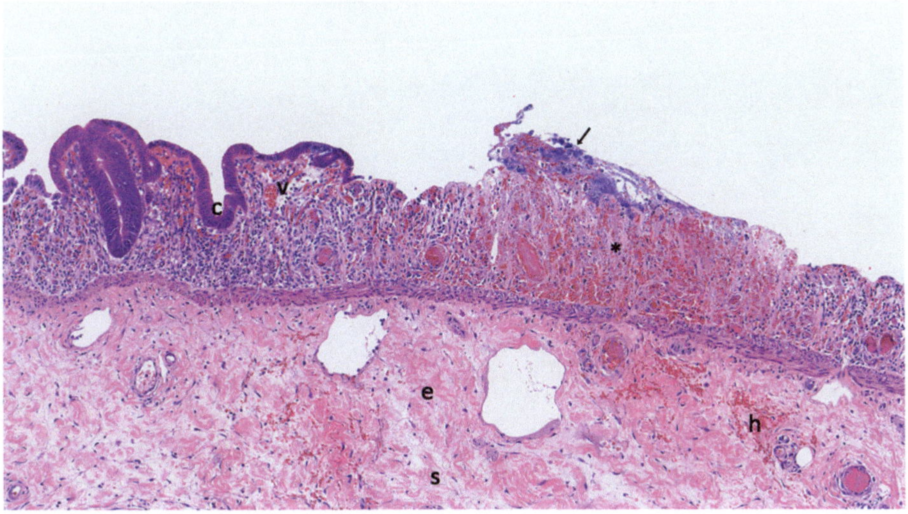

FIGURE 5.3A

Another male rhesus macaque necropsied 8 days after irradiation at 12 Gy has an area of mucosal hemorrhage and necrosis (*) with pronounced superficial bacterial colonization (black arrow). Regenerative crypt epithelium extends from the crypt (c) onto the sides and tip (red arrow) of the villi (v) that are at an early stage of reformation. The thickened submucosa (s) contains edema fluid protein (e) and extravasated erythrocytes (hemorrhage-h). Hematoxylin and eosin stain, 10 × objective magnification.

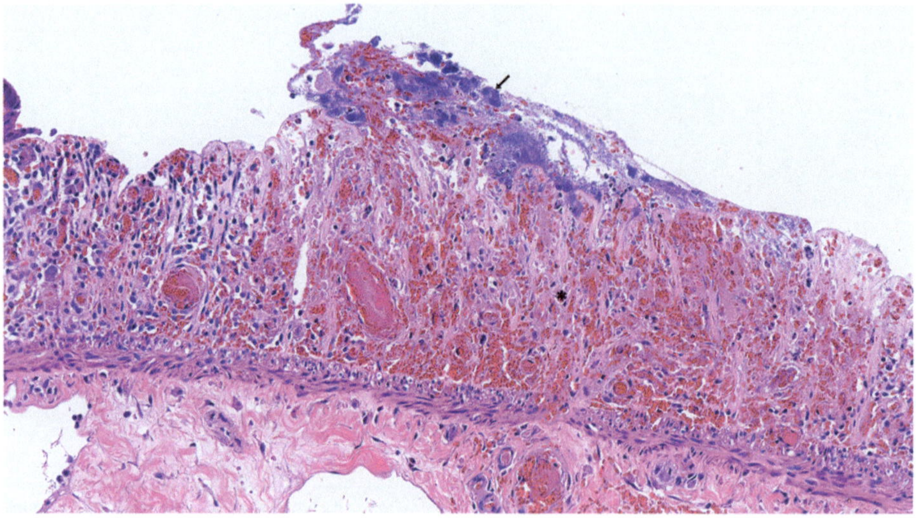

FIGURE 5.3B

Higher magnification of the previous specimen shows bacterial colonization (arrow) overlying a focally extensive area (*) of congestion, hemorrhage, and epithelial necrosis. Note the absence of inflammatory cell infiltration associated with the bacterial colonization. Hematoxylin and eosin stain, 20 × objective magnification.

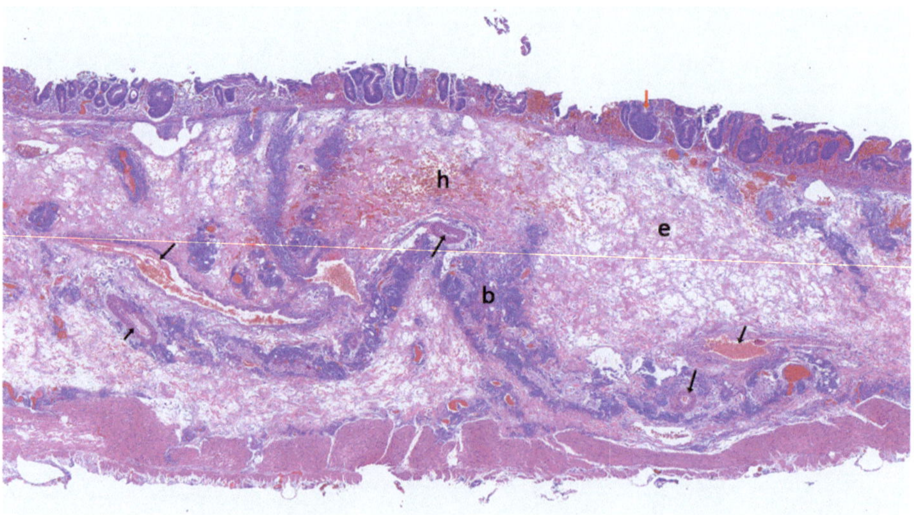

FIGURE 5.4A

The colon of a male rhesus macaque necropsied 9 days after irradiation at 12 Gy has extensive evidence of crypt cell regeneration (red arrow). The thickened submucosa has areas of edema (e) and hemorrhage (h) as well as extensive bacterial colonization (b). The bacterial colonization is associated with the course of major submucosal blood vessels (black arrows). Hematoxylin and eosin stain, 2.5 × objective magnification.

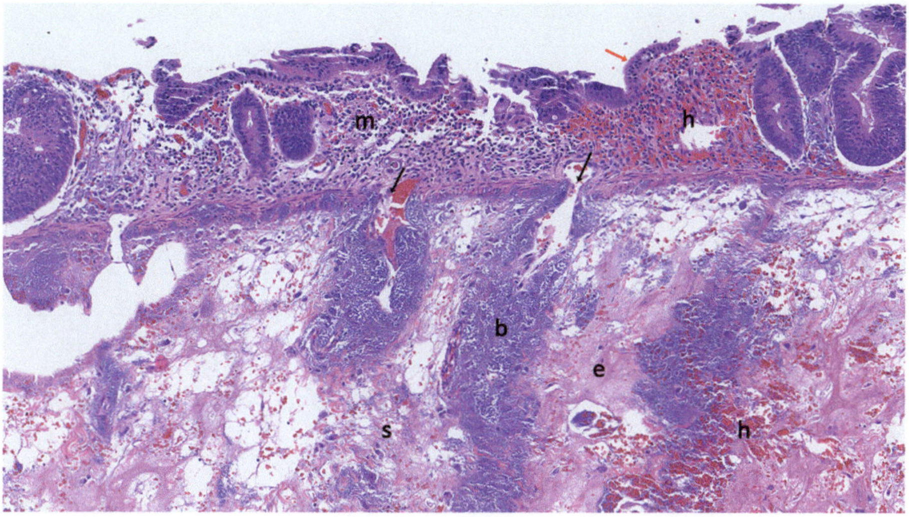

FIGURE 5.4B

Higher magnification of the previous specimen shows prominent bacterial colonization (b) associated with the course of small blood vessels (arrows) that penetrate the muscularis mucosa between the submucosa (s) mucosa (m), suggesting this route of entry of colonic luminal bacteria into the submucosa following irradiation. Extravasated erythrocytes (hemorrhage-h) are present in the mucosa and submucosa, and edema fluid protein (e) is present in the submucosa. Note the absence of inflammatory cell infiltration associated with the bacterial colonization. Hematoxylin and eosin stain, 10 × objective magnification.

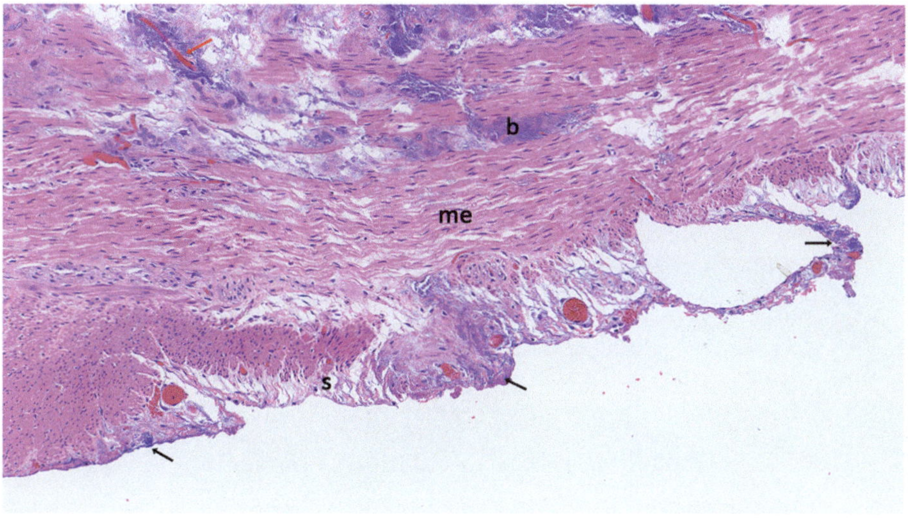

FIGURE 5.4C

A different view of the previous specimen shows bacterial colonies (b) in the external muscular layer (muscularis externa-me). As with bacterial invasion of the submucosa, the bacterial colonization in the external muscular layer is largely associated with blood vessels (red arrow). Multifocal bacterial colonization (black arrows) is present in the serosa (s). Note the absence of inflammatory cell infiltration associated with the bacterial colonization. Hematoxylin and eosin stain, 10 × objective magnification.

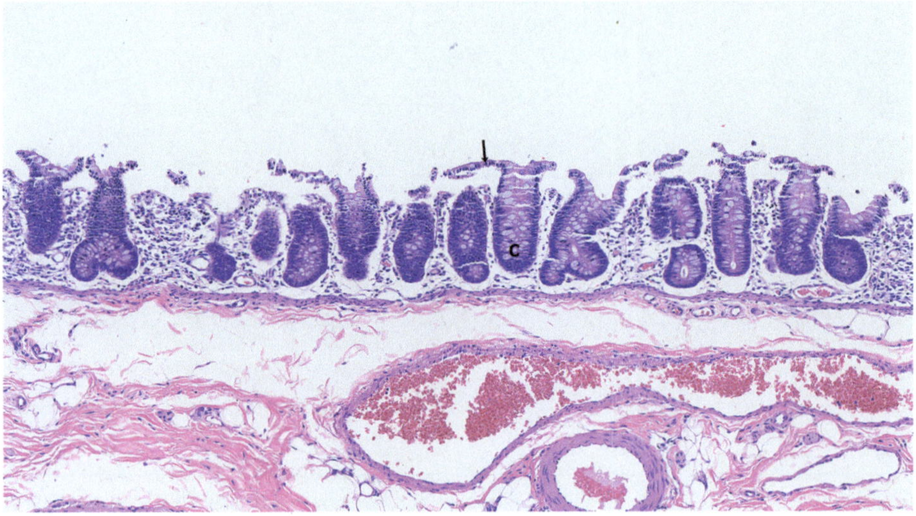

FIGURE 5.5

The colon of a male rhesus macaque necropsied 17 days after irradiation at 11 Gy shows regenerative epithelial cells in virtually all crypts (c). Surface epithelium is present (arrow), though artifactually separated from the mucosal surface by handling during the necropsy procedure. Hematoxylin and eosin stain, 10 × objective magnification.

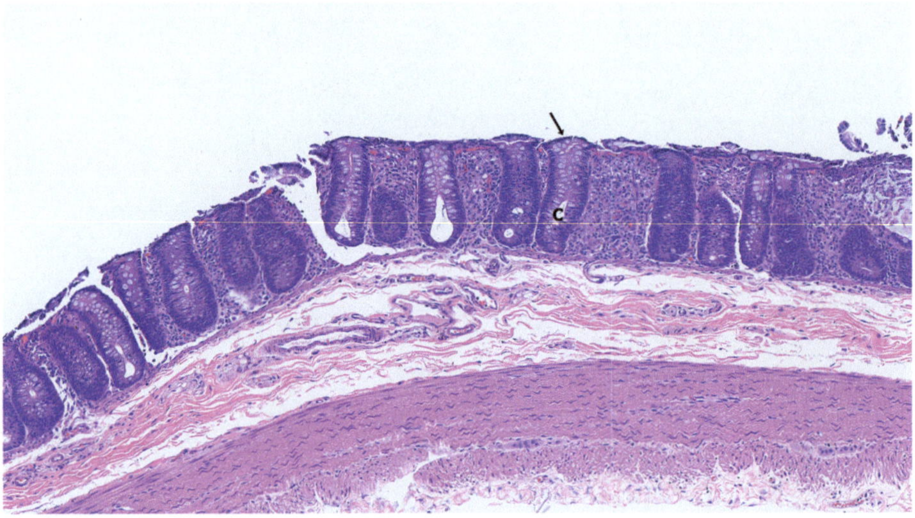

FIGURE 5.6

The colon of a male rhesus macaque necropsied 22 days after irradiation at 11 Gy shows highly proliferative epithelial cells in virtually all crypts (c). Note the presence of regenerating epithelial cells on the luminal surface (arrow). Hematoxylin and eosin stain, 10 × objective magnification.

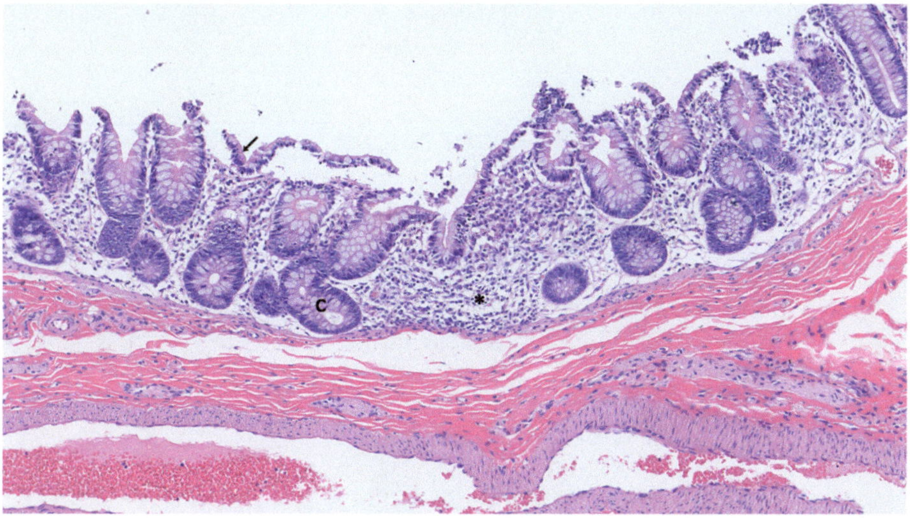

FIGURE 5.7

The colon of a male rhesus macaque necropsied 30 days after irradiation at 10 Gy shows a return to normal histomorphology. Crypts (c) contain the expected number of proliferative cells and the luminal surface is covered by a layer of mature epithelial cells (arrow). The apparent separation of the surface epithelial layer from the underlying mucosa is an artifact resulting from handling procedures during necropsy and subsequent processing. The leukocytic population in the mucosa (*) is within normal limits. Hematoxylin and eosin stain, 10 × objective magnification.

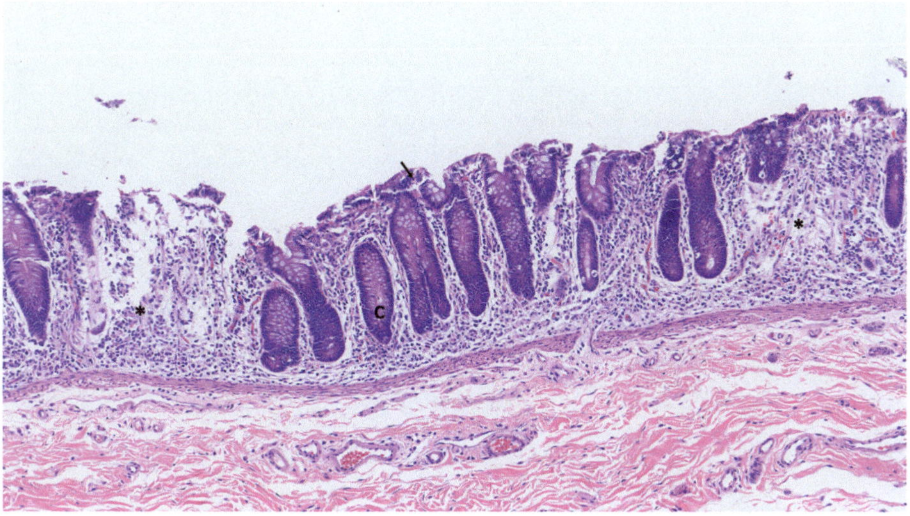

FIGURE 5.8

The colon of a male rhesus macaque necropsied 42 days after irradiation at 11 Gy has an extensive crypt cell population. The presence of goblet cells (arrow) near the luminal aspect of crypts (c) indicates maturation of the epithelial cell population. The luminal surface is covered by a continuous layer of epithelial cells. Some areas of mucosa (*) are devoid of crypts, or have only scant remnants of crypt cells, suggesting repair of the radiation-associated damage is incomplete at this stage. Hematoxylin and eosin stain, 10 × objective magnification.

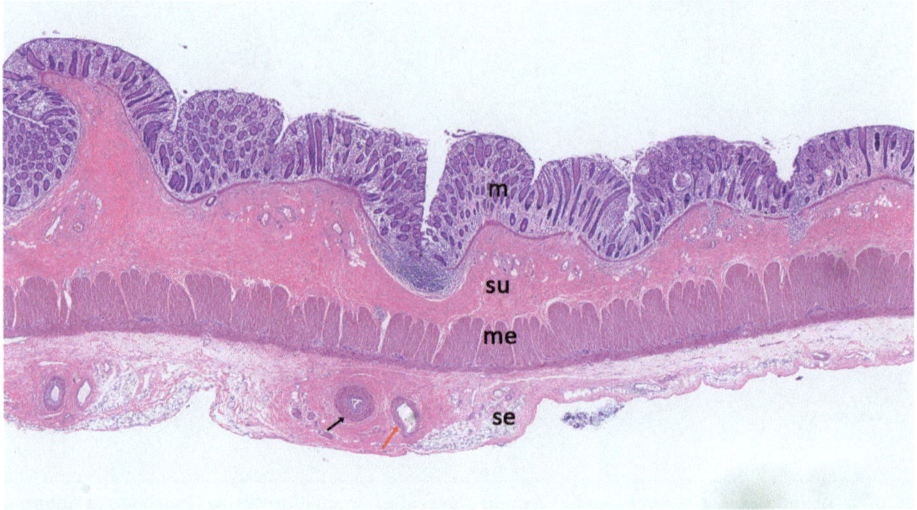

FIGURE 5.9

The colon of a male rhesus macaque necropsied 102 days after irradiation at 12 Gy has a thick layer of dense fibrous connective tissue in the submucosa (su), and similar fibrous connective tissue in the serosa (se) is centered around arteries (black arrow) and veins (red arrow). The mucosa (m) and external muscular layer (muscularis externa-me) have no alterations that are discernible at this magnification. Hematoxylin and eosin stain, 2.5 × objective magnification.

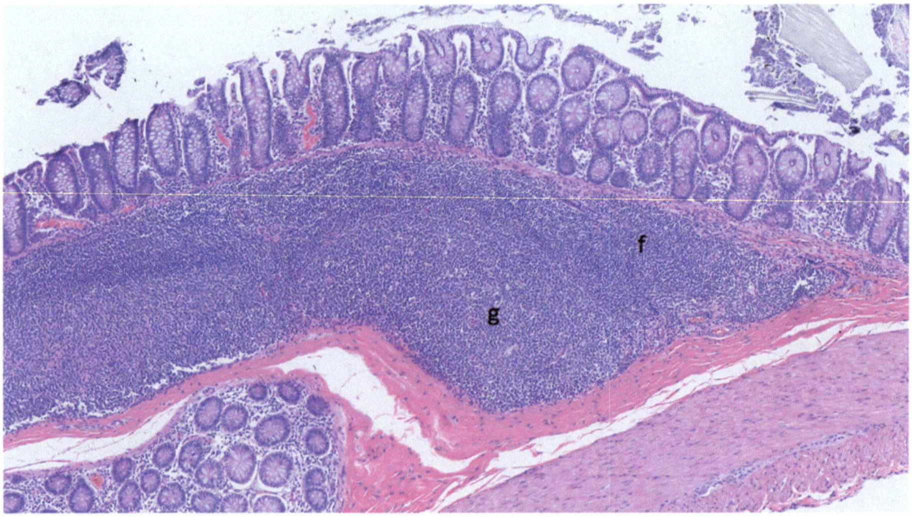

FIGURE 5.10A

The colon of a naive control male rhesus macaque has a prominent submucosal lymphoid follicle (f), also known as a lymphoglandular complex. Note the prominent germinal center (g-circled), indicating active immunological functioning. Hematoxylin and eosin stain, 7.1 × objective magnification.

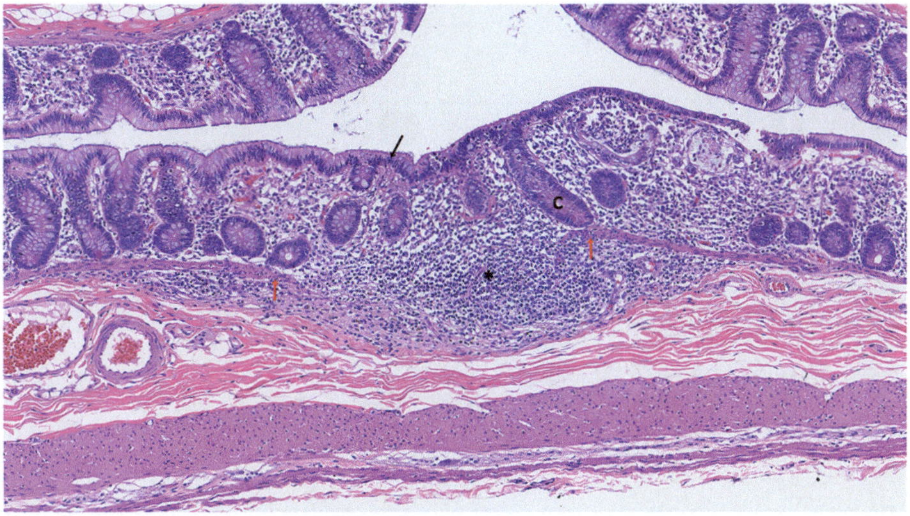

FIGURE 5.10B

The colon of a male rhesus macaque necropsied 37 days after irradiation at 11 Gy has a severely depleted lymphoid follicle (*) that has no evidence of a germinal center. Note the crypts (c) and superficial epithelium (black arrow) are microscopically normal, suggesting the radiation-associated immunological disruption may persist following recovery from the immediate epithelial damage. The discontinuity (between the red arrows) in the muscularis mucosa is normal histomorphology that allows direct communication between the lymphoid follicle and the overlying mucosa. Hematoxylin and eosin stain, 10 × objective magnification.

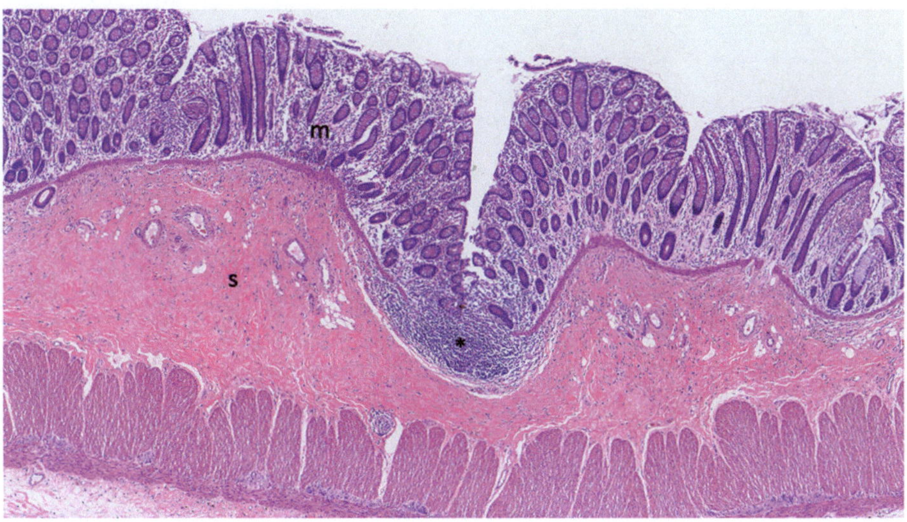

FIGURE 5.10C

The colon of a male rhesus macaque necropsied 102 days after irradiation at 12 Gy has a severely depleted lymphoid follicle (*). The mucosal (M) epithelial structures are normal at this magnification, but the submucosa (s) has a thick layer of fibrous connective tissue. Hematoxylin and eosin stain, 5 × objective magnification.

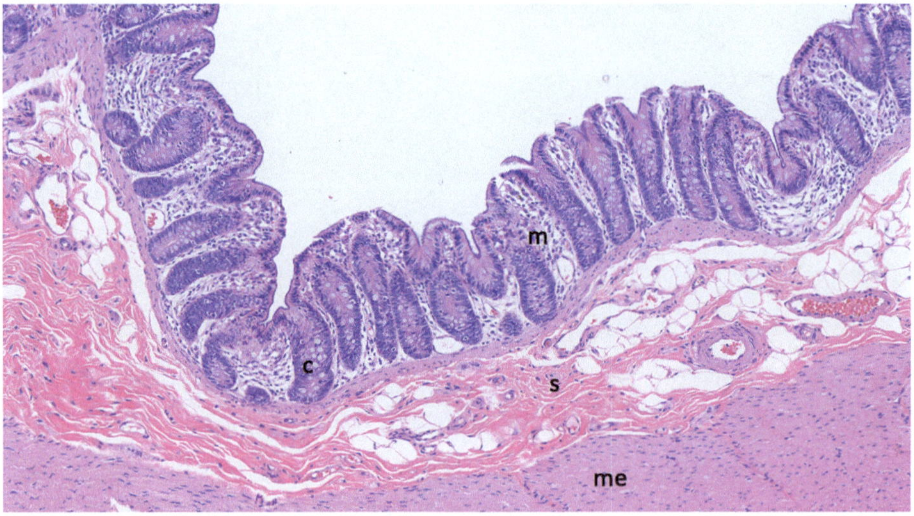

FIGURE 5.11A

The colon of a male rhesus macaque necropsied 111 days after irradiation at 11 Gy has histologically normal mucosa (m) with crypts (c), submucosa (s), and external muscular layer (muscularis externa-me). Hematoxylin and eosin stain, 10 × objective magnification.

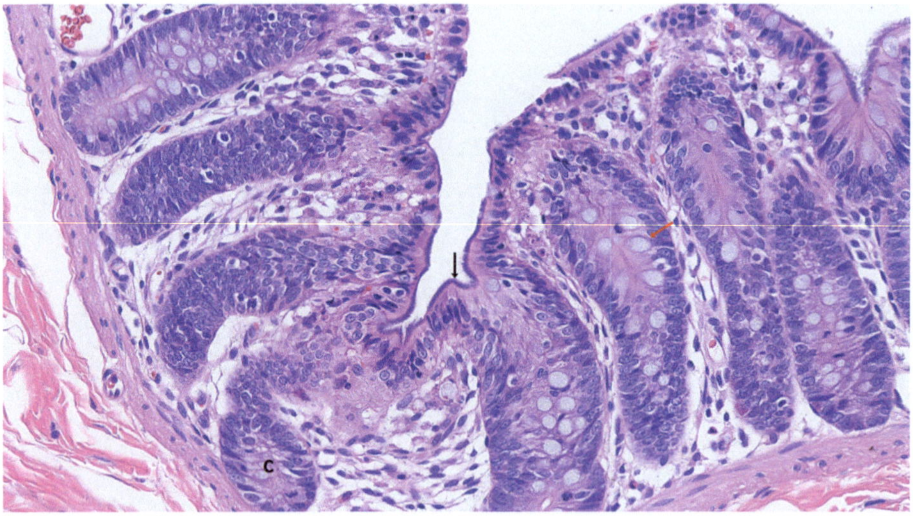

FIGURE 5.11B

Higher magnification of the previous specimen shows crypts (c) with the expected number of goblet cells (red arrow). Note the normal layer of particulate and filamentous bacteria on the luminal surface of the mucosa (black arrow). Hematoxylin and eosin stain, 29 × objective magnification.

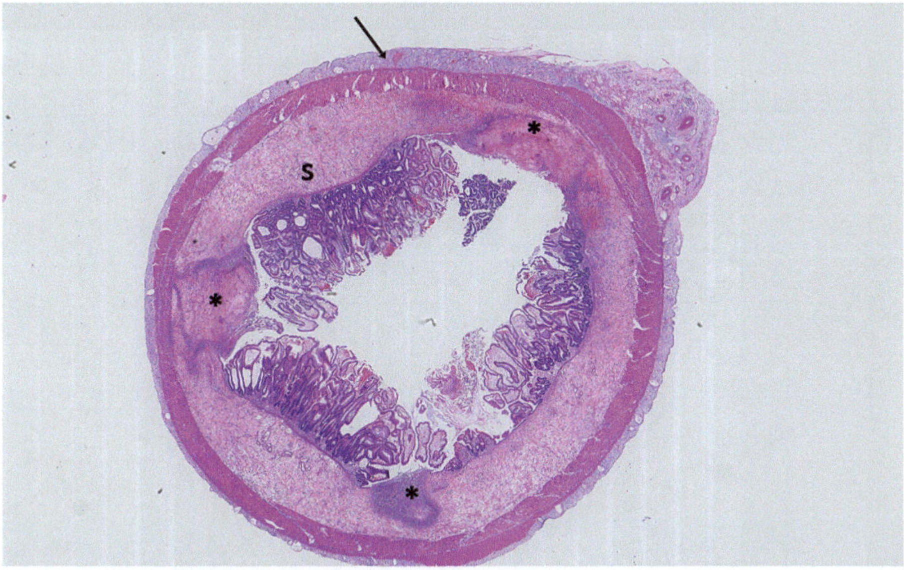

FIGURE 5.12A

Cross-section of the proximal colon from a rhesus macaque that was irradiated at 12 Gy with 3.5% bone marrow sparing and necropsied on Day 22 post-irradiation. Three areas of mucosal ulceration with associated hemorrhage, necrosis, inflammatory infiltration, and bacterial colonization are present (*). The submucosa (sm) is markedly thickened, and moderate edema with myxoid cell populations is present in the serosa (arrow). H&E stain, 0.53 × objective magnification.

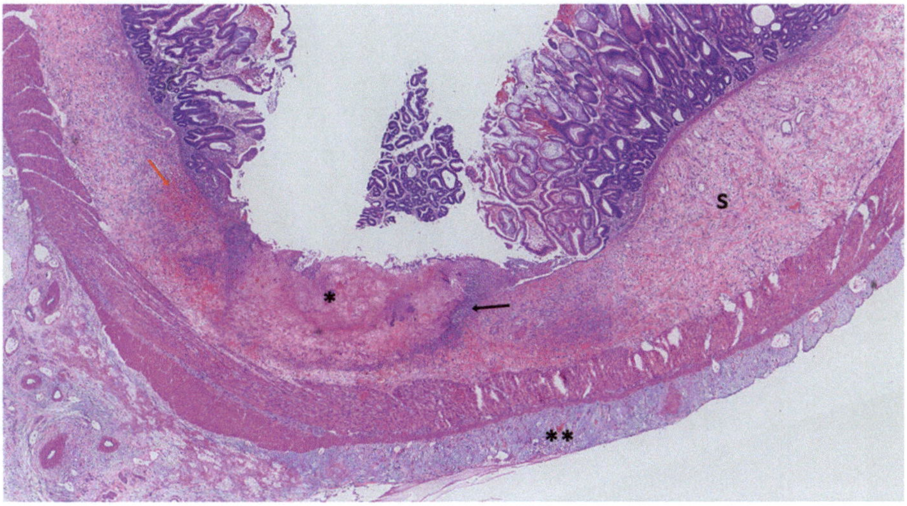

FIGURE 5.12B

Higher magnification of the previous specimen shows a central area of ulceration and necrosis (*) surrounded by a zone of inflammatory cell infiltration and bacterial colonization (black arrow). The submucosa (s) is thickened and edematous, with multifocal hemorrhage (red arrow). The serosa (**) is thickened by the presence of light blue myxoid tissue. Hematoxylin and eosin stain, 1.25 × objective magnification.

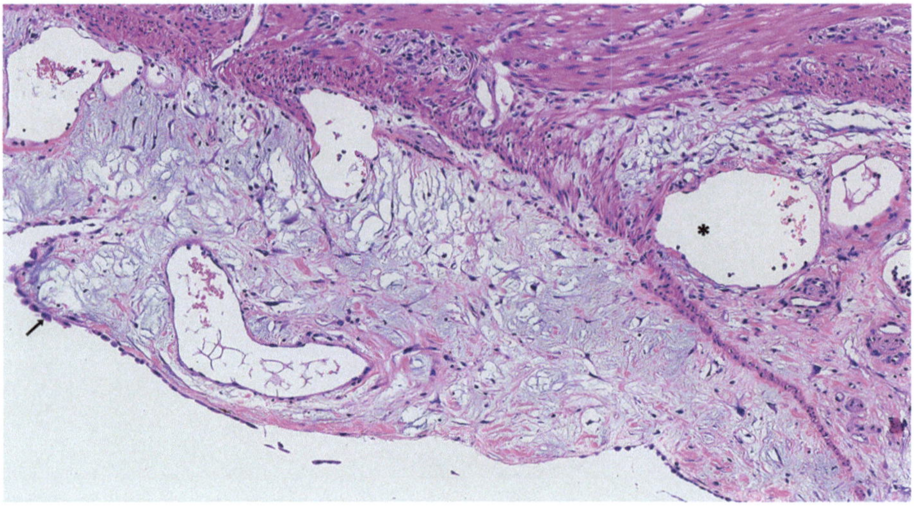

FIGURE 5.12C

A higher magnification image shows the thick serosal layer of light blue-stained myxoid material intermixed with a sparse population of unidentified large mononuclear cells. Note widely dilated lymphatics (*), which suggests obstruction to lymph drainage. Also note the plump "tombstone" appearance of the surface mesothelial cells (arrow), which commonly indicates accumulation of fluid in the peritoneal cavity. H&E stain, 10 × objective magnification.

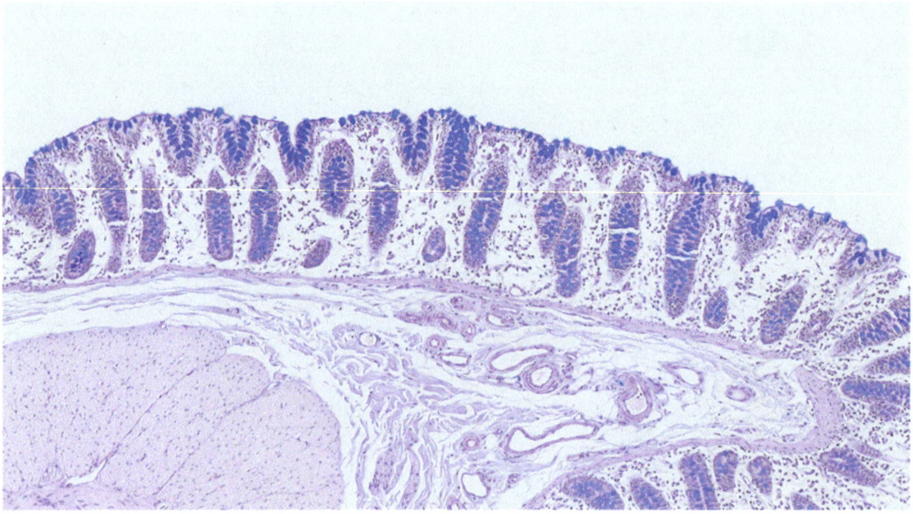

FIGURE 5.13A

Periodic acid-Schiff/Alcian blue stain performed on the colon of a naive male rhesus macaque shows numerous blue-stained goblet cells in the superficial epithelium. 10 × objective magnification 10 × .

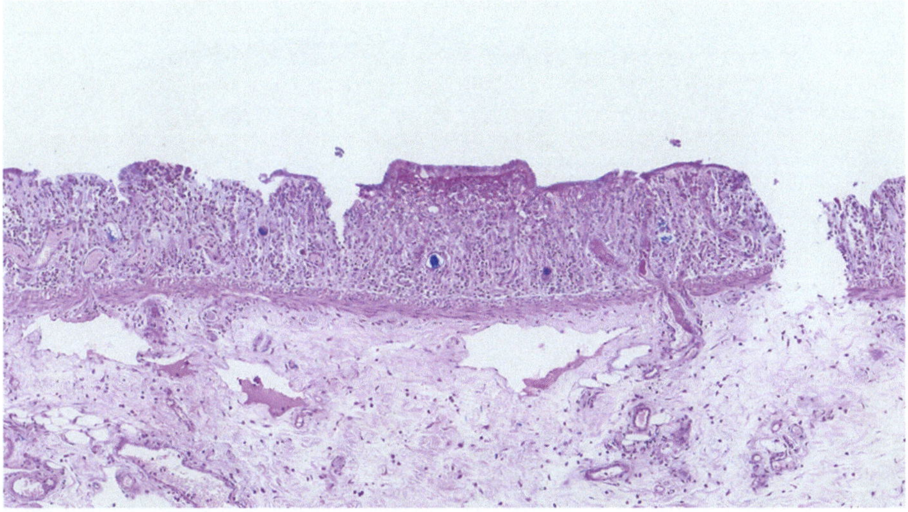

FIGURE 5.13B

Periodic acid-Schiff/Alcian blue stain performed on the colon of a male rhesus macaque necropsied 8 days after irradiation at 12 Gy shows marked reduction of blue-stained goblet cells in the superficial epithelium. 10 × objective magnification.

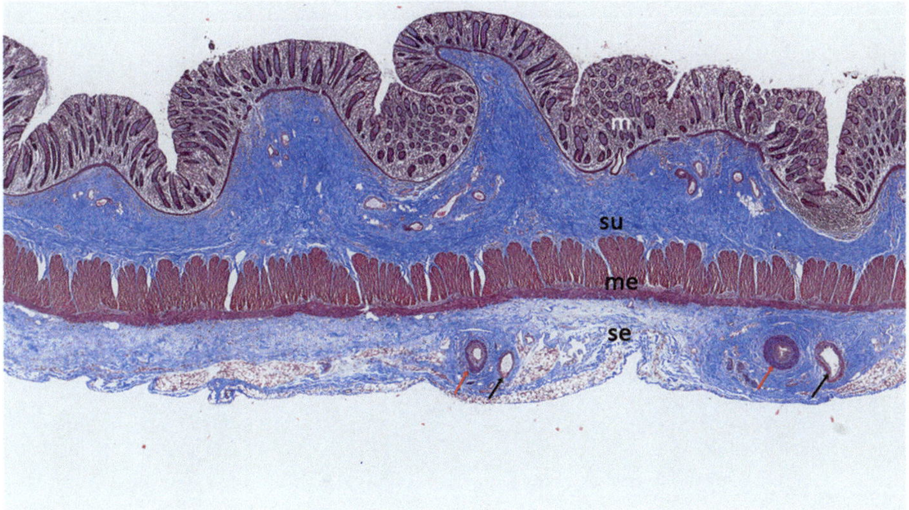

FIGURE 5.14A

Masson's trichrome stain performed on the colon of a male rhesus macaque necropsied 102 days after irradiation at 12 Gy has a thick zone of blue-stained fibrous connective tissue in the submucosa (su), with multifocally intense serosal (se) fibrosis that is concentrated around arteries (red arrows) and veins (black arrows). Increased fibrous connective tissue in the mucosa (m) and muscularis externa (me) is not discernible at this magnification. Objective magnification 2.5 × .

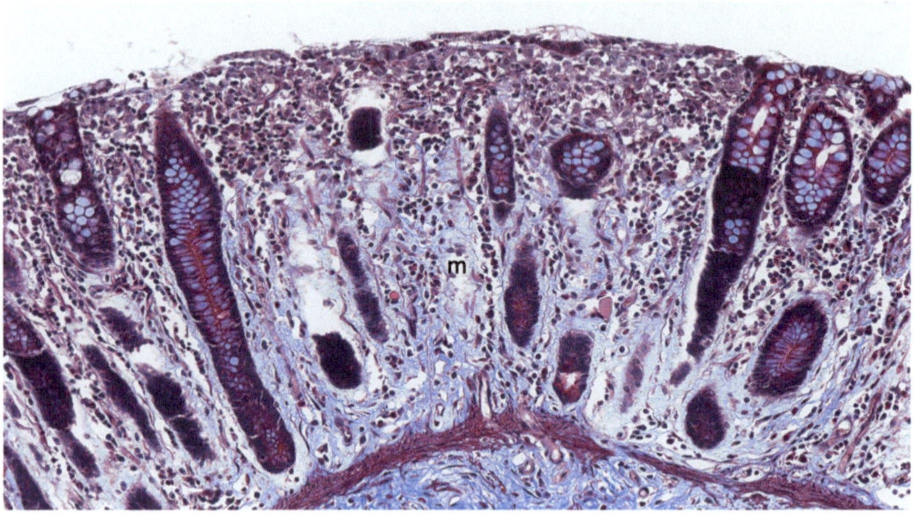

FIGURE 5.14B

Higher magnification of the previous specimen reveals an increased amount of blue-stained fibrous connective tissue that exists as fine fibrils in the mucosa (m). The mucosal fibrosis was multifocally distributed, with sparing of many areas of the colonic mucosa. Masson's trichrome, 20 × objective magnification.

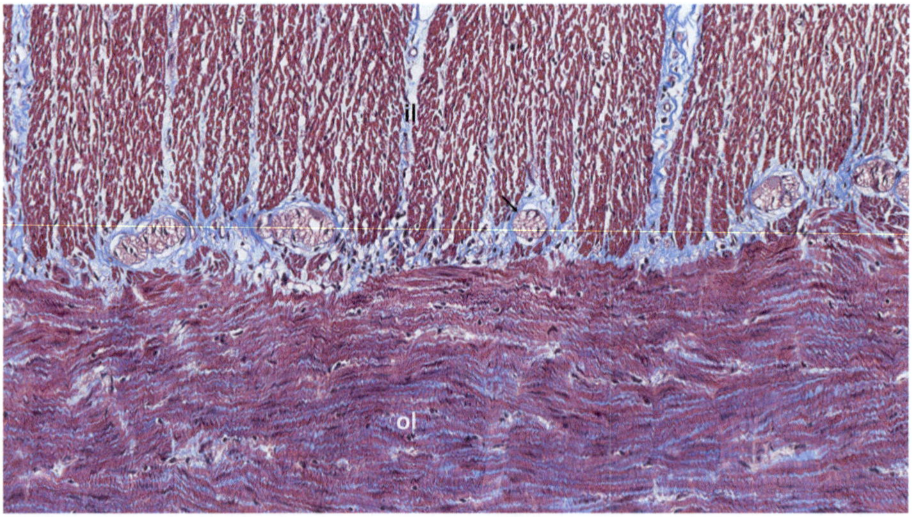

FIGURE 5.14C

Higher magnification of the previous specimen reveals blue-stained fibrous connective tissue fibrils interspersed between smooth muscle fibers in the external layer (ol) of the muscular wall, with relative sparing of the internal layer (il). There is increased fibrous connective tissue in the fascial plane between the two muscular layers surrounds neurons of the myenteric (Auerbach's) plexus (arrow). Masson's trichrome stain, 20 × objective magnification.

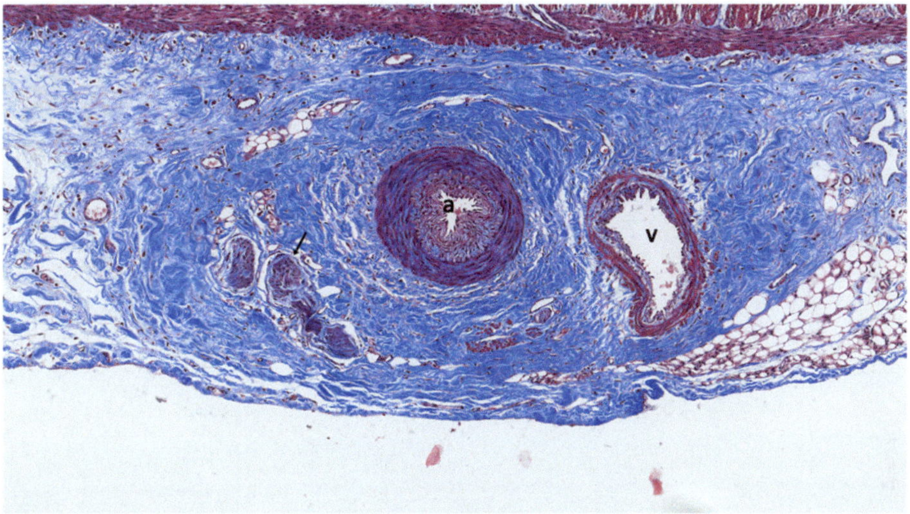

FIGURE 5.14D

Higher magnification of the previous specimen reveals serosal arteries (a), veins (v), and nerves (arrow) are embedded in a zone of dense fibrous connective tissue. Note the wall of the artery also contains blue-stained fibrous connective tissue. Masson's trichrome stain, 10 × objective magnification.

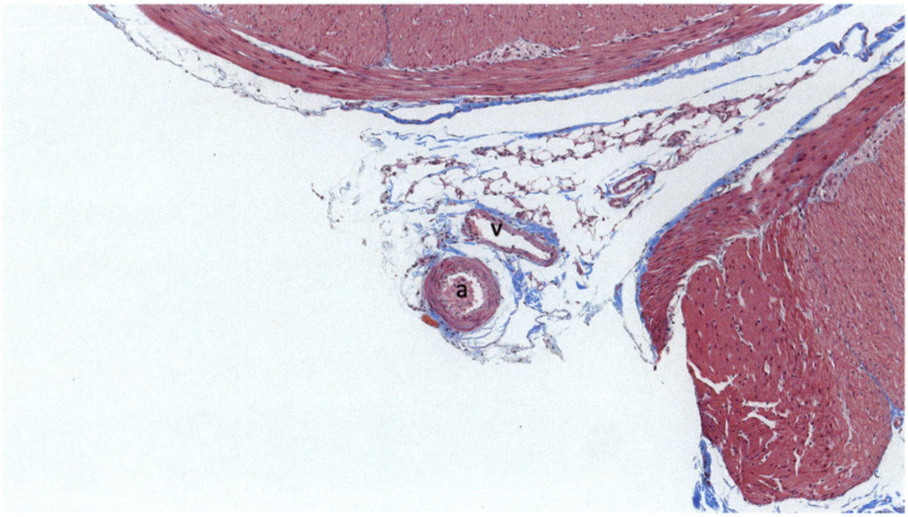

FIGURE 5.14E

For comparison to the previous figure, Masson's trichrome stain performed on the colon of a naive male rhesus macaque reveals a minimal amount of blue-stained fibrous connective tissue surrounding the artery (a) and vein (v) on the serosal surface. 10 × objective magnification.

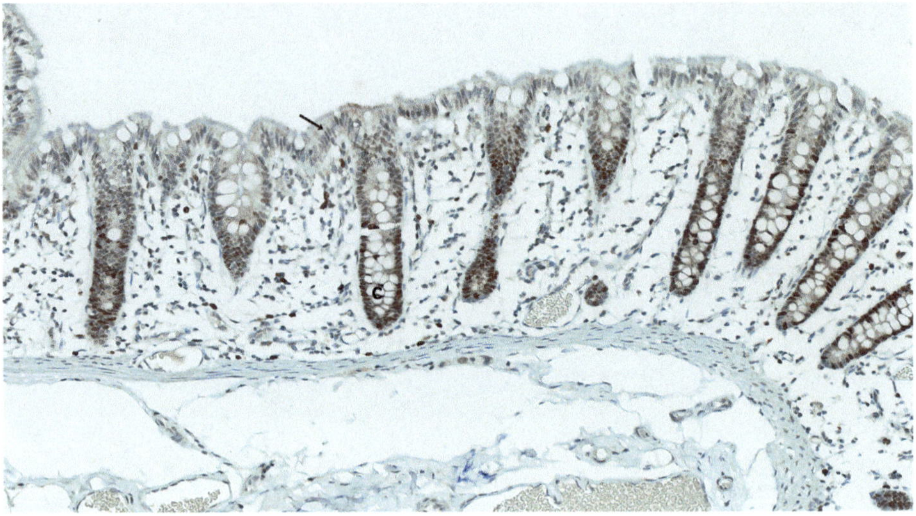

FIGURE 5.15

Immunohistochemical staining for Ki-67 performed on the colon of a naive male rhesus macaque reveals positively (brown) stained cells within crypts (c), with little staining of superficial epithelial cells (arrow). Stain performed with 3, 3′-diaminobenzidine chromogen and hematoxylin counterstain, 20 × objective magnification.

FIGURE 5.16

Immunohistochemical staining for Ki-67 performed on colon from a male rhesus macaque necropsied 6 days after irradiation at 12 Gy reveals essentially no positive staining in the severely disrupted colonic mucosa (m). Stain performed with 3,3′-diaminobenzidine chromogen and hematoxylin counterstain, 10 × objective magnification.

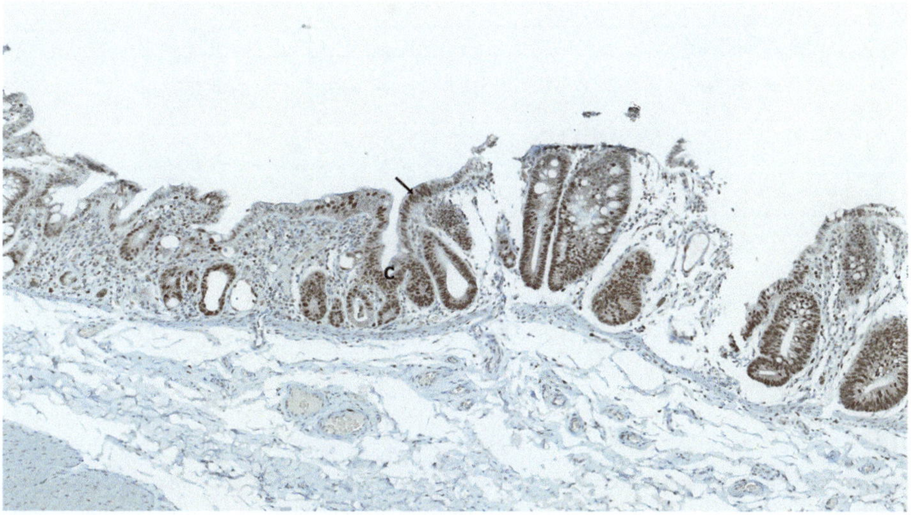

FIGURE 5.17

Immunohistochemical staining for Ki-67 performed on colon from a male rhesus macaque necropsied 8 days after irradiation at 12 Gy reveals positively stained proliferating epithelial cells within crypts (c). Note the positive Ki-67 staining in epithelial cells (arrow) that are extending from crypts to replace the denuded luminal surface. Stain performed with 3,3′-diaminobenzidine chromogen and hematoxylin counterstain, 10 × objective magnification.

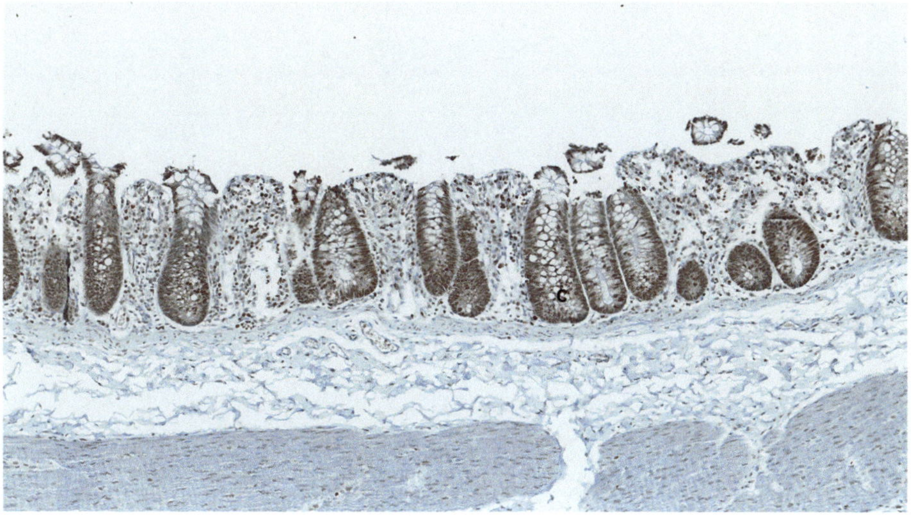

FIGURE 5.18

Immunohistochemical staining for Ki-67 performed on colon from a male rhesus macaque necropsied 22 days after irradiation at 11 Gy reveals numerous positively stained proliferative cells in virtually all crypts (c). Note the enlarged size and rounded contour of crypts as compared to the naive rhesus macaque. Stain performed with 3,3′-diaminobenzidine chromogen and hematoxylin counterstain, 10 × objective magnification.

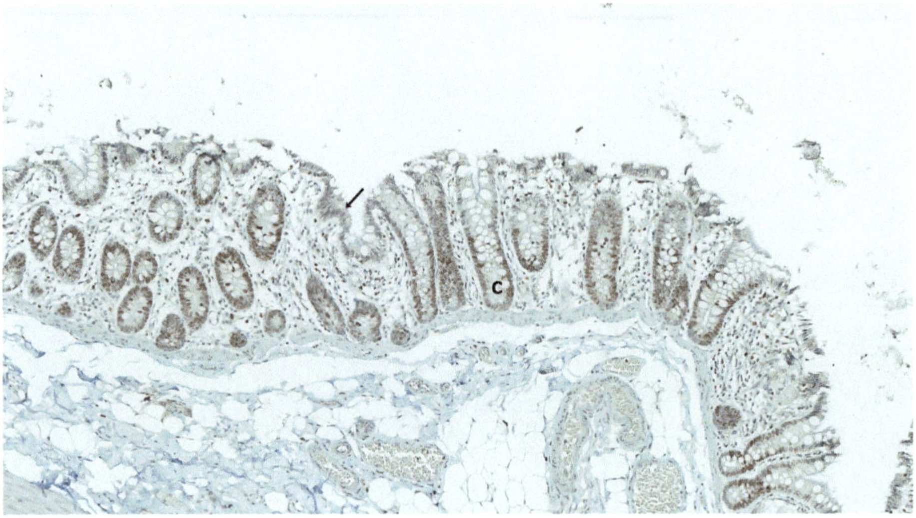

FIGURE 5.19

Immunohistochemical staining for Ki-67 performed on colon from a male rhesus macaque necropsied 84 days after irradiation at 11 Gy reveals the population of positively stained cells in crypts (c) is similar to that seen in the naive control animal. The mature surface epithelial cells (arrow) are not stained, indicating no active cellular proliferation in this mature cell population. Compare the overall contour of the crypts relative to the bulging crypts seen at Day 22 post-irradiation. Stain performed with 3,3′-diaminobenzidine chromogen and hematoxylin counterstain, 10 × objective magnification.

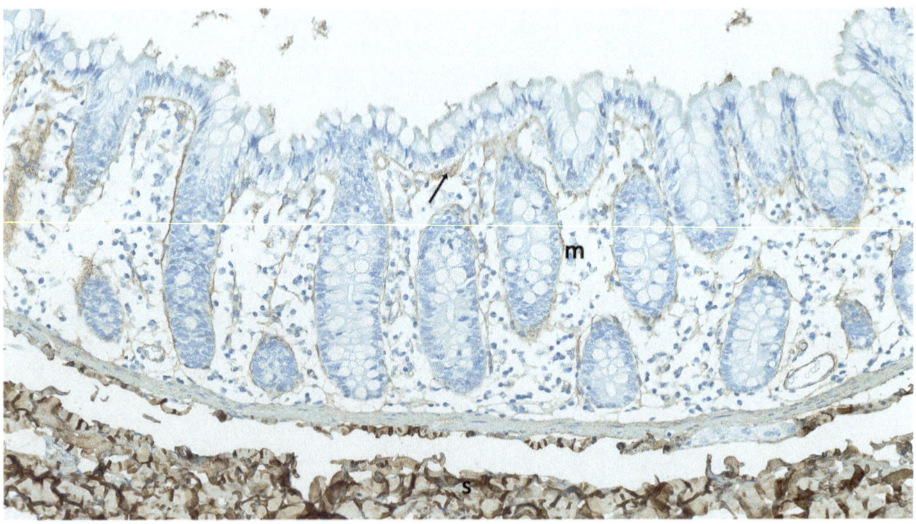

FIGURE 5.20

Immunohistochemical staining for collagen-1 performed on the colon of a naive male rhesus macaque reveals strongly positive staining in the submucosa (s) and fine positively stained collagen fibrils (arrow) in the lamina propria of the mucosa (m). Stain performed with 3, 3′-diaminobenzidine chromogen and hematoxylin counterstain, 20 × objective magnification.

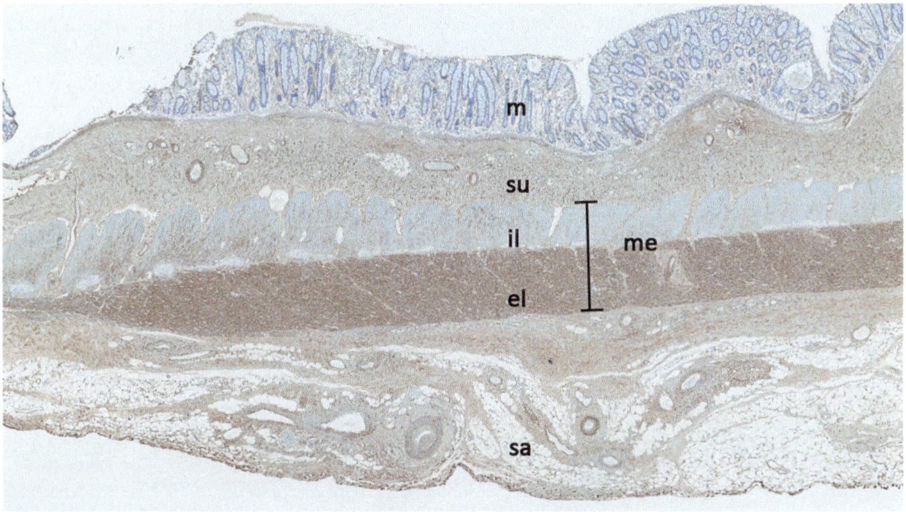

FIGURE 5.21A

Immunohistochemical staining for collagen-1 performed on colon from a male rhesus macaque necropsied 102 days after irradiation at 12 Gy reveals positive staining in the thickened submucosa (su) and the external layer (el) of the muscularis externa (me), but not in the internal layer of the muscularis externa (il). There is a moderate increase in collagen-1 staining in the serosa (sa), and a slight increase in collagen-1 staining is equivocally perceptible in the mucosa (m) at this magnification. Stain performed with 3, 3′-diaminobenzidine chromogen and hematoxylin counterstain, 3.09 × objective magnification.

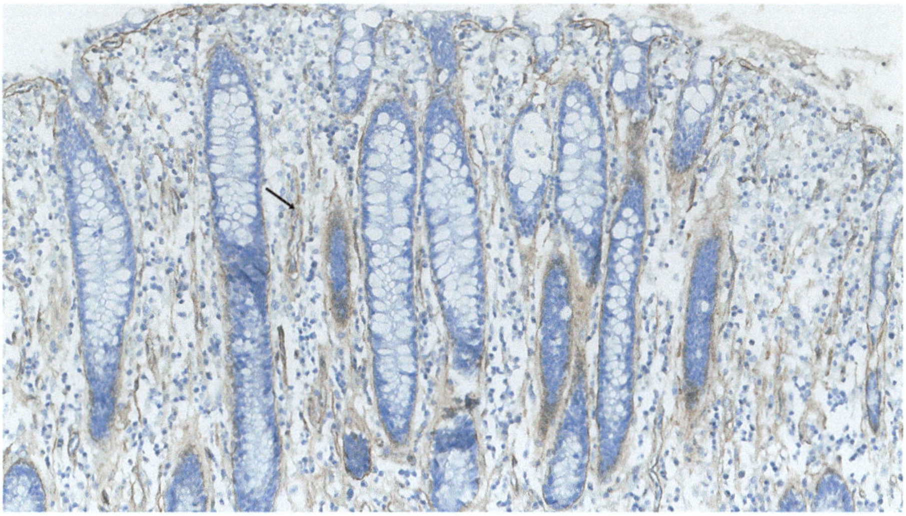

FIGURE 5.21B

Higher magnification of the previous specimen shows an increase in collagen 1^+ fibrils (arrow) in the mucosal lamina propria. Immunohistochemical staining for collagen-1 with 3, 3′-diaminobenzidine chromogen and hematoxylin counterstain, 20 × objective magnification.

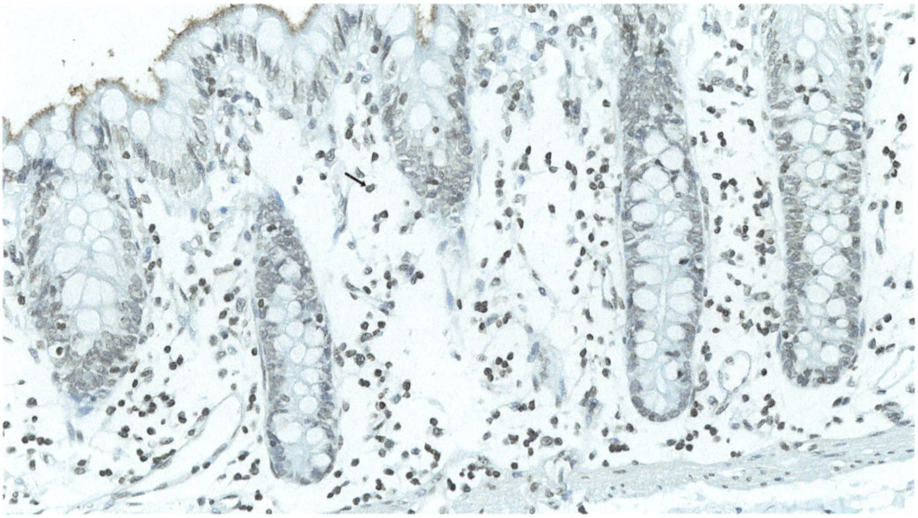

FIGURE 5.22

Immunohistochemical staining for connective tissue growth factor performed on the colon of a naive male rhesus macaque reveals essentially no positive staining in the normal population of mucosal immunocytes (arrow). Stain performed with 3, 3′-diaminobenzidine chromogen and hematoxylin counterstain, 20 × objective magnification.

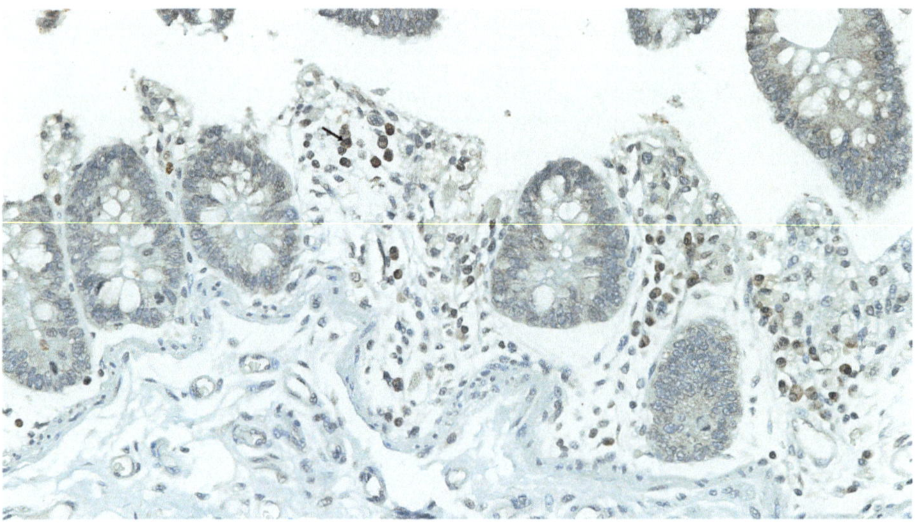

FIGURE 5.23

Immunohistochemical staining for connective tissue growth factor performed on the colon of a male rhesus macaque necropsied 20 days after irradiation at 10 Gy reveals positive cytoplasmic staining in a subpopulation of large mononuclear cells (arrow), consistent with macrophages/dendritic cells. Stain performed with 3, 3′-diaminobenzidine chromogen and hematoxylin counterstain, 20 × objective magnification.

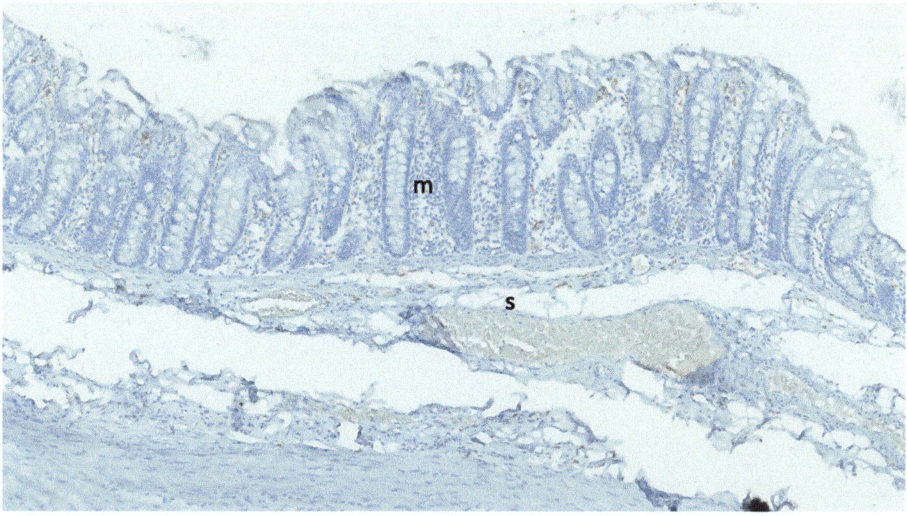

FIGURE 5.24A

Immunohistochemical staining for CD163 performed on the colon of a naive male rhesus macaque shows no positive staining in the mucosa (m) or submucosa (s). Stain performed with 3,3′-diaminobenzidine chromogen and hematoxylin counterstain, 10 × objective magnification.

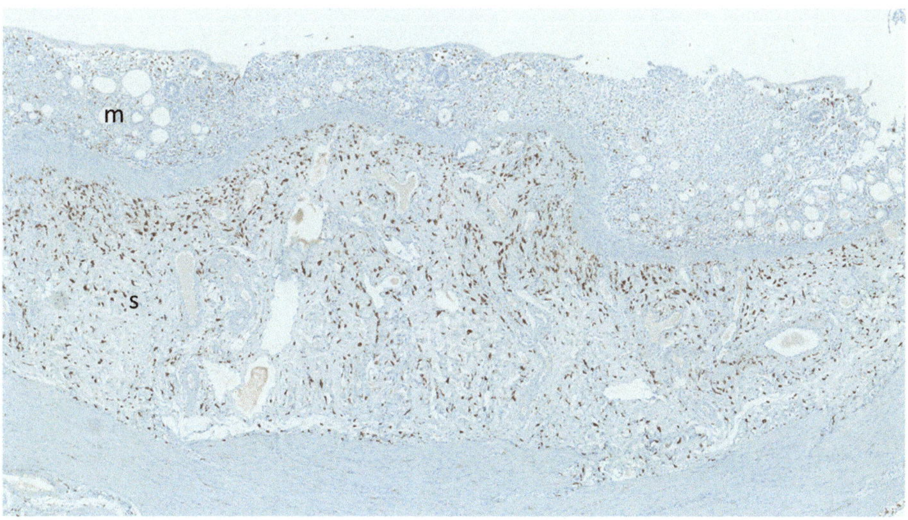

FIGURE 5.24B

Immunohistochemical staining for CD163 performed on the colon of a male rhesus macaque necropsied 5 days after irradiation at 12 Gy shows an abundance of CD163 + cells in the mucosa (m) and submucosa (s). Stain performed with 3, 3′-diaminobenzidine chromogen and hematoxylin counterstain, 4.04 × objective magnification.

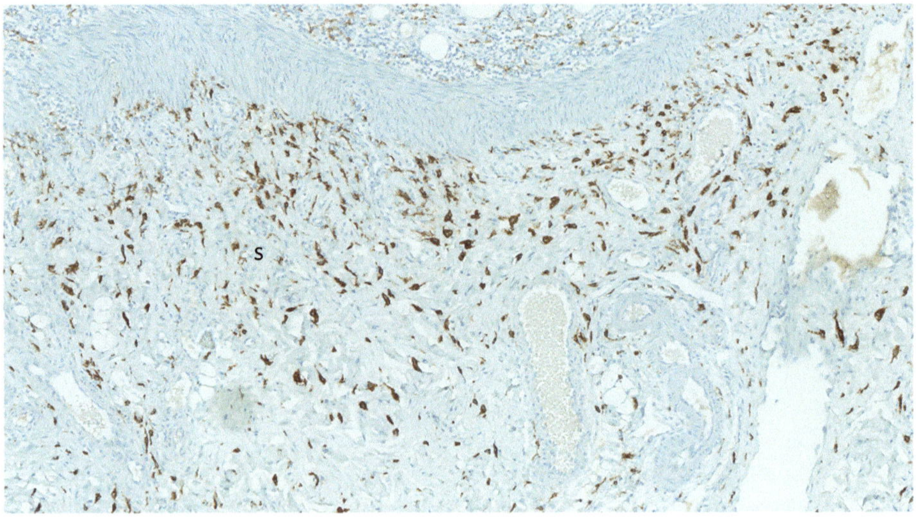

FIGURE 5.24C

Higher magnification of the previous specimen shows the abundant population of CD163 + cells to consist of large cells that are scattered throughout the collagenous submucosa (s). 10 × objective magnification.

FIGURE 5.24D

Higher magnification of the previous specimen shows evenly dispersed $CD163^+$ cells (arrow) within the serosa (s). 10 × objective magnification.

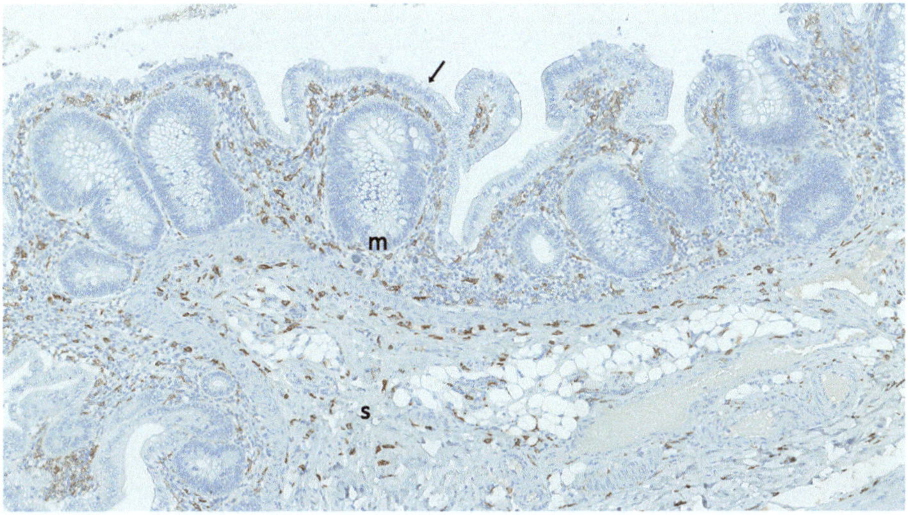

FIGURE 5.24E

Immunohistochemical staining for CD163 performed on the colon of a male rhesus macaque necropsied 9 days after irradiation at 12 Gy shows a pronounced population of $CD163^+$ cells in the mucosa (m) and submucosa (s). Note the continuity of the surface epithelium (arrow) has been reestablished. Stain performed with 3, 3′-diaminobenzidine chromogen and hematoxylin counterstain, 10 × objective magnification.

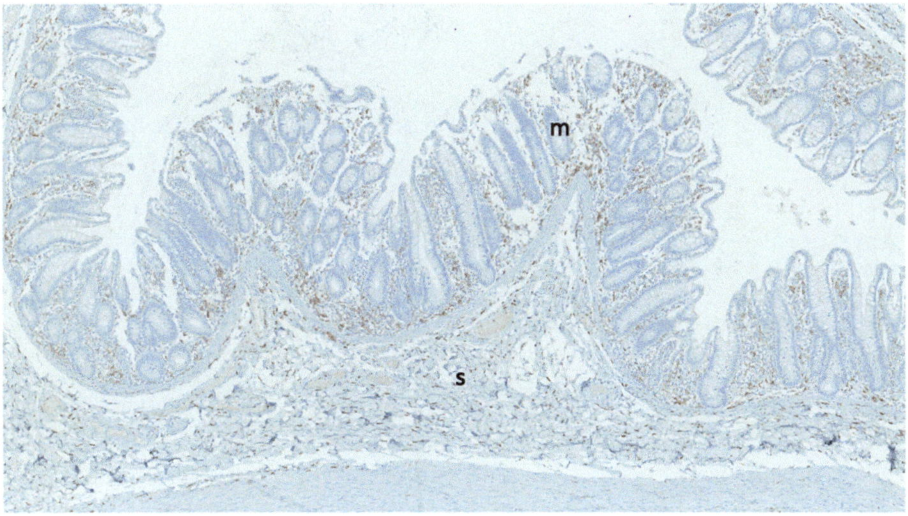

FIGURE 5.24F

Immunohistochemical staining for CD163 performed on the colon of a male rhesus macaque necropsied 25 days after irradiation at 12 Gy shows a pronounced population of $CD163^+$ cells persist in the mucosa (m), with a less pronounced population of $CD163^+$ cells in the submucosa (s). Note the mucosal structure is largely reestablished. Stain performed with 3, 3′-diaminobenzidine chromogen and hematoxylin counterstain, 5 × objective magnification.

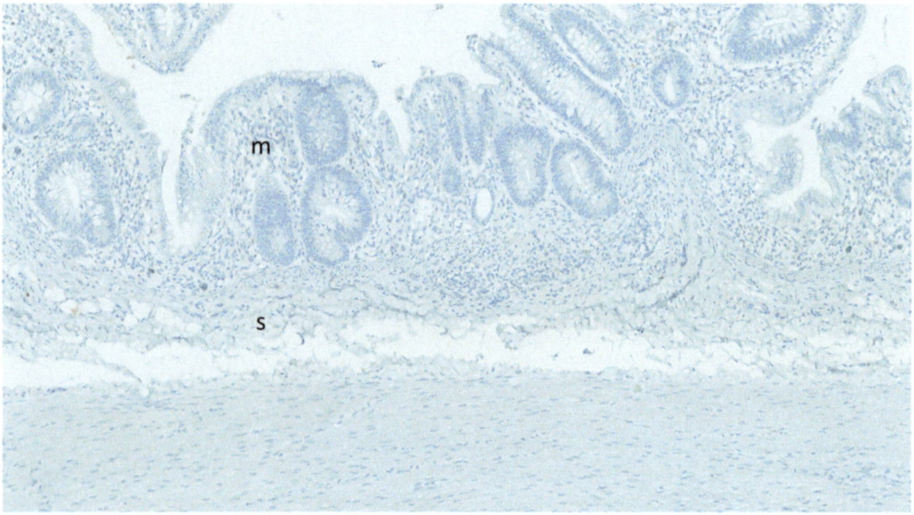

FIGURE 5.25A

Immunohistochemical staining for myeloperoxidase (MPO) performed on a male rhesus macaque necropsied 9 days after irradiation at 12 Gy (same animal as shown for CD163 staining) has no MPO^+ positive cells in the mucosa (m) or submucosa (s). Stain performed with 3, 3′-diaminobenzidine chromogen and hematoxylin counterstain, 10 × objective magnification.

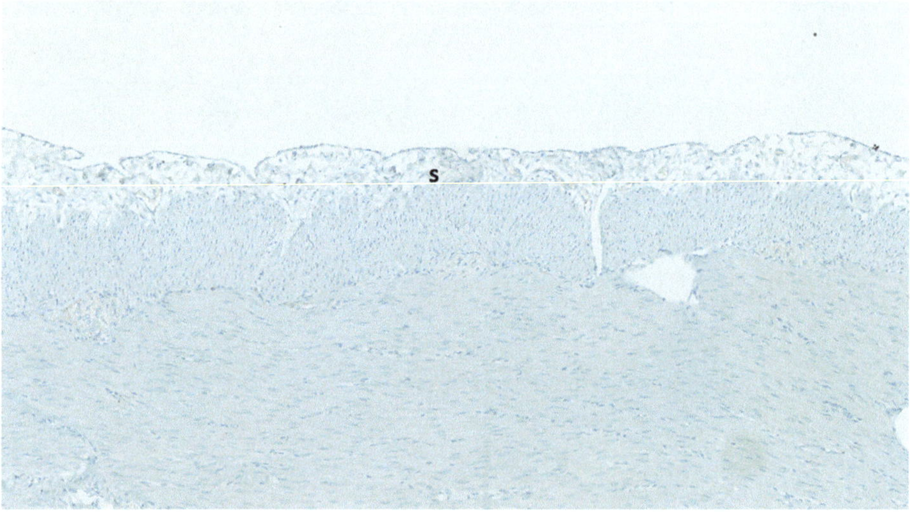

FIGURE 5.25B

Immunohistochemical staining for myeloperoxidase (MPO) performed on a male rhesus macaque necropsied 5 days after irradiation at 12 Gy (same animal as shown for CD163 staining) has no MPO^+ positive cells in the serosa (s). Stain performed with 3, 3′-diaminobenzidine chromogen and hematoxylin counterstain, 10 × objective magnification.

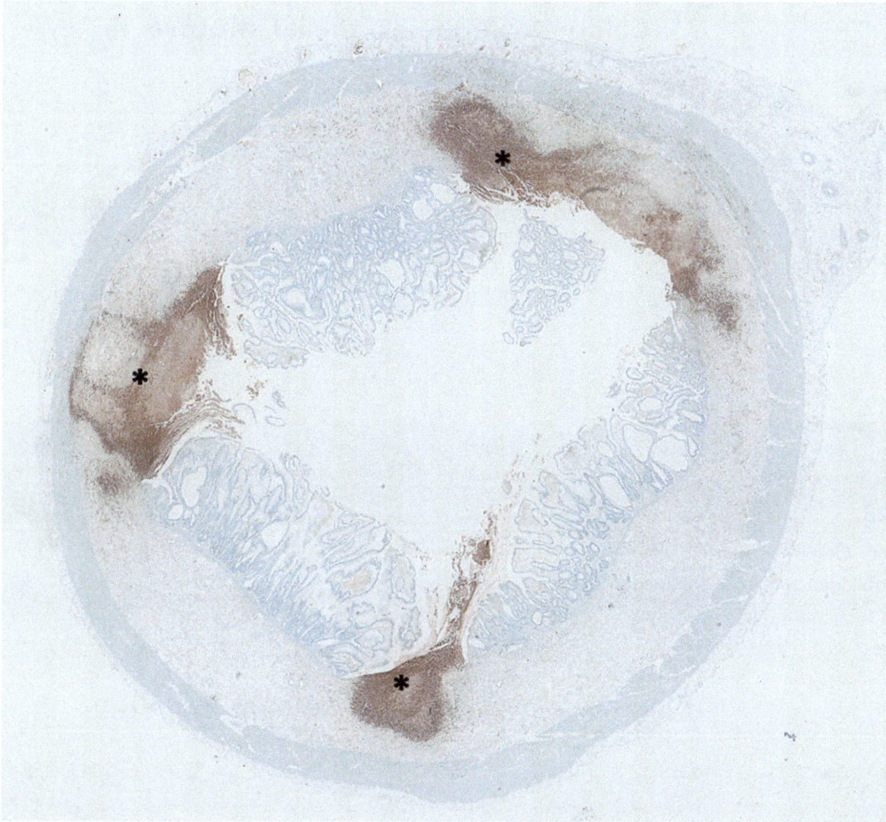

FIGURE 5.25C

Immunohistochemical staining for myeloperoxidase (MPO) performed on a male rhesus macaque necropsied 22 days after irradiation at 12 Gy has pronounced infiltrations of MPO^+ neutrophils and macrophages associated with mucosal ulcerations (*). Stain performed with 3, 3′-diaminobenzidine chromogen and hematoxylin counterstain, 10 × objective magnification.

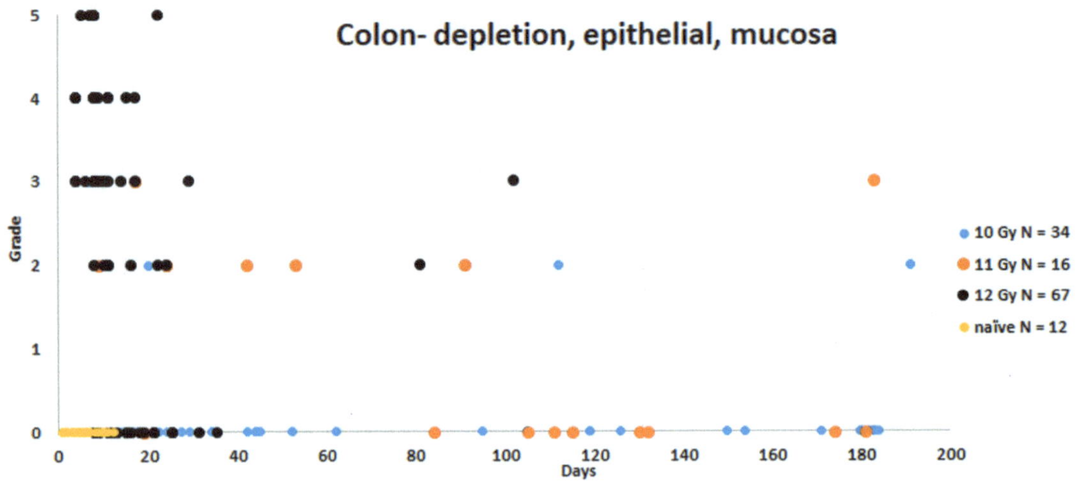

FIGURE 5.26

Scatter plot: Colon—depletion, epithelial, mucosa. Observations based on light microscopic examination of paraffin-embedded specimens stained with hematoxylin and eosin.

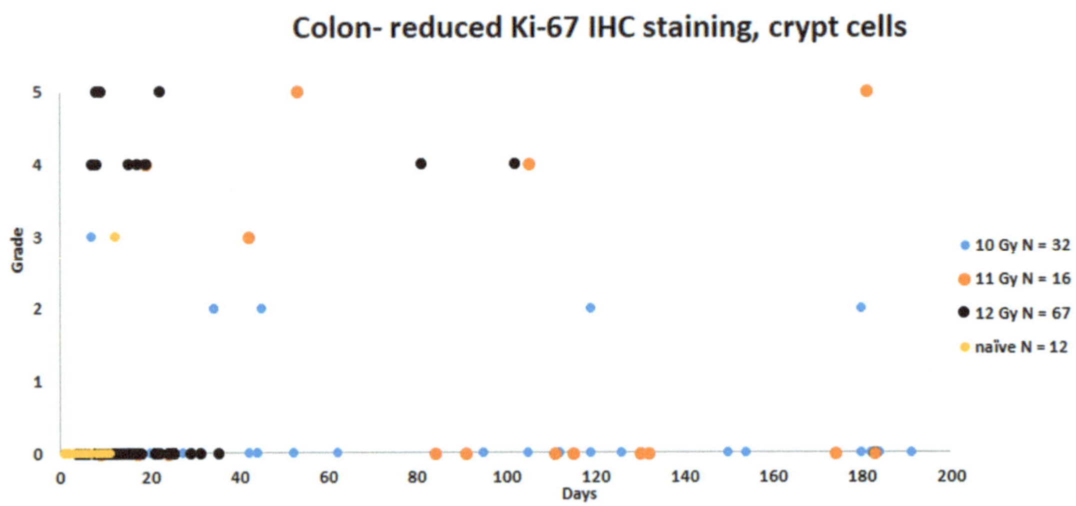

FIGURE 5.27

Scatter plot: Colon—reduced Ki-67^{+} crypt epithelial cells. Observations based on light microscopic examination of paraffin-embedded specimens stained by immunohistochemical procedures for Ki67 with 3,3′-diaminobenzidine chromogen and hematoxylin counterstain.

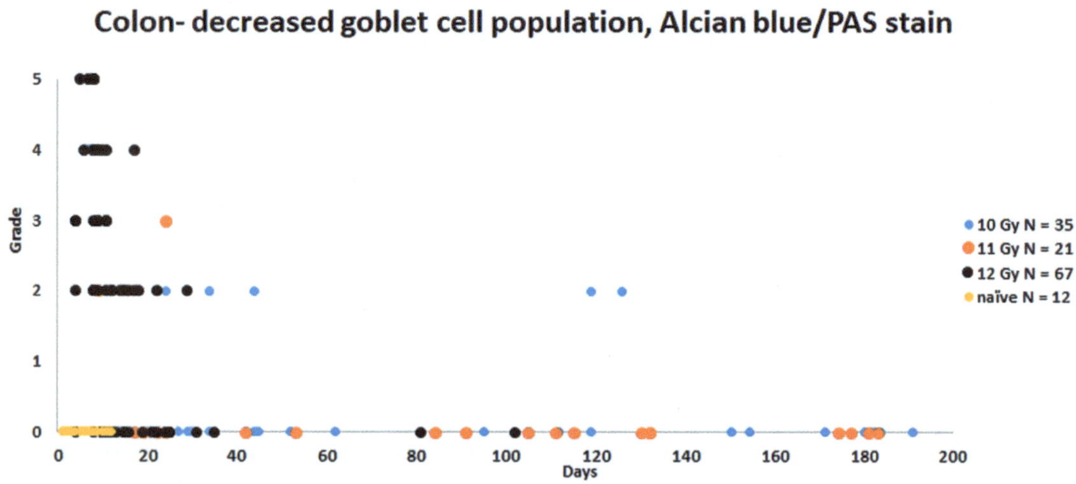

FIGURE 5.28

Scatter plot: Colon—decreased goblet cell population. Observations based on light microscopic examination of paraffin-embedded specimens stained with alcian blue/periodic acid-Schiff stain.

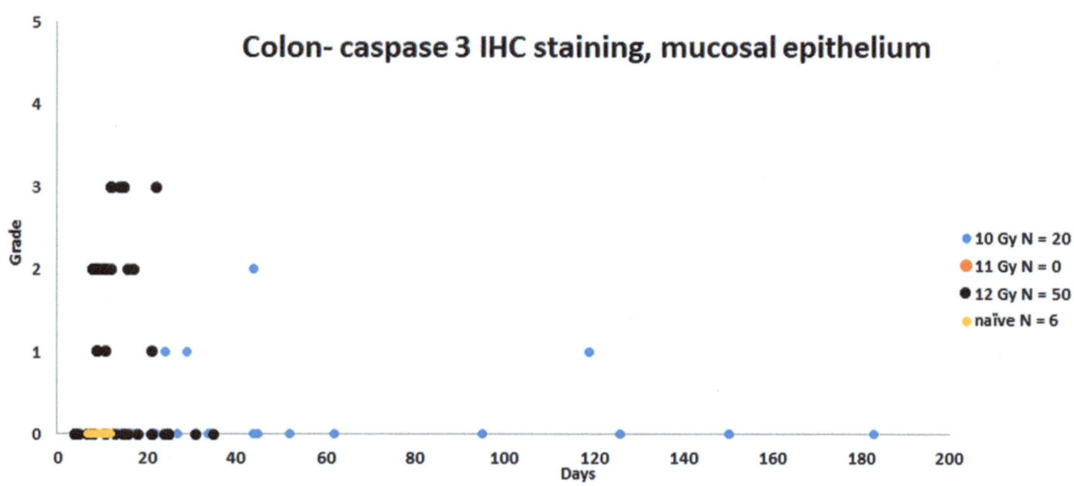

FIGURE 5.29

Scatter plot: Colon—caspase 3 staining, mucosal epithelial cells. Observations based on light microscopic examination of paraffin-embedded specimens stained by immunohistochemical procedures for caspase with 3,3′-diaminobenzidine chromogen and hematoxylin counterstain.

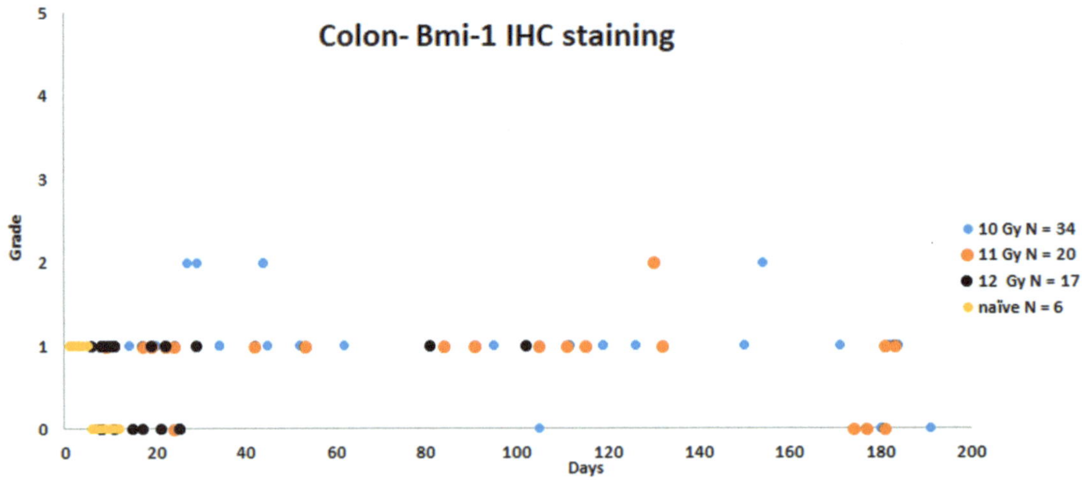

FIGURE 5.30

Scatter plot: Colon—Bmi-1^{+} crypt epithelial cells. Observations based on light microscopic examination of paraffin-embedded specimens stained by immunohistochemical procedures for Bmi-1 with 3,3′-diaminobenzidine chromogen and hematoxylin counterstain.

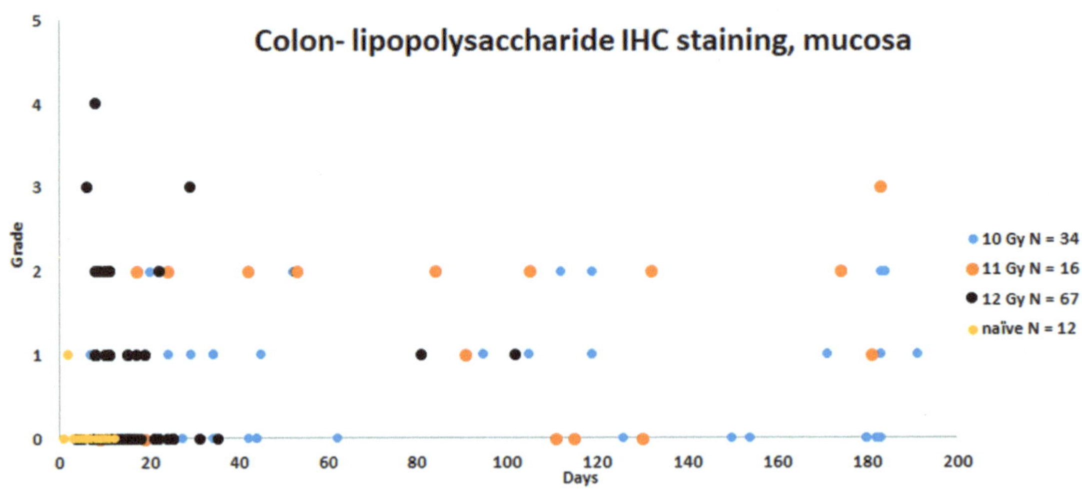

FIGURE 5.31

Scatter plot: Colon—lipopolysaccharide (LPS) core antigen staining, mucosa. Observations based on light microscopic examination of paraffin-embedded specimens stained by immunohistochemical procedures for LPS core antigen with 3,3′-diaminobenzidine chromogen and hematoxylin counterstain.

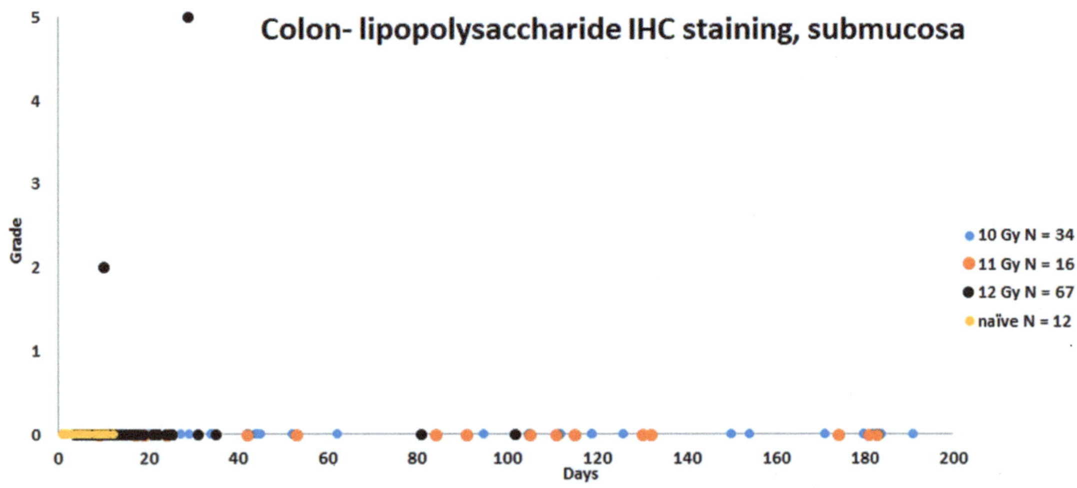

FIGURE 5.32

Scatter plot: Colon—lipopolysaccharide (LPS) core antigen staining, submucosa. Observations based on light microscopic examination of paraffin-embedded specimens stained by immunohistochemical procedures for LPS core antigen with 3,3′-diaminobenzidine chromogen and hematoxylin counterstain.

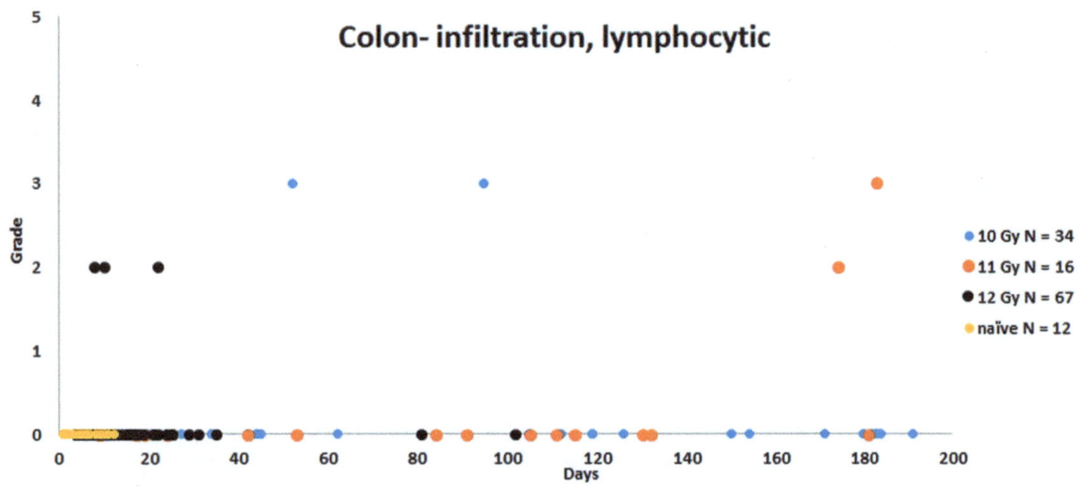

FIGURE 5.33

Scatter plot: Colon—infiltration, lymphocytic. Observations based on light microscopic examination of paraffin-embedded specimens stained with hematoxylin and eosin.

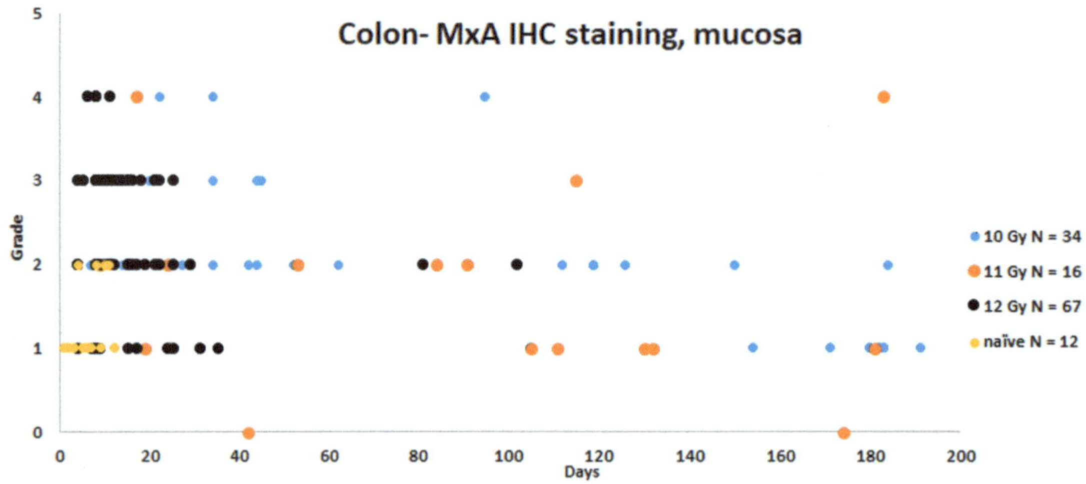

FIGURE 5.34

Scatter plot: Colon-cellular MxA staining, mucosa. Observations based on light microscopic examination of paraffin-embedded specimens stained by immunohistochemical procedures for MxA with 3,3′-diaminobenzidine chromogen and hematoxylin counterstain.

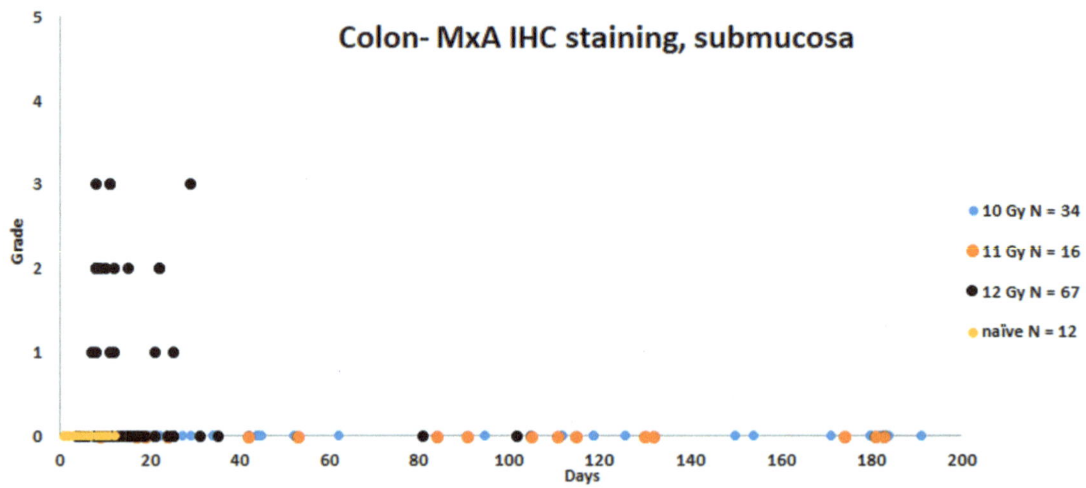

FIGURE 5.35

Scatter plot: Colon—cellular MxA staining, submucosa. Observations based on light microscopic examination of paraffin-embedded specimens stained by immunohistochemical procedures for MxA with 3,3′-diaminobenzidine chromogen and hematoxylin counterstain.

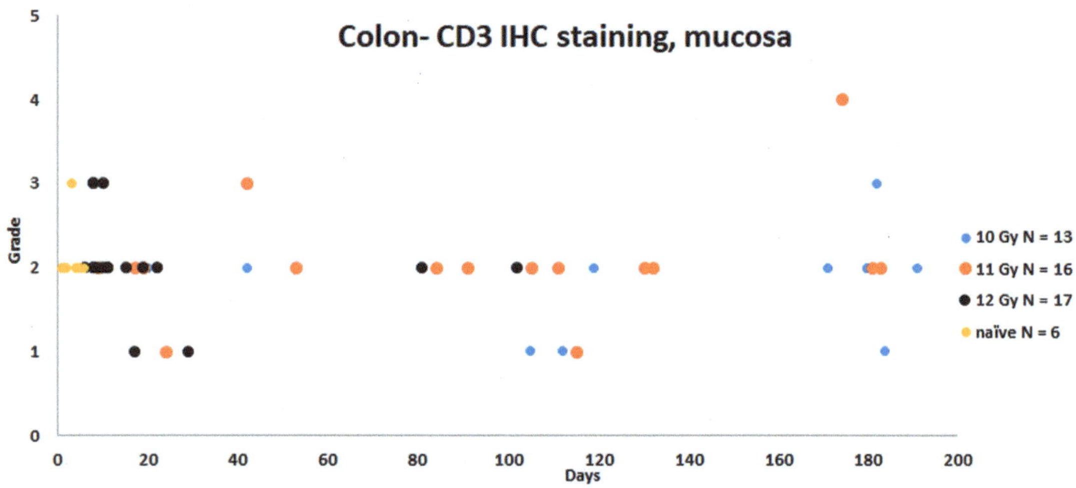

FIGURE 5.36

Scatter plot: Colon—cellular CD3 staining, mucosa. Observations based on light microscopic examination of paraffin-embedded specimens stained by immunohistochemical procedures for CD3 with 3,3′-diaminobenzidine chromogen and hematoxylin counterstain.

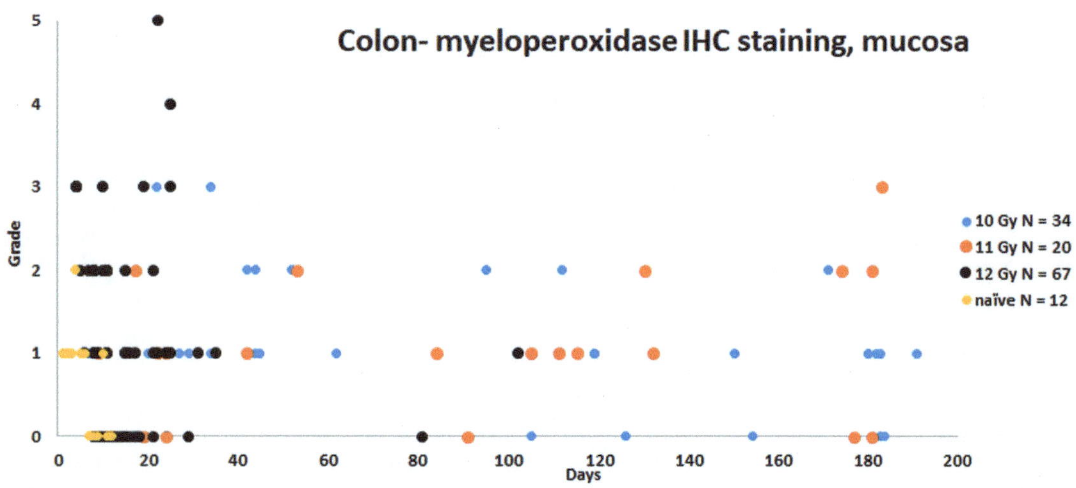

FIGURE 5.37

Scatter plot: Colon—cellular myeloperoxidase staining, mucosa. Observations based on light microscopic examination of paraffin-embedded specimens stained by immunohistochemical procedures for myeloperoxidase with 3,3′-diaminobenzidine chromogen and hematoxylin counterstain.

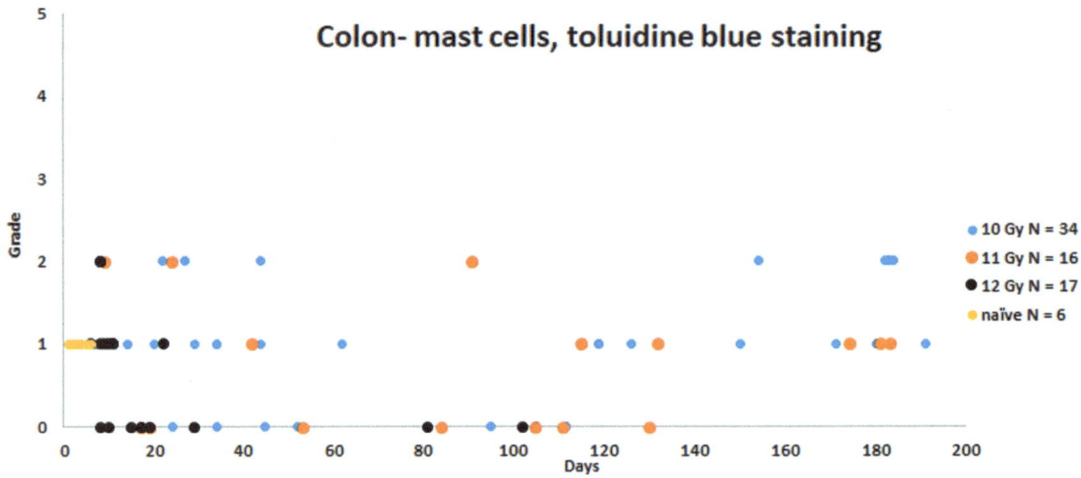

FIGURE 5.38

Scatter plot: Colon—mast cell population, mucosa, toluidine blue staining. Observations based on light microscopic examination of paraffin-embedded specimens stained with toluidine blue.

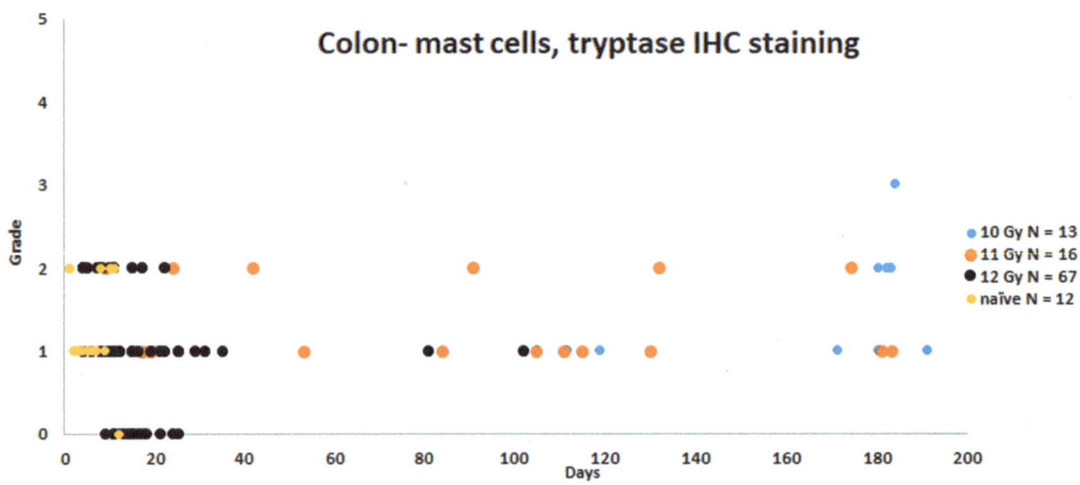

FIGURE 5.39

Scatter plot: Colon—mast cell population, mucosa, tryptase staining. Observations based on light microscopic examination of paraffin-embedded specimens stained by immunohistochemical procedures for tryptase with 3,3′-diaminobenzidine chromogen and hematoxylin counterstain.

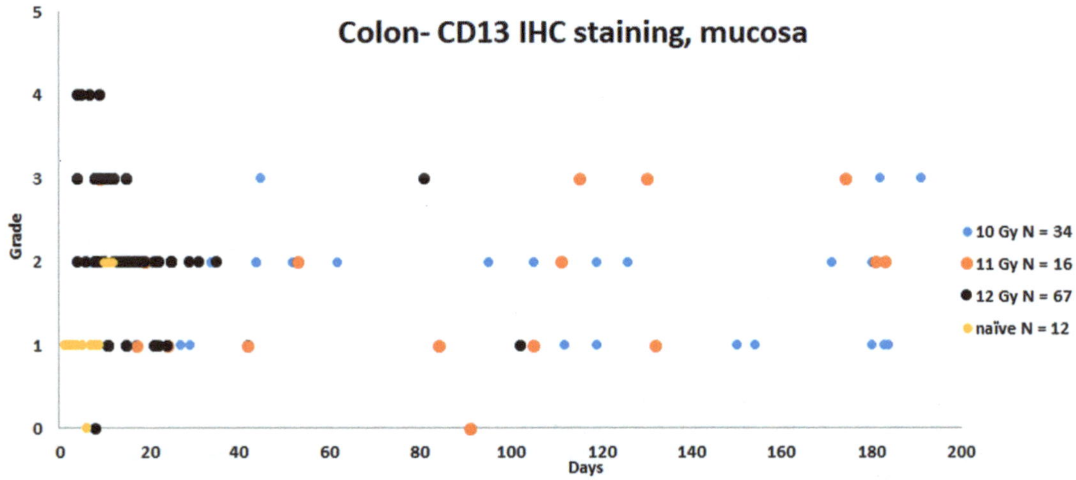

FIGURE 5.40

Scatter plot: Colon—CD13 cellular staining, mucosa. Observations based on light microscopic examination of paraffin-embedded specimens stained by immunohistochemical procedures for CD13 with 3,3′-diaminobenzidine chromogen and hematoxylin counterstain.

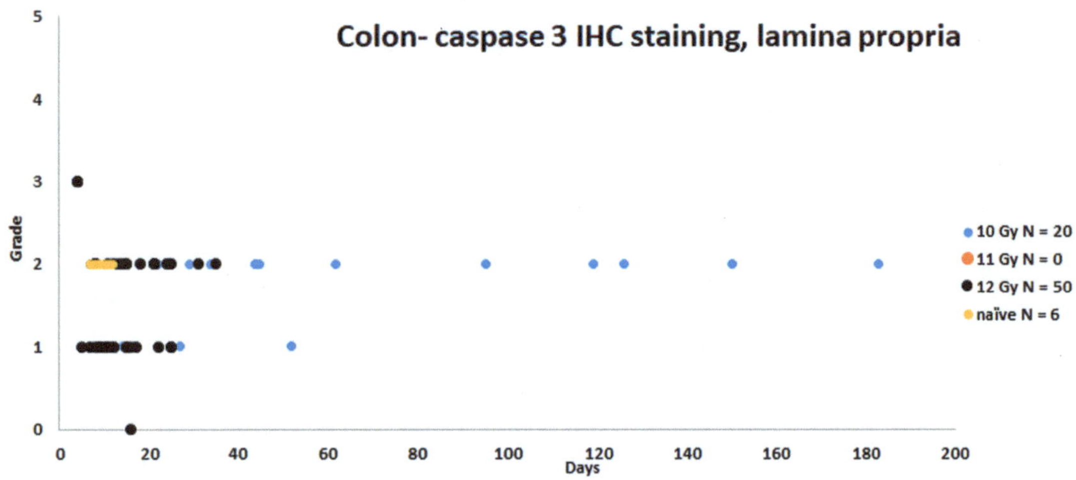

FIGURE 5.41

Scatter plot: Colon—caspase 3 staining, mucosal lamina propria. Observations based on light microscopic examination of paraffin-embedded specimens stained by immunohistochemical procedures for caspase 3 with 3,3′-diaminobenzidine chromogen and hematoxylin counterstain.

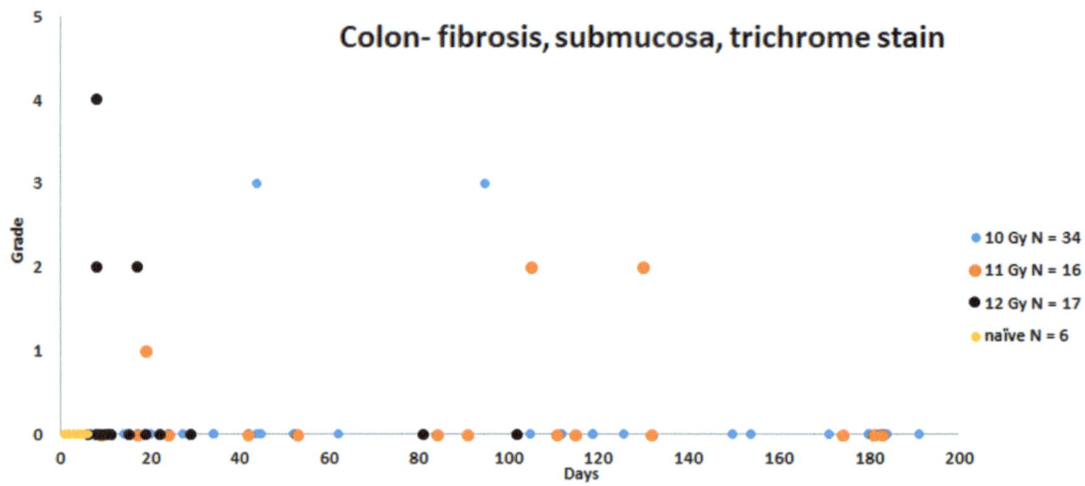

FIGURE 5.42

Scatter plot: Colon—fibrosis, serosa. Observations based on light microscopic examination of paraffin-embedded specimens stained with Masson's trichrome stain.

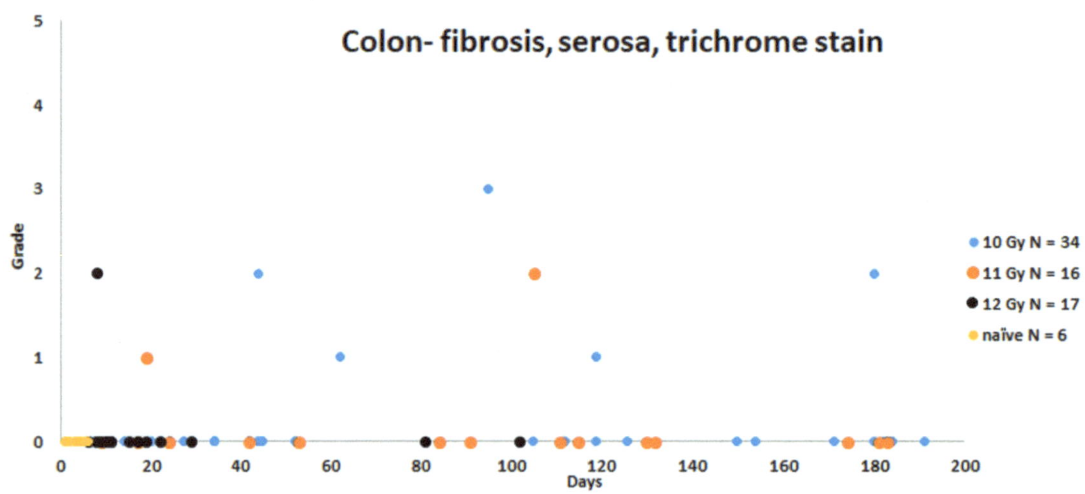

FIGURE 5.43

Scatter plot: Colon—fibrosis, submucosa. Observations based on light microscopic examination of paraffin-embedded specimens stained with Masson's trichrome stain.

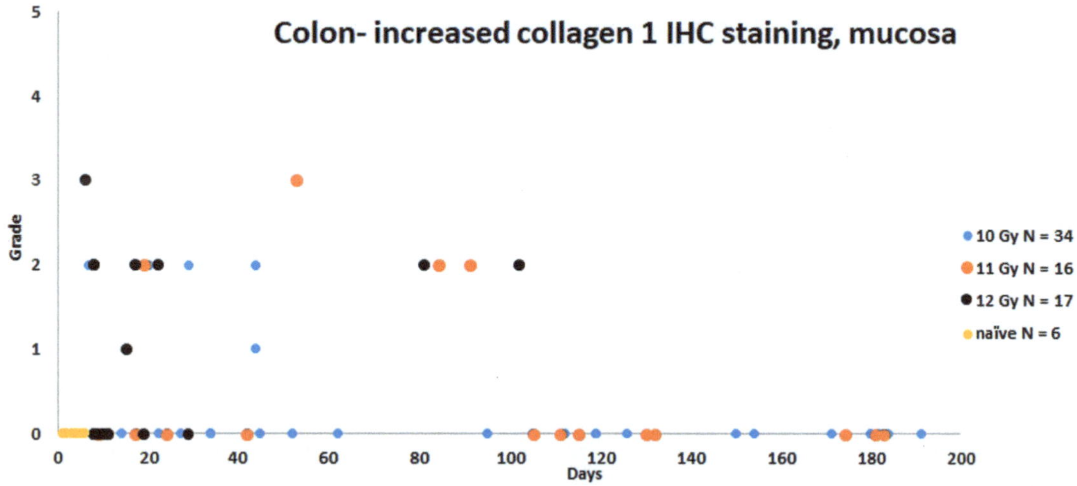

FIGURE 5.44

Scatter plot: Colon—increased collagen 1 staining, mucosa. Observations based on light microscopic examination of paraffin-embedded specimens stained by immunohistochemical procedures for collagen 1 with 3,3′-diaminobenzidine chromogen and hematoxylin counterstain.

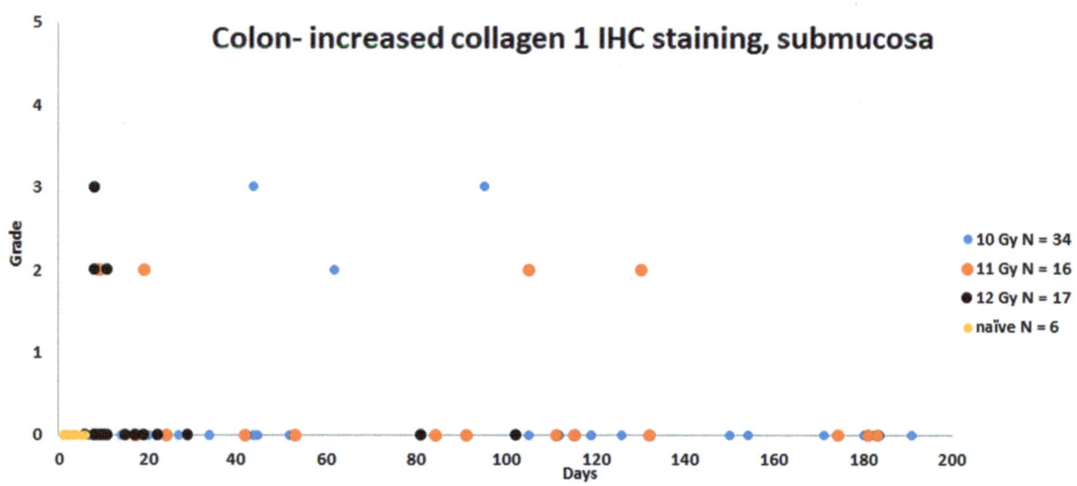

FIGURE 5.45

Scatter plot: Colon—increased collagen 1 staining, submucosa. Observations based on light microscopic examination of paraffin-embedded specimens stained by immunohistochemical procedures for collagen 1 with 3,3′-diaminobenzidine chromogen and hematoxylin counterstain.

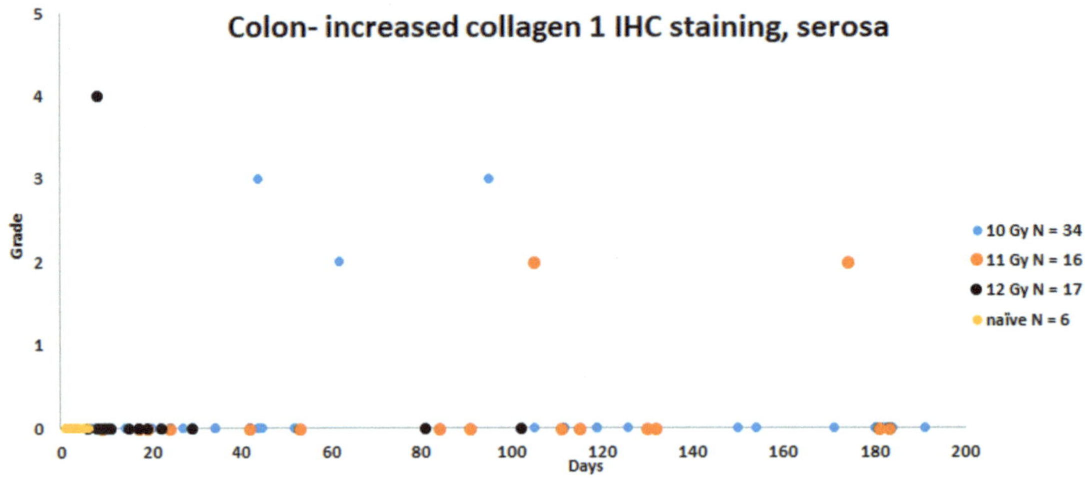

FIGURE 5.46

Scatter plot: Colon—increased collagen 1 staining, serosa. Observations based on light microscopic examination of paraffin-embedded specimens stained by immunohistochemical procedures for collagen 1 with 3,3′-diaminobenzidine chromogen and hematoxylin counterstain.

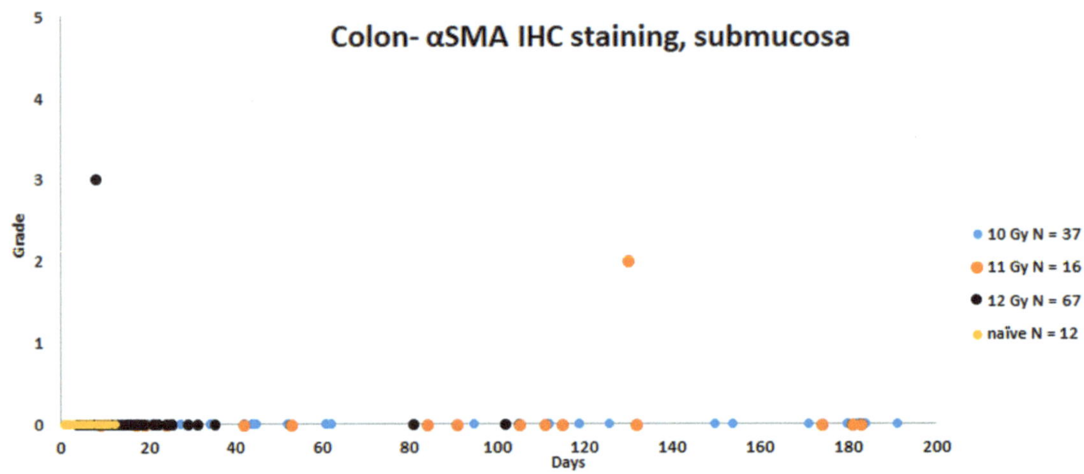

FIGURE 5.47

Scatter plot: Colon—increased alpha-smooth muscle actin (αSMA) staining, submucosa. Observations based on light microscopic examination of paraffin-embedded specimens stained by immunohistochemical procedures for αSMA with 3,3′-diaminobenzidine chromogen and hematoxylin counterstain.

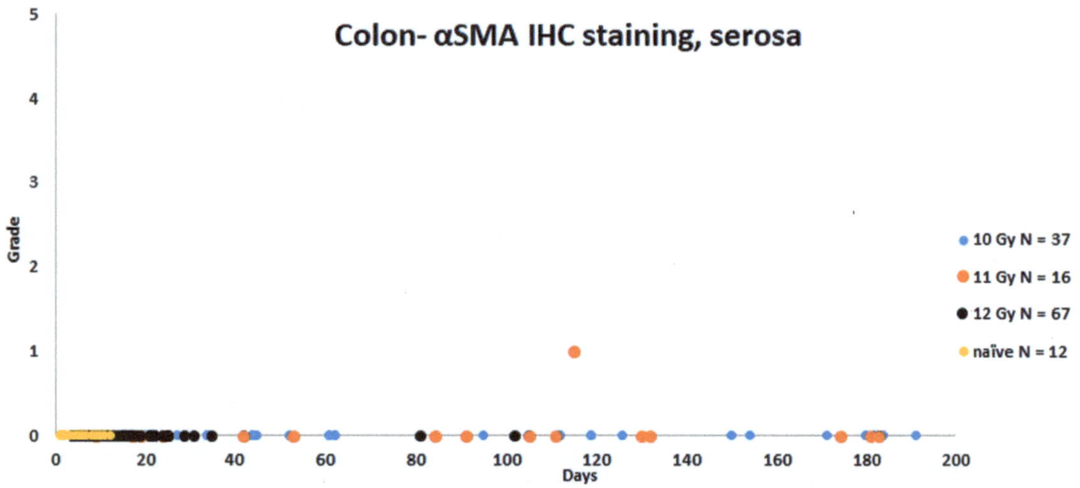

FIGURE 5.48

Scatter plot: Colon—increased alpha-smooth muscle actin (αSMA) staining, serosa. Observations based on light microscopic examination of paraffin-embedded specimens stained by immunohistochemical procedures for αSMA with 3,3′-diaminobenzidine chromogen and hematoxylin counterstain.

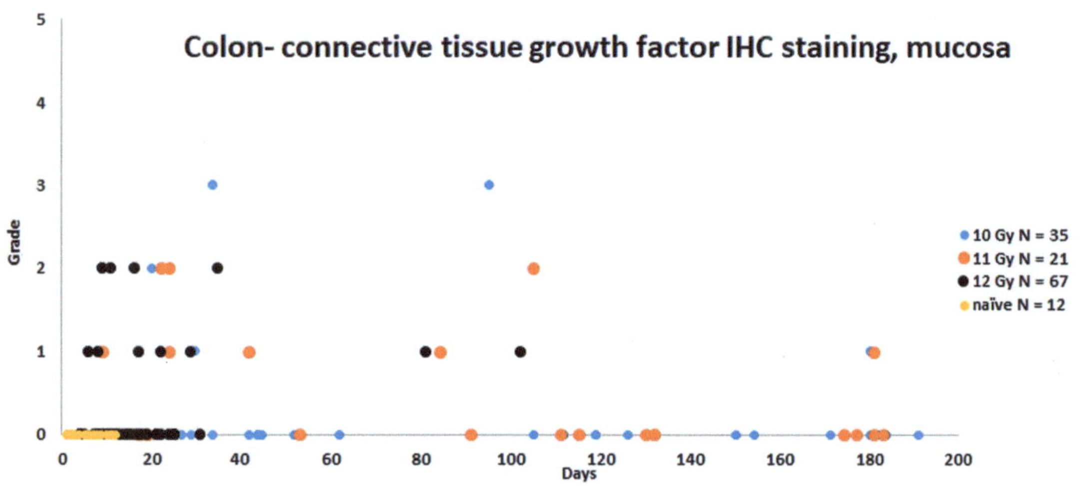

FIGURE 5.49

Scatter plot: Colon—connective tissue growth factor (CTGF) staining, mucosa. Observations based on light microscopic examination of paraffin-embedded specimens stained by immunohistochemical procedures for CTGF with 3,3′-diaminobenzidine chromogen and hematoxylin counterstain.

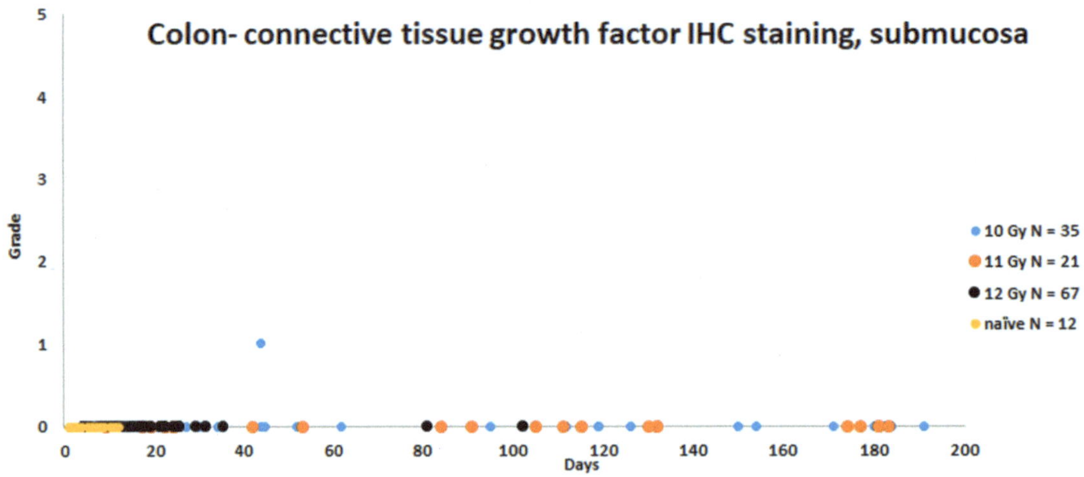

FIGURE 5.50

Scatter plot: Colon—connective tissue growth factor (CTGF) staining, submucosa. Observations based on light microscopic examination of paraffin-embedded specimens stained by immunohistochemical procedures for CTGF with 3,3′-diaminobenzidine chromogen and hematoxylin counterstain.

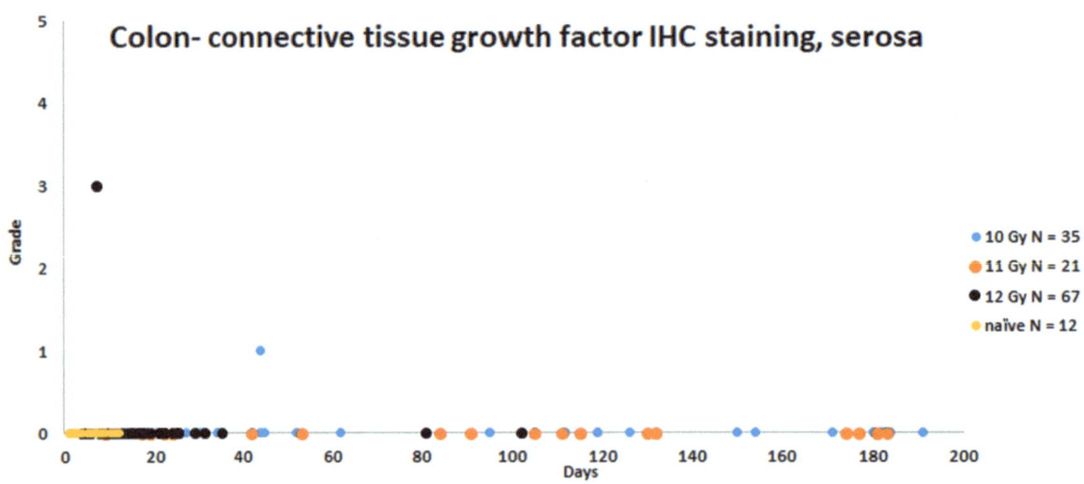

FIGURE 5.51

Scatter plot: Colon—connective tissue growth factor (CTGF) staining, serosa. Observations based on light microscopic examination of paraffin-embedded specimens stained by immunohistochemical procedures for CTGF with 3,3′-diaminobenzidine chromogen and hematoxylin counterstain.

CHAPTER

Radiation-associated changes in the mesenteric lymph node of rhesus macaques

6

Overview

Radiation-associated histologic alterations in the mesenteric lymph node consisted primarily of lymphoid changes, fibrosis, and changes secondary to upstream radiation-associated alterations in the intestine. Lymphoid depletion was noted in all animals necropsied prior to day 50 postirradiation, and in numerous animals throughout the remaining observation period. Potentially immunosuppressive corticosteroids were included in the medical management of animals that met specified clinical parameters; thus it is possible that medical management contributed to the incidence and/or severity of lymphoid depletion in the later stages of the observation period.

Fibrosis, manifested primarily as an increase in collagen 1 as demonstrated by immunohistochemical staining, consisted of an increase in the amount of fibrous connective tissue that normally surrounds blood vessel the medulla and a slight increase in collagen fibrils in the cortex. It should be noted that the Masson's trichrome stains used to define fibrosis in other organs were generally inadequate for demonstration of the minimal increase in collagen fibrils in the lymph node cortex.

The mesenteric lymph nodes also served a sentinel function by revealing pathologic alterations in regions of the intestine drained by the lymph nodes. Populations of engorged macrophages (histiocytes) within subcortical sinuses of mesenteric lymph nodes indicated migration of macrophages from the intestinal tract to the draining mesenteric lymph nodes. In this sense, the mesenteric lymph nodes served a "filter" function and maybe a better indicator of intestinal injury than the histologic specimens of jejunum and colon.

Even though the intestine had little direct histologic evidence of increased inflammatory cell populations, mesenteric lymph nodes of irradiated animals contained an increased number of activated immunocytes, as demonstrated by positive MxA immunohistochemical staining. The MxA^+ cells were commonly present in subcapsular, paracortical, and medullary sinuses of mesenteric lymph nodes, suggesting the cells were migrating from the intestine. These observations suggested a radiation-associated shift in the activation status of intestinal leukocytes that was not apparent as an overt increase in the number of intestinal leukocytes in routinely stained histologic sections.

Immunohistochemical staining for myeloperoxidase (MPO) revealed an increased population of macrophages in mesenteric lymph nodes of irradiated animals. Localization of the macrophages in subcapsular, paracortical, and medullary sinuses suggests the MPO^+ cell population originated from the radiation-injured intestine. The presence of MPO^+ cells in the mesenteric lymph node was related to level of irradiation, reflecting the greater degree of intestinal mucosal injury associated with higher levels of irradiation. Irradiated animals had a pronounced increase in profibrotic $CD163^+$ macrophages in mesenteric lymph nodes, presumably a reflection of the increased

Histopathology Atlas of Acute Radiation Syndrome and Delayed Effects in Rhesus Macaques. DOI: https://doi.org/10.1016/B978-0-323-91393-5.00006-2

population of $CD163^+$ macrophages seen in the intestinal sections. The mesenteric lymph nodes of irradiated animals also had an increased population of $CD38^+$ (pro-inflammatory) macrophages, suggesting radiation-associated intestinal reactive changes that were not manifested as a microscopically discernible increase in macrophage or leukocyte populations in the intestine.

Increased populations of lipopolysaccharide $(LPS)^+$ macrophages in mesenteric lymph nodes indicated passage of enteric microbes or microbe fragments through the intestinal barrier and translocation to the mesenteric lymph nodes.

Tryptase staining of lymph nodes from irradiated animals, but not naive animals, commonly revealed $tryptase^+$ histiocytic cells in subcapsular sinuses. The tryptase staining was typically located at the cell surface, suggesting accumulation of $tryptase^+$ material on the surface of the cells. Degranulation of intestinal mast cells was considered a likely source of the $tryptase^+$ material in/on subcapsular sinus cells in the mesenteric lymph nodes.

Bacterial colonization was occasionally noted in mesenteric lymph nodes, but bacterial invasion was not a common feature of the mesenteric lymph node response. Likewise, there was little evidence of overt inflammation in the mesenteric lymph nodes.

Staining and figures list

Routine histology stain-

Hematoxylin & eosin (H&E)—Figs. 6.1A, 6.1B, 6.1C, 6.2A, 6.2B, 6.2C, 6.3A, 6.3B, 6.4A, 6.4B, 6.5, 6.6A, 6.6B, 6.6C, 6.7A, 6.7B

Histochemical stains-

Masson's trichrome—Figs. 6.8A, 6.8B, 6.9A, 6.9B, 6.10

Immunohistochemical stains-

CD38—Fig. 6.25
CD163—Figs. 6.23A, 6.23B, 6.24A, 6.24B
Collagen 1—Figs. 6.11A, 6.11B, 6.12A, 6.12B
Connective tissue growth factor—Fig. 6.19
Lipopolysaccharide—Figs. 6.18A, 6.18B
MxA—Figs. 6.16, 6.17A, 6.17B, 6.17C, 6.17D
Myeloperoxidase—Figs. 6.13A, 6.13B, 6.14A, 6.14B, 6.15A, 6.15B, 6.15C, 6.17C
Transforming growth factor-beta—Fig. 6.20
Tryptase—Figs. 6.21, 6.22A, 6.22B

Graphic presentations (scatter plots) of individual histologic observations in the jejunum are presented in the following figures:

Fig. 6.26. Mesenteric lymph node—depletion, lymphoid
Fig. 6.27. Mesenteric lymph node—histiocytosis, sinus
Fig. 6.28. Mesenteric lymph node—myxoid cells
Fig. 6.29. Mesenteric lymph node—lipopolysaccharide core antigen staining, subcapsular sinus
Fig. 6.30. Mesenteric lymph node—lipopolysaccharide core antigen staining, cortex

Detailed microscopic observations

Routine histologic stain

Marked bacterial invasion was noted in the mesenteric lymph node of one irradiated animal. The identity of the bacterial species was not apparent from routine histopathologic examination, but the general features suggested the organisms were coccoid rather than the rod-shaped organisms typically encountered with coliform bacteria. Tissue gram stains revealed a mixed bacterial population, with a preponderance of cocci.

The connective tissue adjacent to mesenteric lymph nodes of two irradiated animals had aggregates of myxoid cells that were morphologically similar to those seen in the lung. This may be meaningful with regard to the genesis and significance of similar myxoid cells in other sites.

Eosinophil infiltrations in mesenteric lymph nodes of a few animals, including naive controls, were assigned no particular importance, as they were suspected to be associated with intestinal parasitism.

Trichrome stain

The trichrome stains revealed minimal evidence of fibrosis in the mesenteric lymph nodes. See collagen stains below, which resulted in a more precise identification of fibrous connective tissue in the lymph nodes.

Toluidine blue stain

Mast cells as revealed by toluidine blue staining were essentially absent from the mesenteric lymph nodes of most animals. Mast cell populations were recorded only in the parenchyma of the mesenteric lymph nodes and did not include the normal minor mast cell populations in connective tissue surrounding the lymph nodes.

Tryptase immunohistochemical stain

Tryptase staining revealed a slightly greater population of mast cells within mesenteric lymph nodes than were revealed by toluidine blue staining. This could reflect a population of intestinal mucosal mast cells that translocated to the mesenteric lymph nodes and were not revealed by toluidine blue staining.

Apparent tryptase reactivity was noted on the cell membrane surfaces of sinusoidal macrophages in irradiated animals. The genesis and significance of this observation are unknown. Macrophage tryptase staining was not observed in mesenteric lymph nodes of the naive control animals; thus, it is likely the macrophage tryptase reactivity was in some way related to irradiation. Given the evidence of radiation effects on mast cell populations in the current study and reported in the literature, it is possible the tryptase reactivity on the surface of sinusoidal macrophages represents sequestration of tryptase enzymes released from degranulated intestinal mast cells.

Alpha-smooth muscle actin immunohistochemical stain

There was little evidence of alpha-smooth muscle actin (α-SMA) staining in the mesenteric lymph nodes.

MxA immunohistochemical stain

Staining for MxA was topographically subcategorized as germinal center, medullary, or subcapsular sinus staining. Low-level MxA^+ cell populations staining was present in the medulla of all naive control animals and MxA^+ cell populations were present in lymph node germinal centers of two of six naive control animals, indicating the presence of activated immune cells in these locations as part of normal immune system homeostasis. Irradiated animals had more prominent populations of MxA^+ medullary cells. Subcapsular sinuses of mesenteric lymph nodes from naive control animals had no discernible MxA^+ cells. By contrast, subcapsular sinuses of many of the irradiated animals contained MxA^+ cells that were histologically consistent with macrophages or dendritic cells. The presence of the MxA^+ cells in subcapsular sinuses was evidence of immunologic reactions in the intestines, with subsequent translocation of MxA^+ cells to the mesenteric lymph nodes.

LPS core immunohistochemical stain

Staining for LPS core antigen was topographically subcategorized as cortical, germinal center, medullary, or subcapsular sinus staining. The naive control animals had low-level populations of LPS^+ cells in each of these sublocations. Irradiated animals had more pronounced populations of LPS^+ cells in the various sublocations of the mesenteric lymph nodes. The mesenteric lymph node appeared to serve as a collecting vessel for LPS^+ cells that were migrating from upstream sites in the intestine.

Myeloperoxidase immunohistochemical stain

MPO staining in the mesenteric lymph node was topographically subcategorized as medullary or subcapsular sinus. Positively stained cells appeared to be primarily macrophages/histiocytes or neutrophils, based on the size and nuclear profile of the cells.

Low-level populations of MPO^+ cells were noted in the medulla of mesenteric lymph nodes of most animals, including all of the naive control animals. In six irradiated animals, the medullary population of MPO^+ cells was somewhat more pronounced than that seen in naive controls. MPO^+ cells were not observed in the subcapsular sinuses of naive control animals but were noted in low incidence among irradiated animals. MPO-positive cells in subcapsular sinuses were particularly common in animals irradiated at 12 Gy, the great majority of which did not survive beyond 30 days. MPO-positive cells in subcapsular sinuses were noted at a lower incidence rate in animals irradiated at 11 Gy and were less commonly noted in animals irradiated at 10 Gy. Cells in subcapsular sinuses of mesenteric lymph nodes typically originate from pathologic processes in the intestine, thus it is presumed the subcapsular cell population observed in the mesenteric lymph nodes reflected radiation-associated intestinal injury. While that may be the case, it should be noted that animals irradiated at 12 Gy typically survived less than 30 days, while those irradiated at 10 or 11 Gy tended to survive longer. As a result, the apparent radiation dose association of MPO + cells in the mesenteric lymph nodes may reflect survival time rather than a direct effect of radiation dose.

Collagen 1 immunohistochemical stain

Data entries of positive collagen staining the mesenteric lymph nodes refer to collagen deposits other than the normal collagenous tissue that surrounds blood vessels in the hilus and medulla of the lymph nodes. In essence, this was collagen deposition in the cortex of the lymph nodes. This additional collagen deposition was observed in low incidence in animals irradiated at 10 or 11 Gy, but not in animals irradiated at 12 Gy. Collagen generation and deposition are typically associated with chronic pathologic processes, thus the absence of lymph node collagen deposition in animals irradiated at 12 Gy may reflect the shorter survival time of animals in that exposure group.

Connective tissue growth factor immunohistochemical stain

Staining for connective tissue growth factor (CTGF) was not noted in mesenteric lymph nodes of the naive control animals and was uncommonly noted in mesenteric lymph nodes of irradiated animals.

Transforming growth factor β immunohistochemical stain

Low-level medullary transforming growth factor β (TGF-β) staining was noted in one naive control animal, but the mesenteric lymph nodes of the remaining naive control animals were devoid of TGF-β staining. Medullary TGF-β staining was more commonly noted in irradiated animals, particularly in animals from the 12 Gy irradiation group. Subcapsular sinuses of naive control animals had no TGF-β^+ cells, but TGF-β^+ cells were present in a substantial subset of irradiated animals that were necropsied within 22 days of irradiation, and sporadically thereafter.

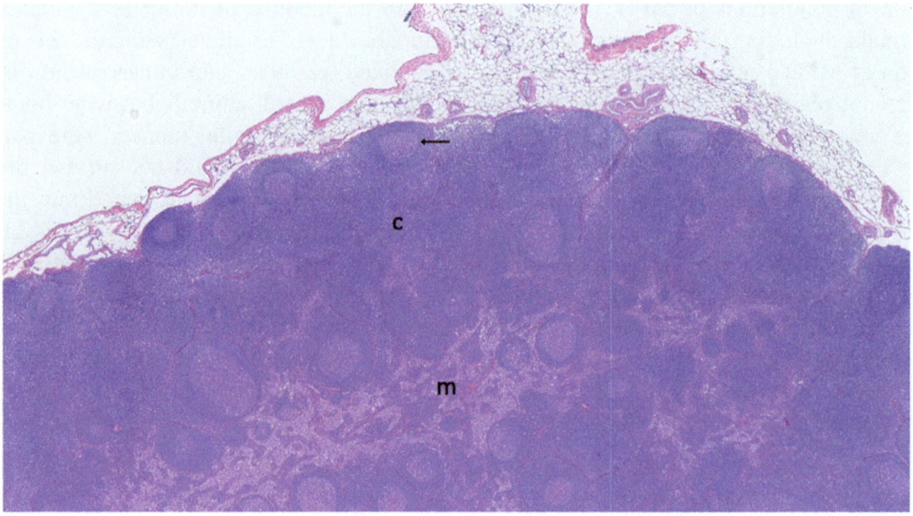

FIGURE 6.1A

The mesenteric lymph node from a naive male rhesus macaque has highly cellular cortex (c) and less cellular medulla (m). Numerous follicles with germinal centers (arrow) indicate active immunological functioning. Hematoxylin and eosin stain, 2.5 × objective magnification.

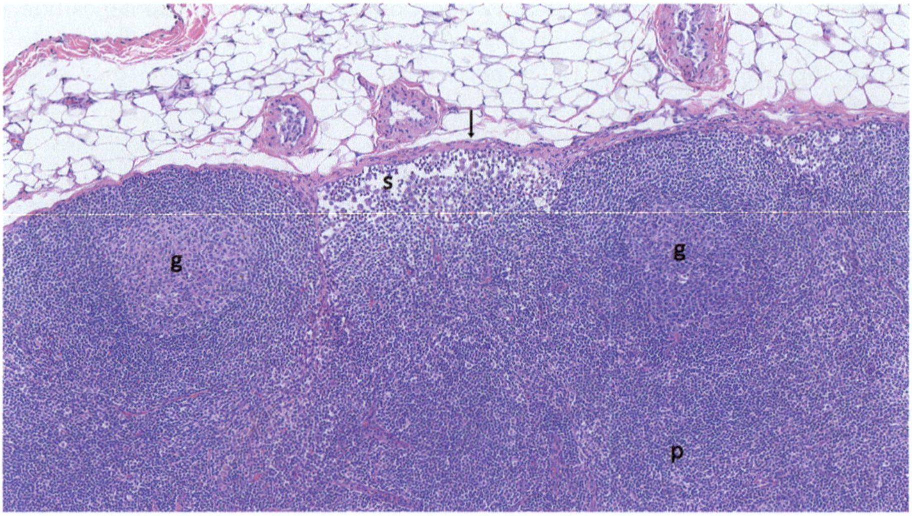

FIGURE 6.1B

A higher magnification of the previous specimen shows a follicle with germinal centers (g), highly cellular paracortex (p), and subcapsular sinus (s) containing a mixed population of lymphocytes and histiocytes. Hematoxylin and eosin stain, 10 × objective magnification.

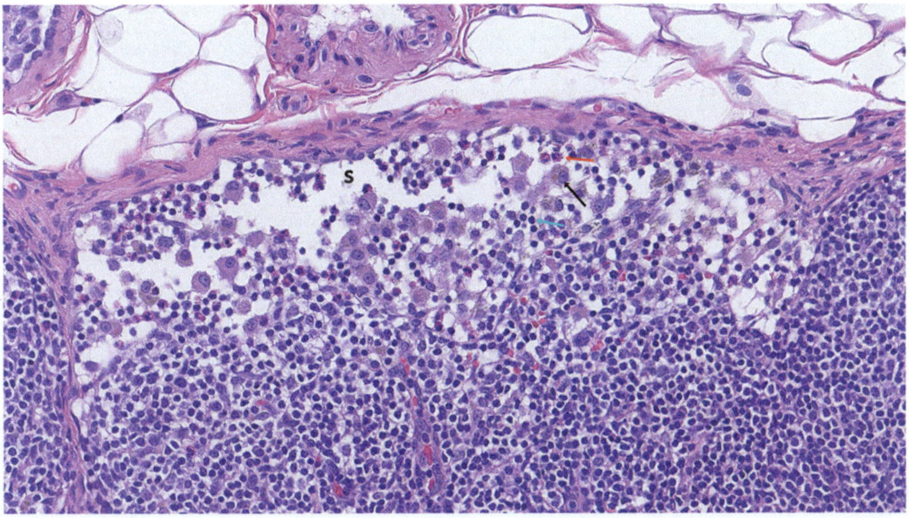

FIGURE 6.1C

Higher magnification of the previous specimen shows a subcapsular sinus (s) containing histiocytes/dendritic cells (black arrow), lymphocytes (blue arrow), and eosinophils (red arrow). The cell populations in the subcapsular sinuses represent transmigration from the region of the intestine that is drained by the lymph node, thus reflect pathological processes in the intestine. Lymphocytes and histiocytes/dendritic cells in subcapsular sinuses are part of normal immunological functioning, but the presence of a substantial eosinophil population suggests intestinal parasitism or allergic reaction in this animal. Hematoxylin and eosin stain, 30.3 × objective magnification.

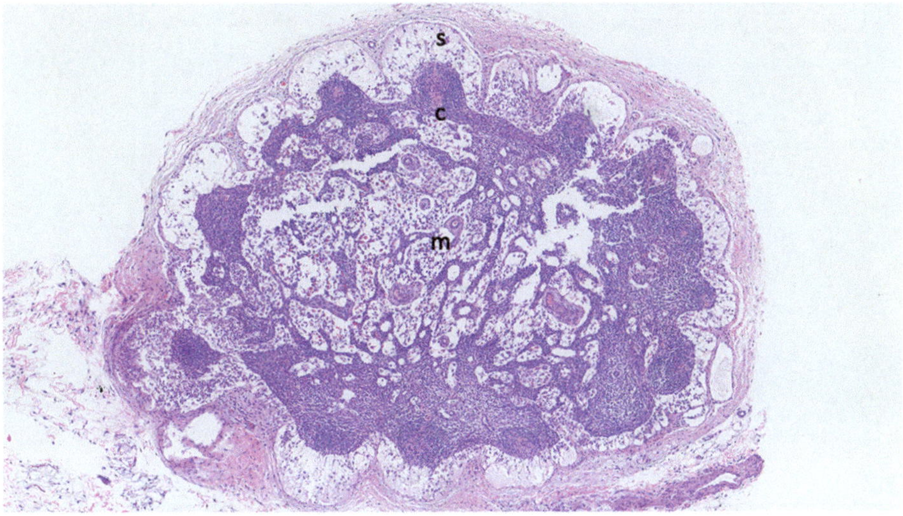

FIGURE 6.2A

The mesenteric lymph node of a male rhesus macaque necropsied 8 days after irradiation at 12 Gy has marked dilation of subcapsular sinuses (s) and reduction of cortical (c) cellularity. Medullary (m) sinuses are dilated, and cellularity is somewhat reduced. Hematoxylin and eosin stain, 4.62 × objective magnification.

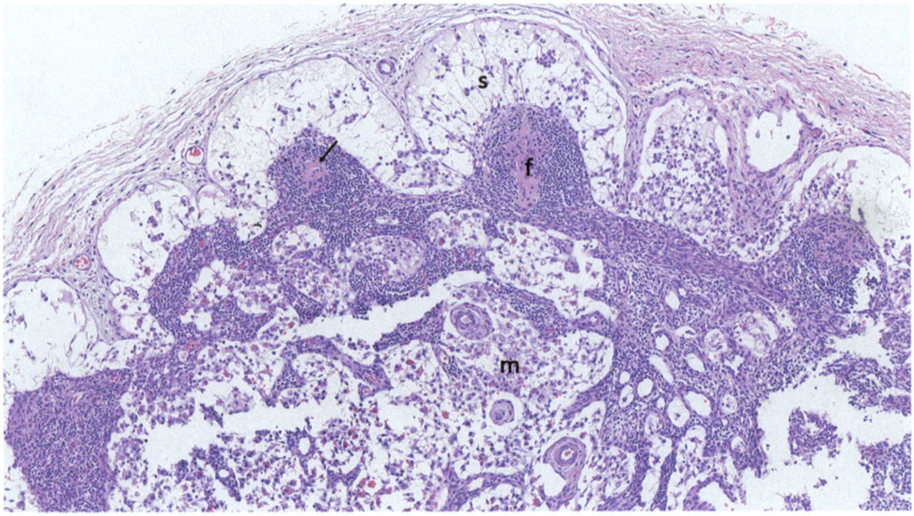

FIGURE 6.2B

Higher magnification of the previous specimen shows marked dilation of subcapsular sinuses (s) and reduced cellularity of cortical follicles (f). Dilated medullary (m) sinuses are populated primarily by histiocytes/dendritic cells, with a reduced population of lymphocytes. Accumulations of homogeneous proteinic material (arrow) within involuted follicles is a remnant of a species characteristic of rhesus macaques rather than a radiation-associated alteration. Hematoxylin and eosin stain, 10 × objective magnification.

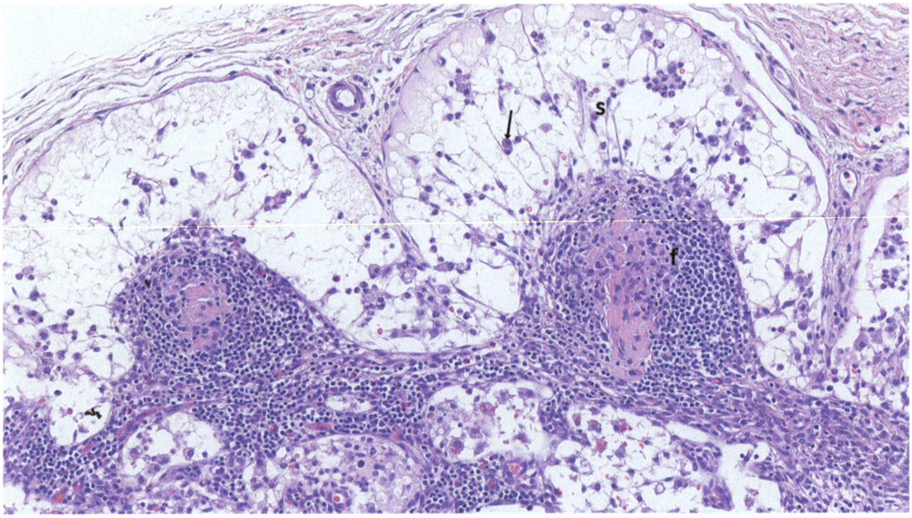

FIGURE 6.2C

Higher magnification of the previous specimen shows dilation of subcapsular sinuses (s) containing histiocytes/dendritic cells (arrow) and lightly eosinophilic lymph proteins. Follicles (f) have markedly reduced lymphocytic populations, and few lymphocytes are present in the subcapsular sinuses. The reduced lymphocyte populations reflect the known sensitivity of lymphocytes to radiation injury. Hematoxylin and eosin stain, 20 × objective magnification.

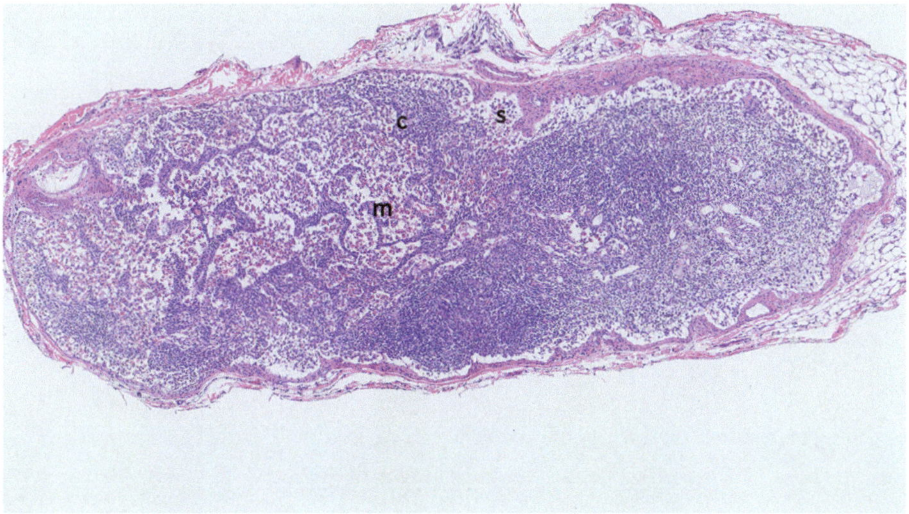

FIGURE 6.3A

The mesenteric lymph node of a male rhesus macaque necropsied 15 days after irradiation at 12 Gy shows persistent reduction of cellularity of the cortex (c) and medulla (m), with mild dilation of subcapsular sinuses (s). Hematoxylin and eosin stain, 5 × objective magnification.

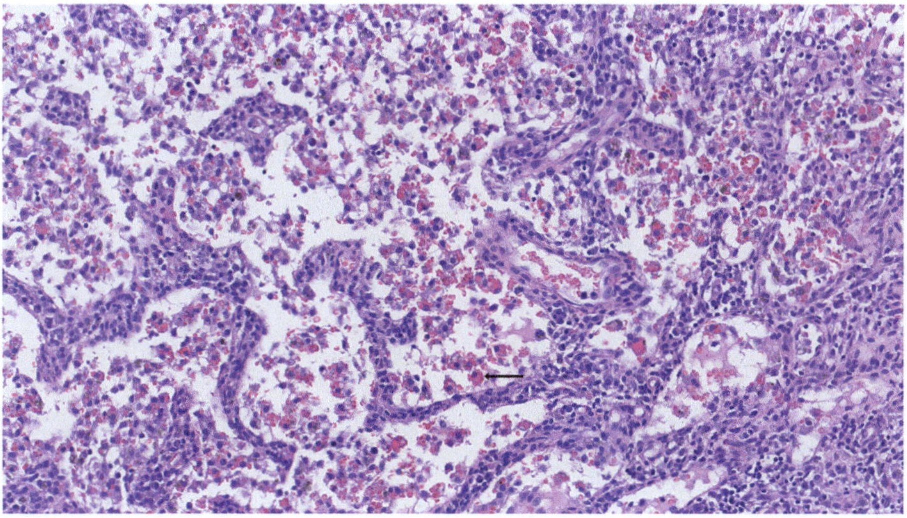

FIGURE 6.3B

Higher magnification of the previous specimen reveals numerous brightly eosinophilic erythrocytes within medullary sinuses. Many of the extravasated erythrocytes are contained within phagocytic cells (arrow). Accumulation of blood within mesenteric lymph nodes reflects hemorrhage in the intestinal segment associated with the lymph node. Hematoxylin and eosin stain, 20 × objective magnification.

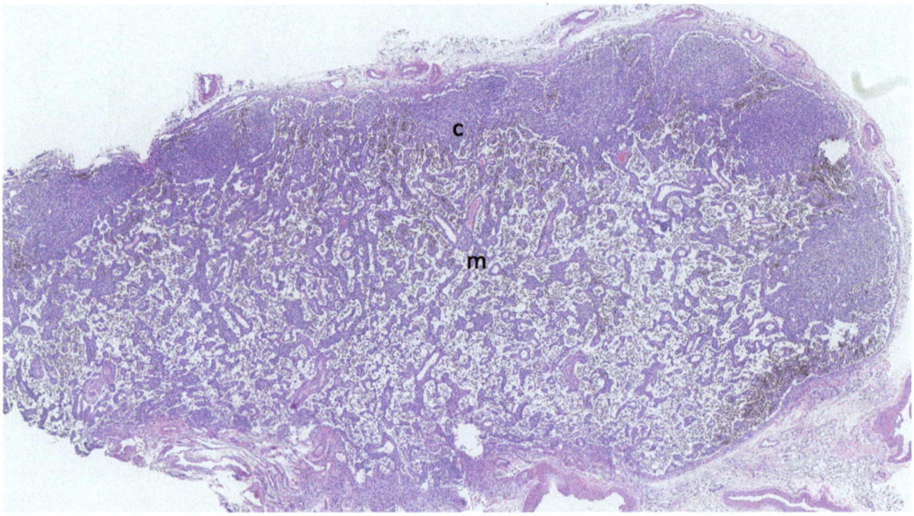

FIGURE 6.4A

The mesenteric lymph node of a male rhesus macaque necropsied 22 days following irradiation at 12 Gy shows some increase in cellularity of the cortex (c) and medulla (m) relative to the previous images, but the overall cellularity remains markedly reduced in comparison to the naive control animal. Hematoxylin and eosin stain, 2.5 × objective magnification.

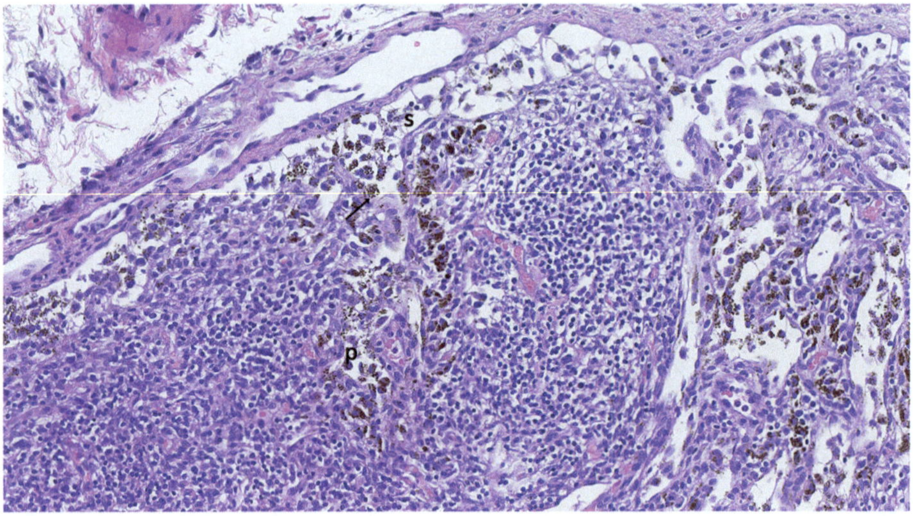

FIGURE 6.4B

Higher magnification of the previous specimen shows numerous macrophages (arrow) laden with golden-brown pigment within subcapsular (s) and paracortical (p) sinuses. The golden-brown pigment represents hemosiderin accumulation resulting from degradation of extravasated blood. Hematoxylin and eosin stain, 20 × objective magnification.

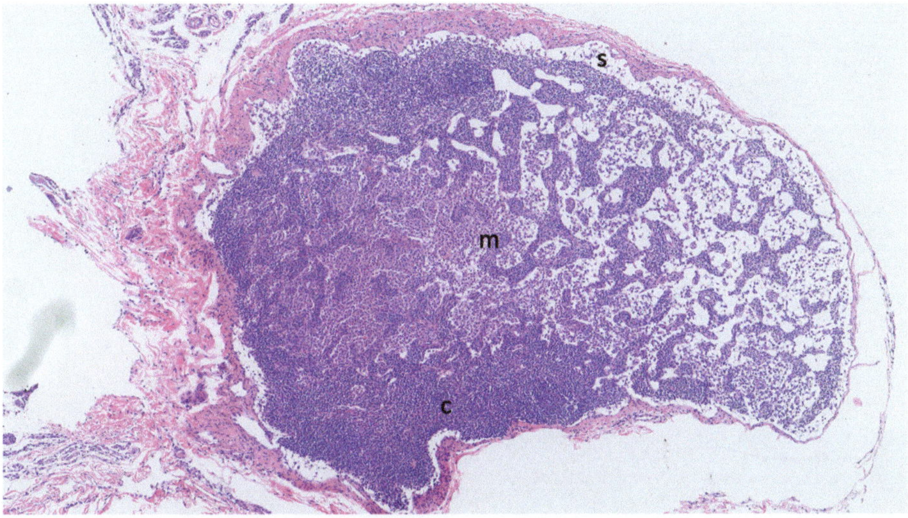

FIGURE 6.5

The mesenteric lymph node of a male rhesus macaque necropsied 42 days after irradiation at 11 Gy. The cortex (c) and medulla (m) have regained some level of cellularity, but the cellularity remains reduced in comparison to the naive controls. Subcapsular sinuses (s) are moderately dilated. Hematoxylin and eosin stain, 5 × objective magnification.

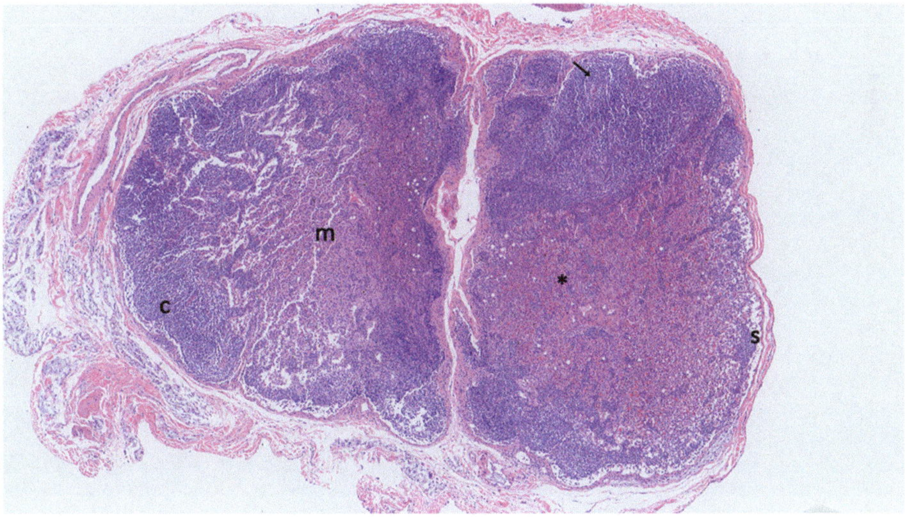

FIGURE 6.6A

The mesenteric lymph node of a male rhesus macaque necropsied 53 days after irradiation at 11 Gy. The cortex (c) and medulla (m) have regained a substantial level of cellularity, and there is evidence of follicle formation (arrow) in the cortex. Subcapsular sinuses (s) are mildly dilated. An extensive area of the medulla (*) has red coloration associated with accumulations of extravasated erythrocytes, indicating hemorrhage in the associated intestinal segment. Hematoxylin and eosin stain, 5 × objective magnification.

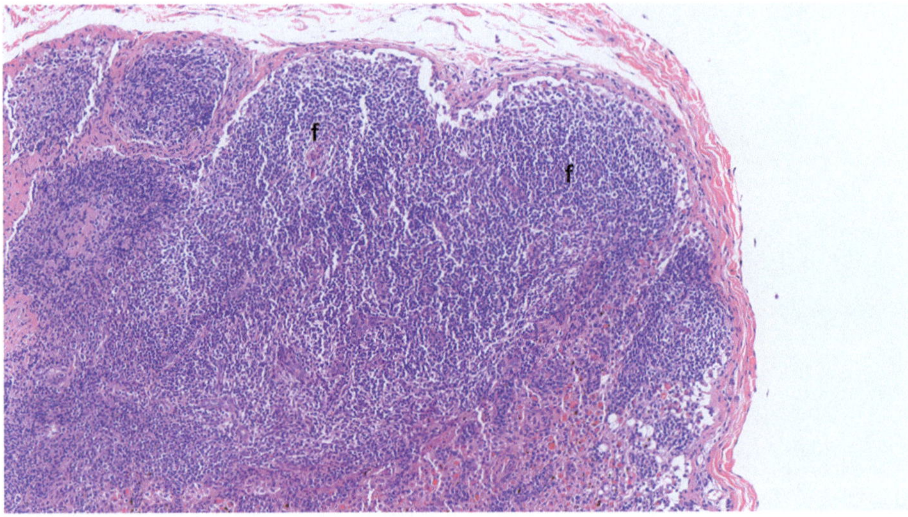

FIGURE 6.6B

Higher magnification of the previous specimen reveals follicular structures (f) but no evidence of active germinal centers, indicating lack of full immunological function. Hematoxylin and eosin stain, 10 × objective magnification.

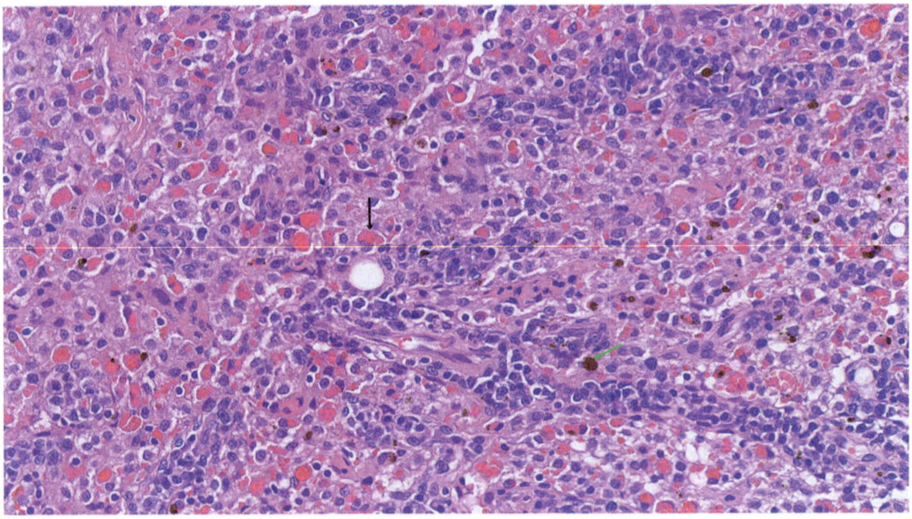

FIGURE 6.6C

Higher magnification of the previous specimen reveals numerous macrophages (black arrow) engorged with erythrocytes and erythrocytic debris, reflecting hemorrhage in the associated intestinal region. Golden-brown pigment within macrophages (green arrow) indicates degradation of hemoglobin to hemosiderin within macrophages. Hematoxylin and eosin stain, 30.5 × objective magnification.

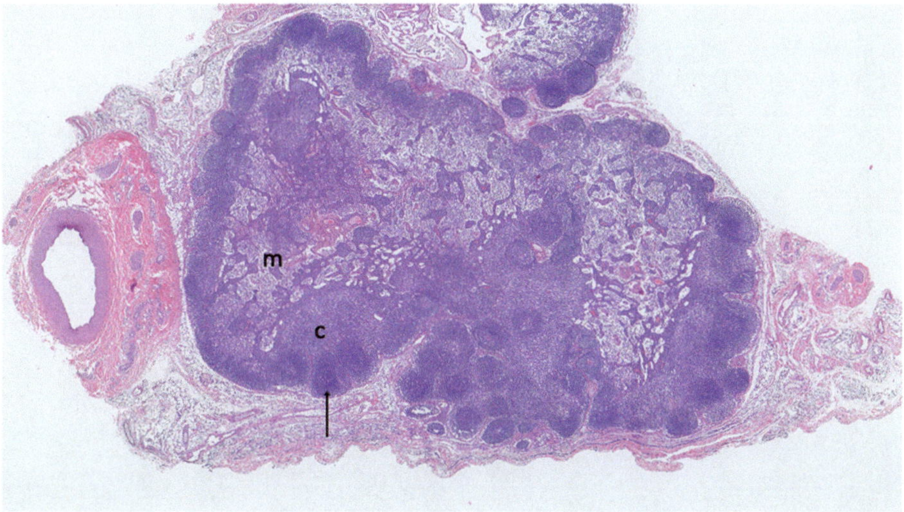

FIGURE 6.7A

The mesenteric lymph node of a male rhesus macaque necropsied 91 days after irradiation at 11 Gy has moderately cellular cortex (c) and medulla (m). Numerous follicles (arrow) are present in the cortex. Hematoxylin and eosin stain, 1.59 × objective magnification.

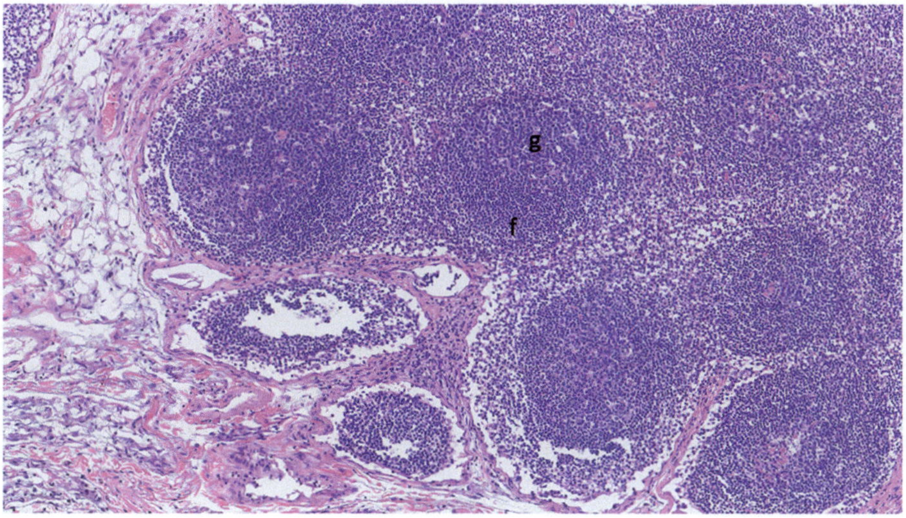

FIGURE 6.7B

Higher magnification of the previous specimen shows mature follicles (f) with active germinal centers (g), indicating recovery of this major facet of immunological function. Hematoxylin and eosin stain, 10 × objective magnification.

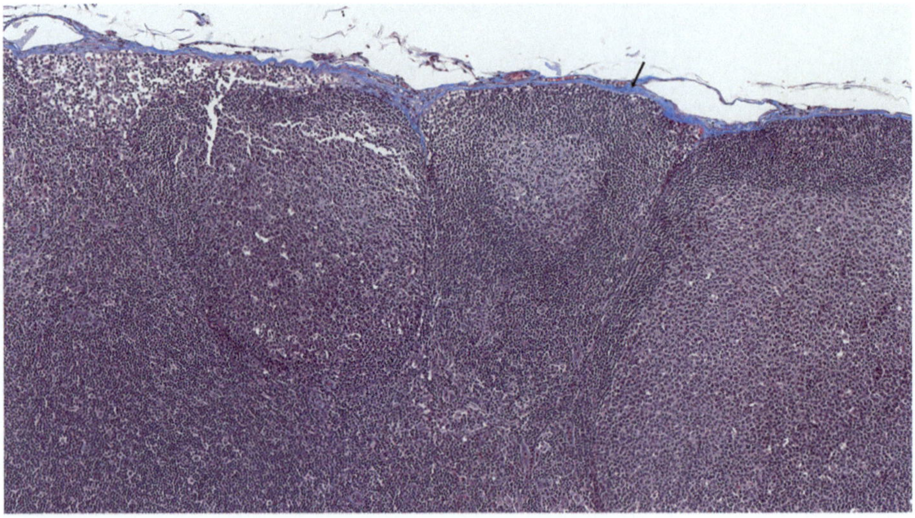

FIGURE 6.8A

Masson's trichrome stain performed on the mesenteric lymph node of a naive male rhesus macaque shows blue-stained fibrous connective tissue in the capsule (arrow). 10 × objective magnification.

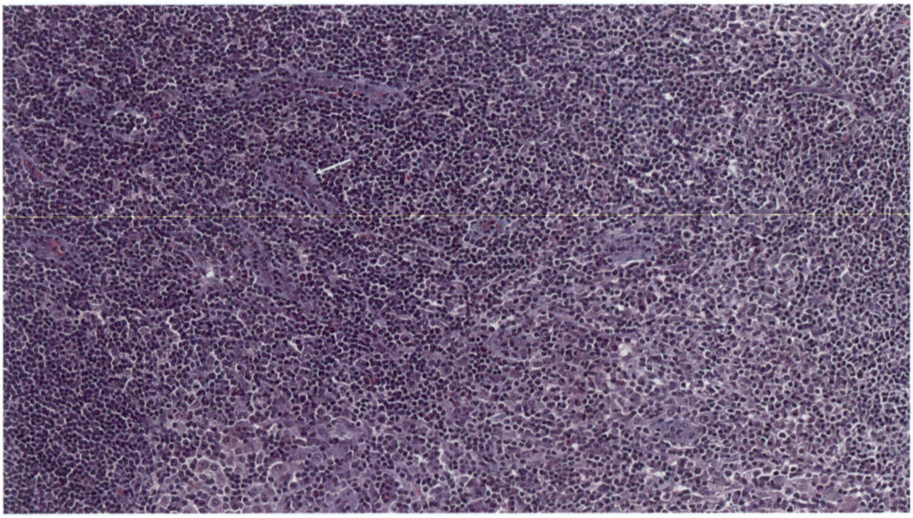

FIGURE 6.8B

Higher magnification of the previous specimen shows a minimal amount of blue-stained fibrous connective tissue (arrow) that is primarily associated with vascular structures in the cortex of the mesenteric lymph node. Masson's trichrome stain, 20 × objective magnification.

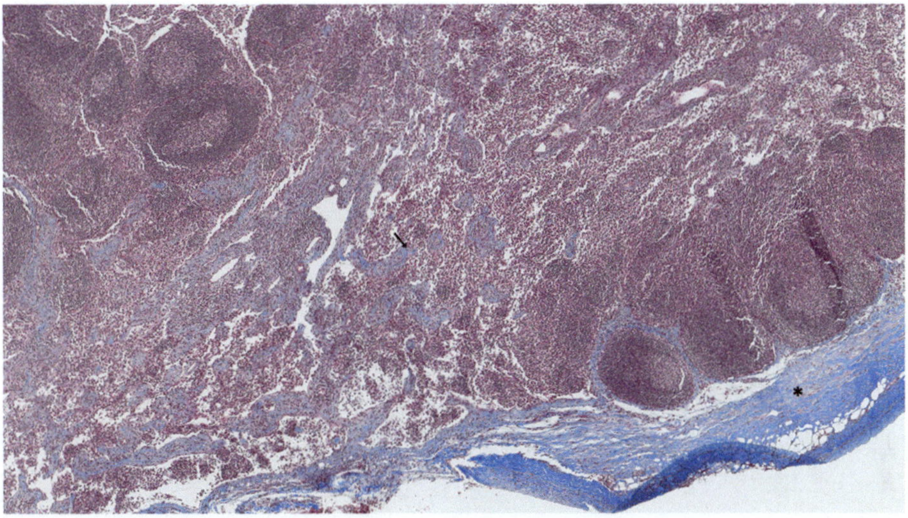

FIGURE 6.9A

The Masson's trichrome stain performed on the mesenteric lymph node of a male rhesus macaque necropsied 132 days after irradiation at 11 Gy shows a thick layer of fibrous connective tissue in the capsule (*) and prominent fibrous connective tissue (arrow) associated with medullary blood vessels, 5 × objective magnification.

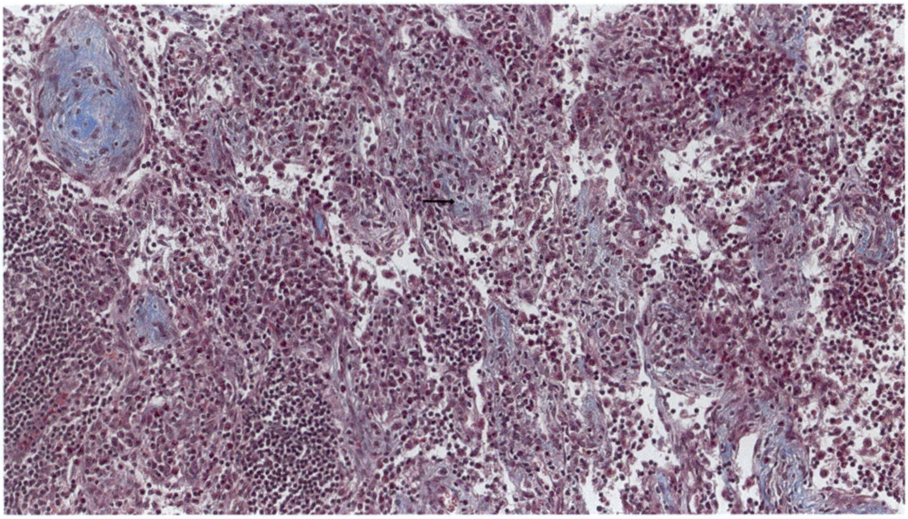

FIGURE 6.9B

Higher magnification of the previous specimen shows an increased amount of blue-stained fibrous connective tissue (arrow) within the medullary cords of the mesenteric lymph node. Masson's trichrome stain, 20 × objective magnification.

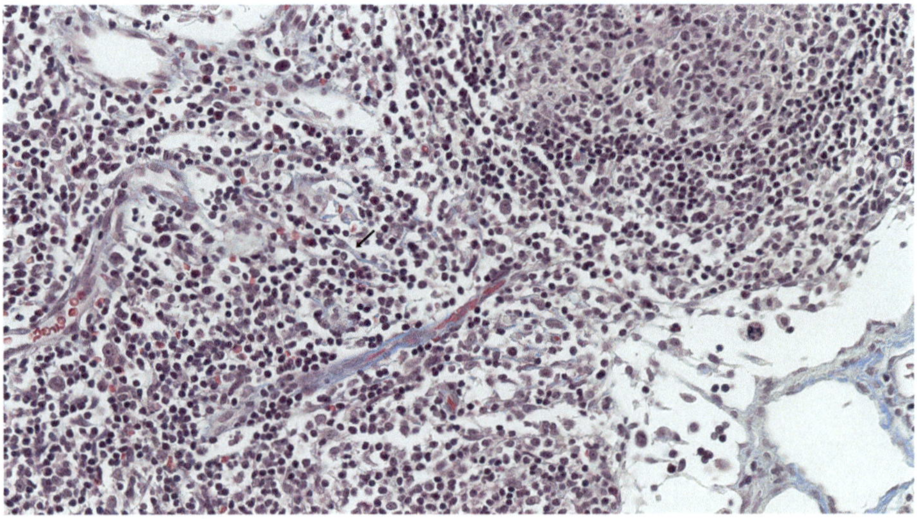

FIGURE 6.10

The Masson's trichrome stain performed on the mesenteric lymph node of a male rhesus macaque that was necropsied 130 days after irradiation at 11 Gy shows numerous fine blue-stained collagen fibrils (arrow) throughout the cortex. Though not visually prominent, the presence of this collagenous tissue could present an impediment to the normal trafficking that is part of lymphocytic maturation and functioning, 30.7 × objective magnification.

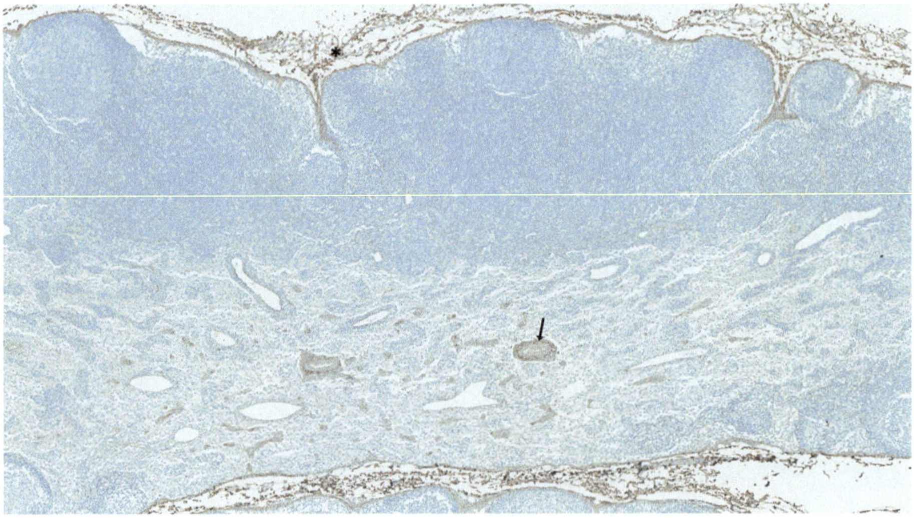

FIGURE 6.11A

Immunohistochemical staining for collagen-1 performed on the mesenteric lymph node of a naive male rhesus macaque shows a large amount of collagen-1 staining in the fibrous capsure (*), and a moderate amount of staining in association with medullary blood vessels (arrow) of the mesenteric lymph node. Stain performed with 3, 3′-diaminobenzidine chromogen and hematoxylin counterstain, 5 × objective magnification.

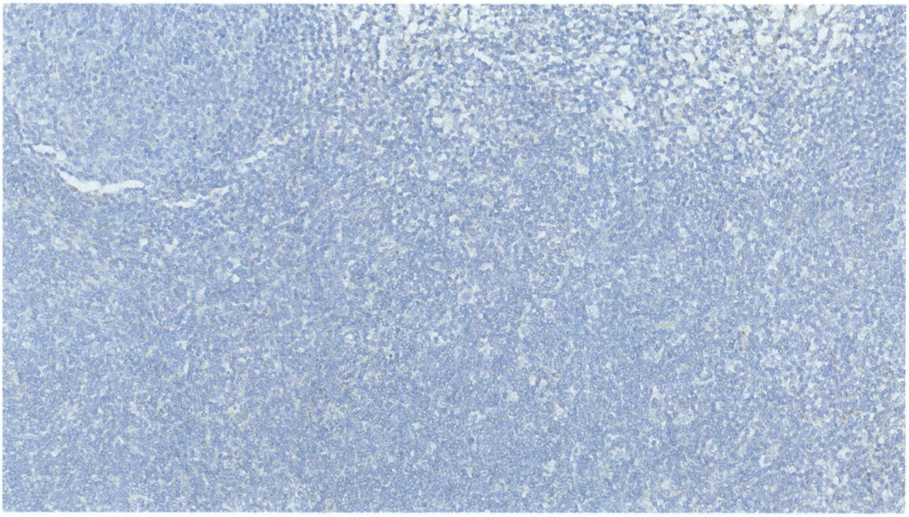

FIGURE 6.11B

Higher magnification of the previous specimen shows the mesenteric lymph node cortex to be virtually devoid of collagen-1 immunostaining. Immunohistochemical stain for collagen-1 with 3, 3′-diaminobenzidine chromogen and hematoxylin counterstain, 20 × objective magnification.

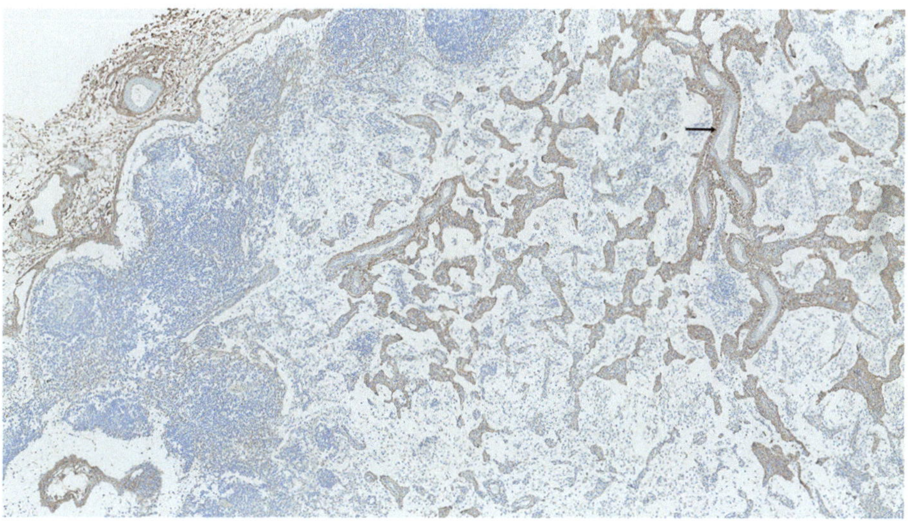

FIGURE 6.12A

Immunohistochemical staining for collagen-1 performed on the mesenteric lymph node of a male rhesus macaque necropsied 130 days after irradiation at 11 Gy shows a distinct increase in the collagen-1 staining associated with medullary vascular structures (arrow). Stain performed with 3, 3′-diaminobenzidine chromogen and hematoxylin counterstain, 5 × objective magnification.

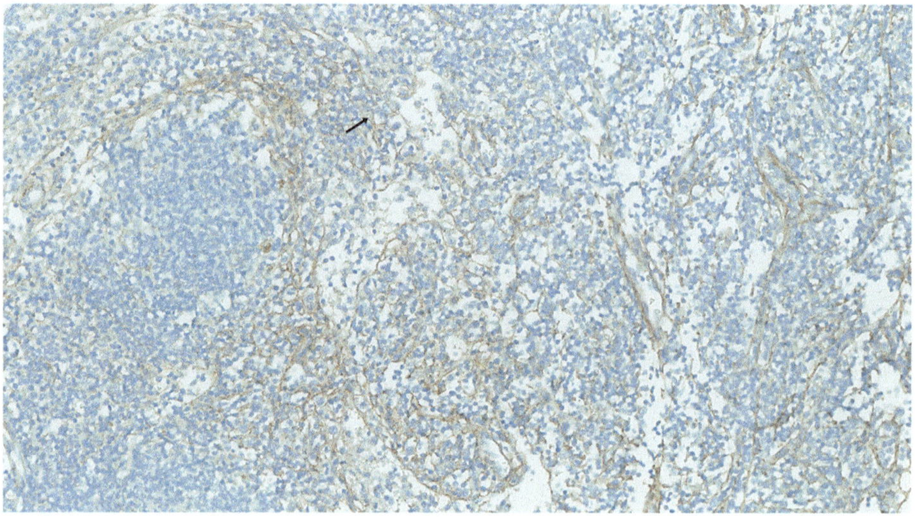

FIGURE 6.12B

Higher magnification of the previous specimen shows a delicate network of collagen 1 staining (arrow) surrounding lymphocytes in the cortex of the mesenteric lymph node. Collagen-1 immunohistochemical staining provided greater detail than trichrome staining in revealing the presence of fibrous connective tissue in mesenteric lymph nodes. Immunohistochemical stain for collagen-1 with 3, 3′-diaminobenzidine chromogen and hematoxylin counterstain, 20 × objective magnification.

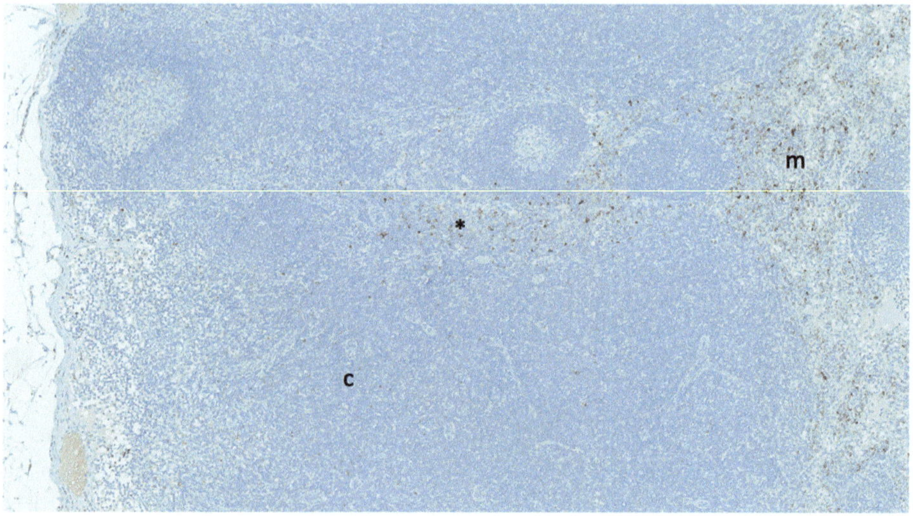

FIGURE 6.13A

Immunohistochemical staining for myeloperoxidase performed on the mesenteric lymph node of a naive male rhesus macaque shows a small number of MPO + cells in the medulla (m) and paracortical sinuses (*), with relative sparing of the cortex (c). Stain performed with 3,3′-diaminobenzidine chromogen and hematoxylin counterstain, 8.43 × objective magnification.

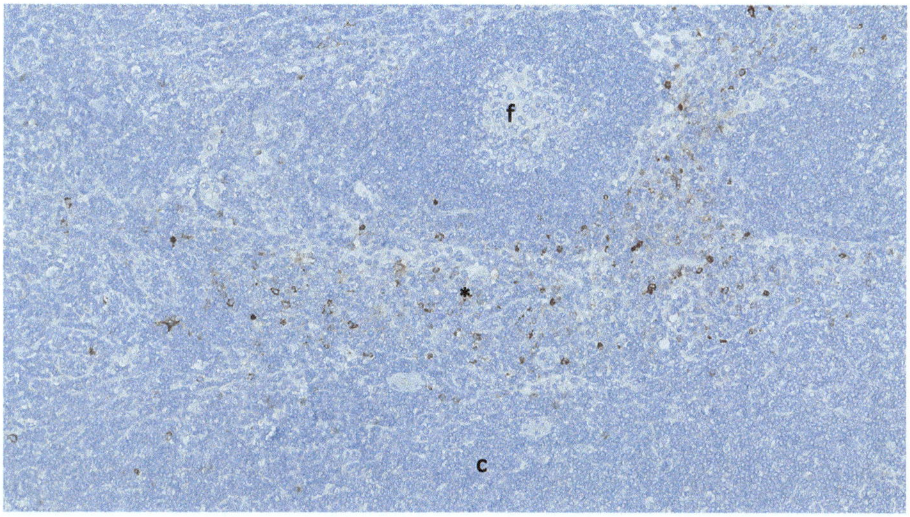

FIGURE 6.13B

Higher magnification of the previous specimen shows the myeloperoxidase (MPO) staining is localized in large mononuclear cells, consistent with cells of the macrophage/dendritic cell line. C = cortex, f = follicle, * = cortical sinus. Stain performed with 3, 3′-diaminobenzidine chromogen and hematoxylin counterstain, 20 × objective magnification.

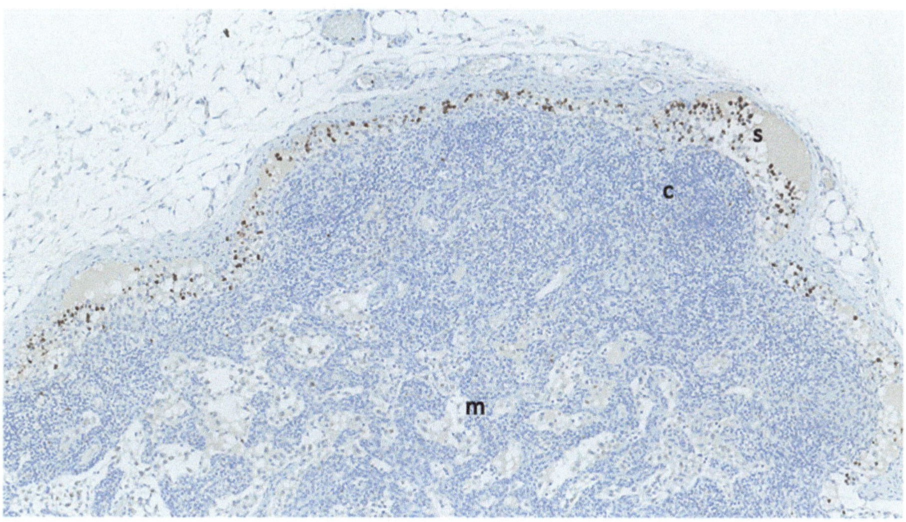

FIGURE 6.14A

Immunohistochemical staining for myeloperoxidase performed on the mesenteric lymph node of a male rhesus macaque necropsied 8 days after irradiation at 12 Gy shows a moderate number of MPO^+ cells in the subcapsular sinus (s), with relative sparing of the cortex (c) and medulla (m). The MPO^+ cells have translocated from the intestine to the mesenteric lymph node. Stain performed with 3,3′ diaminobenzidine chromogen and hematoxylin counterstain, 10 × objective magnification.

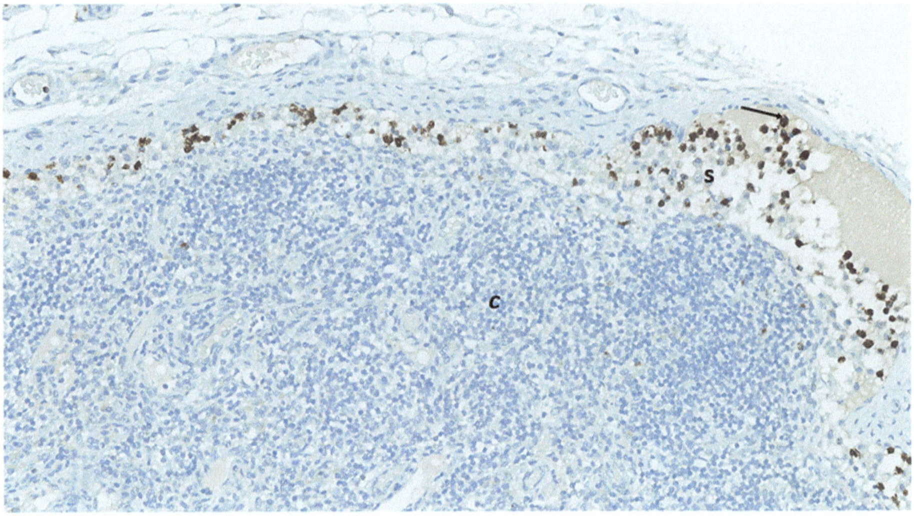

FIGURE 6.14B

Higher magnification of the previous specimen reveals the myeloperoxidase-positive cells in the subcapsular sinus (s) to be primarily large mononuclear cells (arrow), consistent with cells of macrophage/dendritic cell lineage. Note the underlying cortex (c) is devoid of MPO^+ cells. Immunohistochemical stain for myeloperoxidase with 3, 3′-diaminobenzidine chromogen and hematoxylin counterstain, 20 × objective magnification.

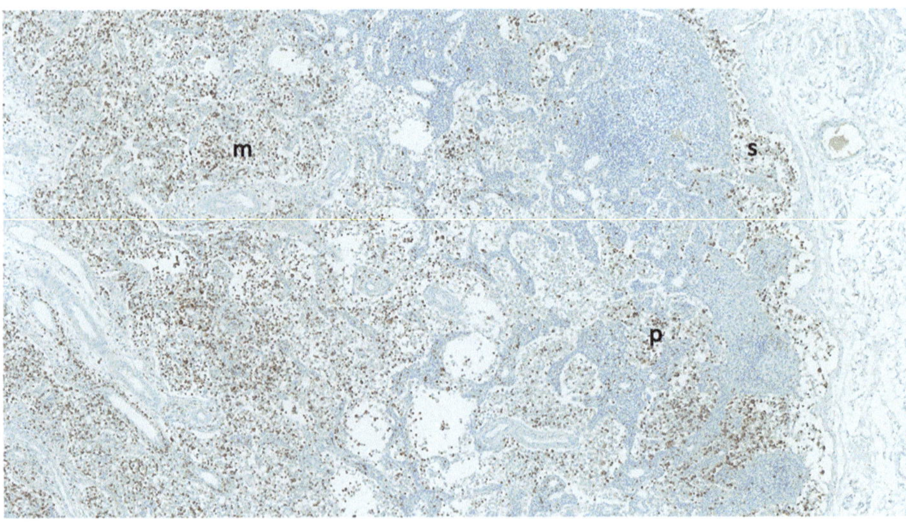

FIGURE 6.15A

Immunohistochemical staining for myeloperoxidase performed on the mesenteric lymph node of a male rhesus macaque necropsied 11 days after irradiation at 10 Gy shows a large number of MPO^+ cells in the subcapsular (s), paracortica (p) and medullary (m) sinuses. The increased population of MPO^+ cells is derived primarily from an influx of cells from the intestinal region served by the lymph node. Stain performed with 3, 3′-diaminobenzidine chromogen and hematoxylin counterstain, 5 × objective magnification.

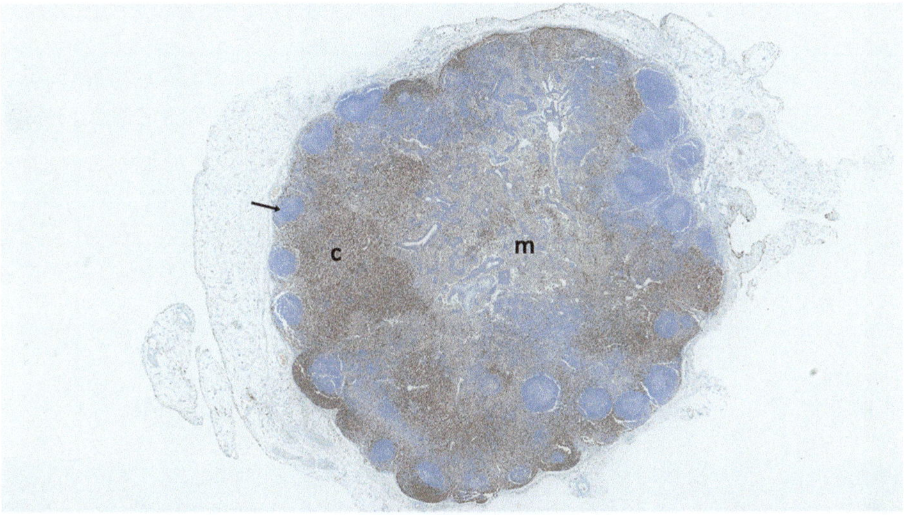

FIGURE 6.15B

Immunohistochemical staining for myeloperoxidase performed on the mesenteric lymph node of a male rhesus macaque necropsied 177 days after irradiation at 11 Gy shows a large number of MPO + cells in the cortex (c), with fewer MPO + cells in the medullar (m). Cortical follicles (arrow) are essentially devoid of MPO + cells. Stain performed with 3,3-diaminobenzidine chromogen and hematoxylin counterstain, 1.56 × objective magnification.

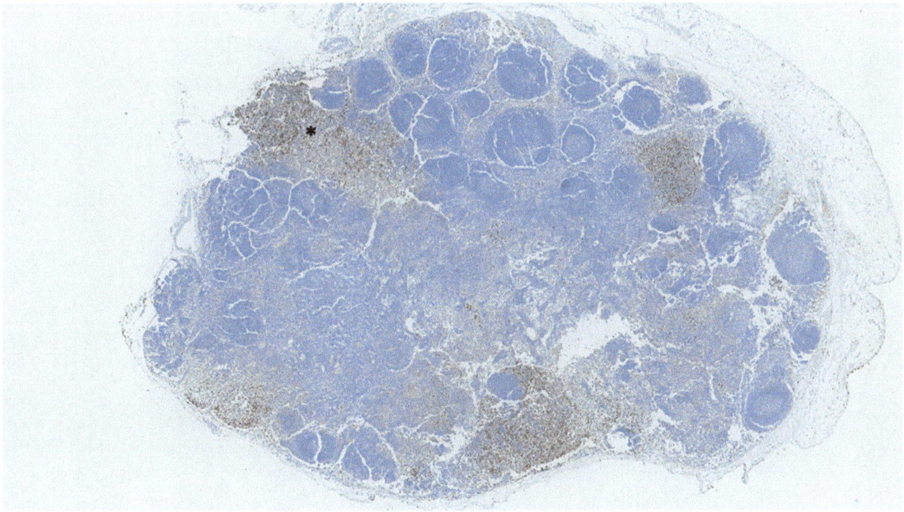

FIGURE 6.15C

An additional lymph node specimen from the same animal shown in the previous has multifocally intense populations of MPO^+ cells (*). The variation in MPO^+ cell populations is associated with variation in the radiation-associated pathological alterations in the intestinal regions drained by the different mesenteric lymph nodes. Immunohistochemical stain for myeloperoxidase with 3, 3′-diaminobenzidine chromogen and hematoxylin counterstain, 2.16 × objective magnification.

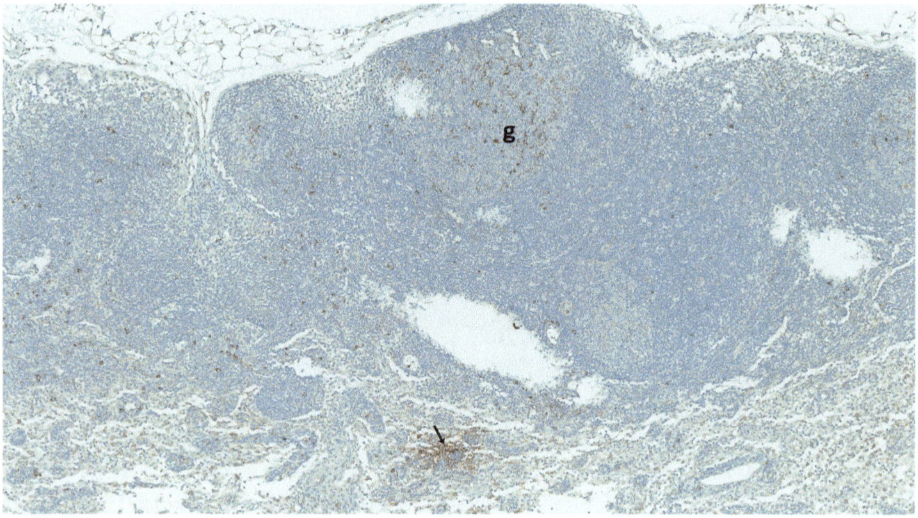

FIGURE 6.16

An immunohistochemical stain for MxA, a marker for activated immunocytes, performed on the mesenteric lymph node of a naive male rhesus macaque shows a scattered population of positively stained cells throughout the lymph node, with concentrations in the germinal center (g) of an active follicle and in a cluster (arrow) in the medulla. Stain performed with 3, 3′-diaminobenzidine chromogen and hematoxylin counterstain, 7.97 × objective magnification.

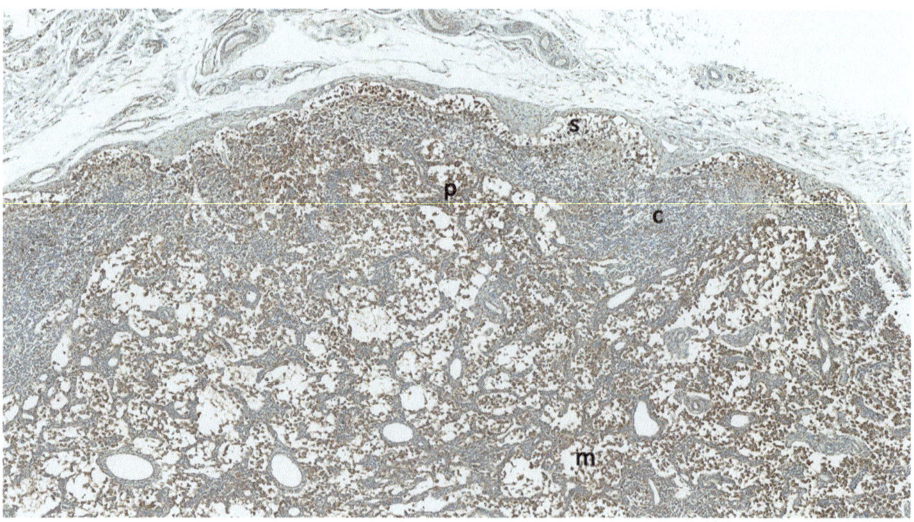

FIGURE 6.17A

An immunohistochemical stain for MxA performed on a mesenteric lymph node of a male rhesus macaque necropsied 22 days after irradiation at 11 Gy has a massive population of positively stained cells in subcapsular (s), paracortical (P), and medullary (m) sinuses. The small amount of remaining cortical tissue (c) has little MxA staining. Stain performed with 3, 3′-diaminobenzidine chromogen and hematoxylin counterstain, 5 × objective magnification.

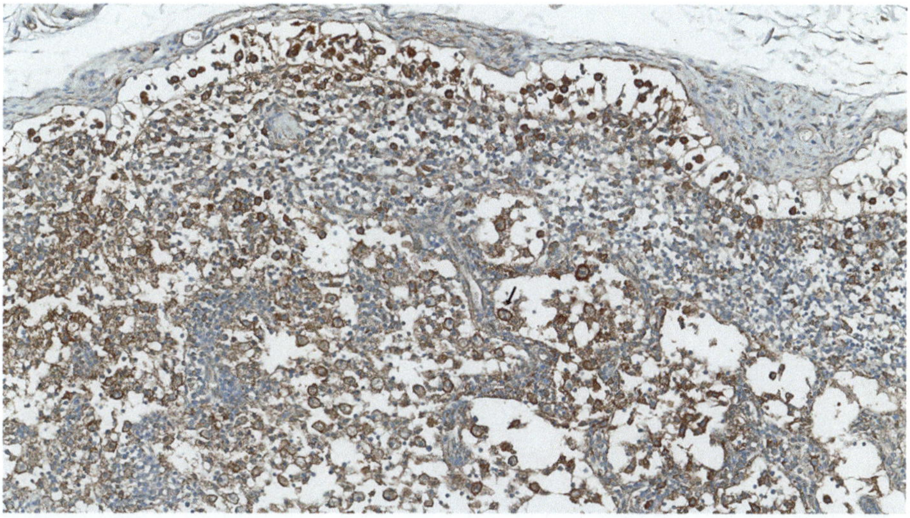

FIGURE 6.17B

Higher magnification of the previous specimen reveals the MxA$^+$ cell population consists primarily of large mononuclear cells (arrow), consistent with cells of the macrophage/dendritic cell lineage. Immunohistochemical stain for MxA with 3, 3′-diaminobenzidine chromogen and hematoxylin counterstain, 20 × objective magnification.

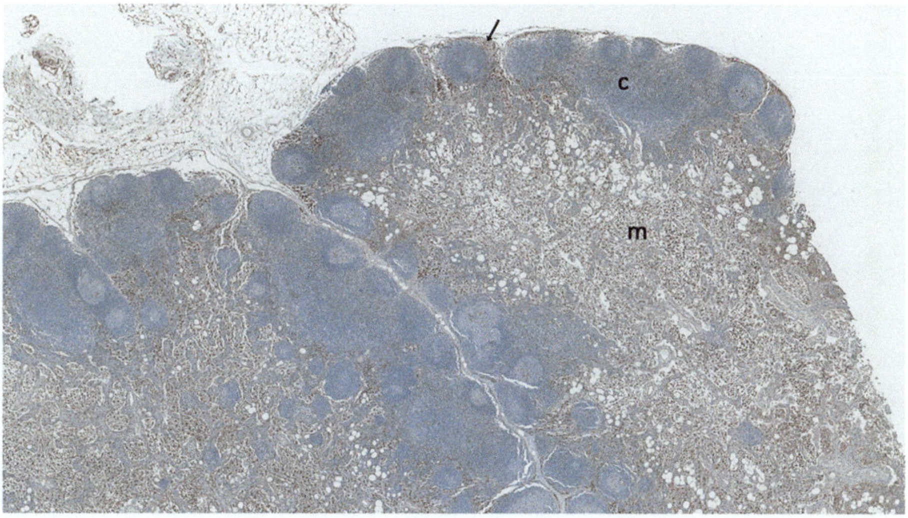

FIGURE 6.17C

An immunohistochemical stain for MxA performed on a mesenteric lymph node of a male rhesus macaque necropsied 183 days after irradiation at 10 Gy has a massive population of positively stained cells in the medulla (m) and subcapsular sinus (arrow), with fewer positively stained cells in the cortex (c). Mesenteric lymph nodes with the most pronounced populations of MxA$^+$ cells tended to have the most pronounced IHC staining for lipopolysaccharide (LPS) core antigen, suggesting the increased population of activated immunocytes in the lymph node was associated with leakage of LPS core antigen from the intestine. Stain performed with 3, 3′-diaminobenzidine chromogen and hematoxylin counterstain, 2.5 × objective magnification.

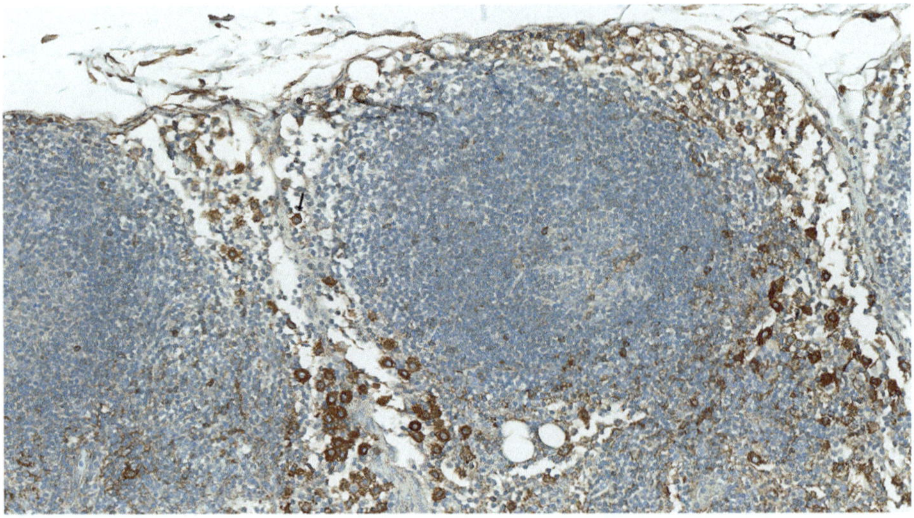

FIGURE 6.17D

Higher magnification of the previous specimen shows the MxA$^+$ cell population to consist primarily of large mononuclear cells (arrow), consistent with cells of the macrophage/dendritic cell line. Immunohistochemical stain for MxA with 3, 3′-diaminobenzidine chromogen and hematoxylin counterstain, 20 × objective magnification.

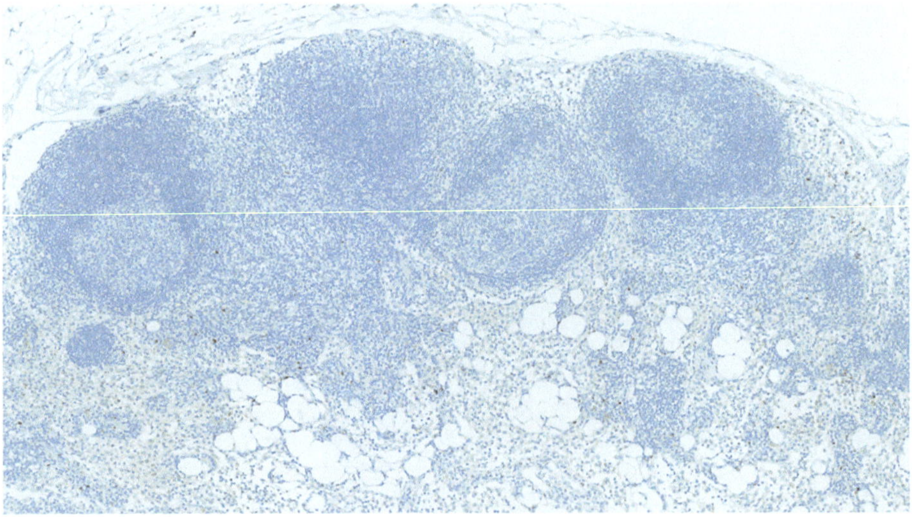

FIGURE 6.17E

An immunohistochemical stain for myeloperoxidase (MPO) performed on the same mesenteric lymph node as shown in the previous figures reveals a minimal population of MPO$^+$ macrophages. Stain performed with 3, 3′-diaminobenzidine chromogen and hematoxylin counterstain, 10 × objective magnification.

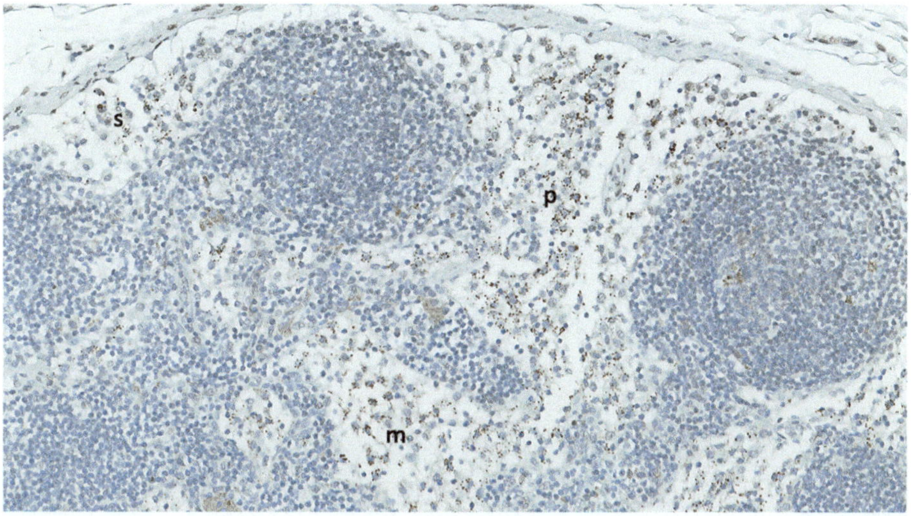

FIGURE 6.18A

An immunohistochemical stain for lipopolysaccharide (LPS) core antigen performed on the mesenteric lymph node of a male rhesus macaque necropsied 183 days after irradiation at 10 Gy reveals substantial LPS$^+$ particulate staining in subcapsular (s), paracortical (p), and medullary (m) sinuses. This is the same mesenteric lymph node specimen used to demonstrate MxA staining. Stain performed with 3, 3′-diaminobenzidine chromogen and hematoxylin counterstain, 20 × objective magnification.

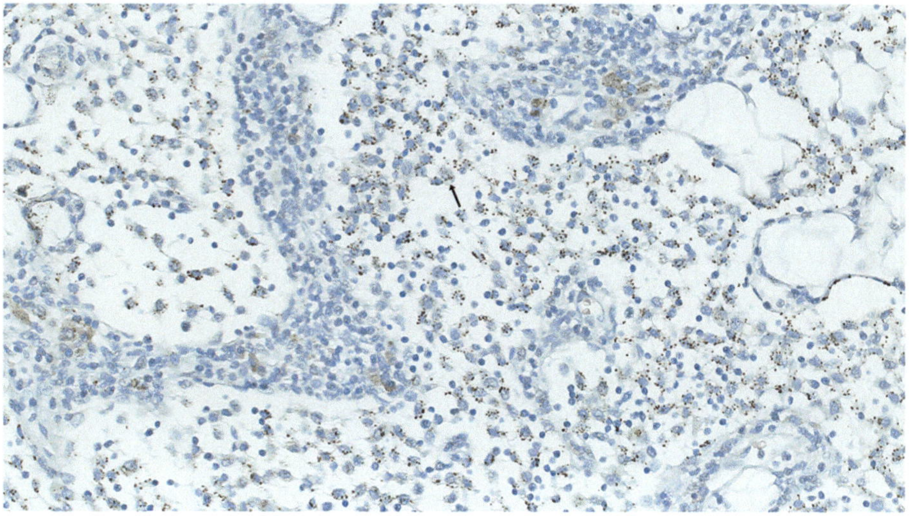

FIGURE 6.18B

A higher magnification of the previous specimen shows particulate LPS$^+$ staining that is primarily associated with large mononuclear cells (arrow) in medullary sinuses. Immunohistochemical stain for lipopolysaccharide core antigen with 3, 3′-diaminobenzidine chromogen and hematoxylin counterstain, 30.3 × objective magnification.

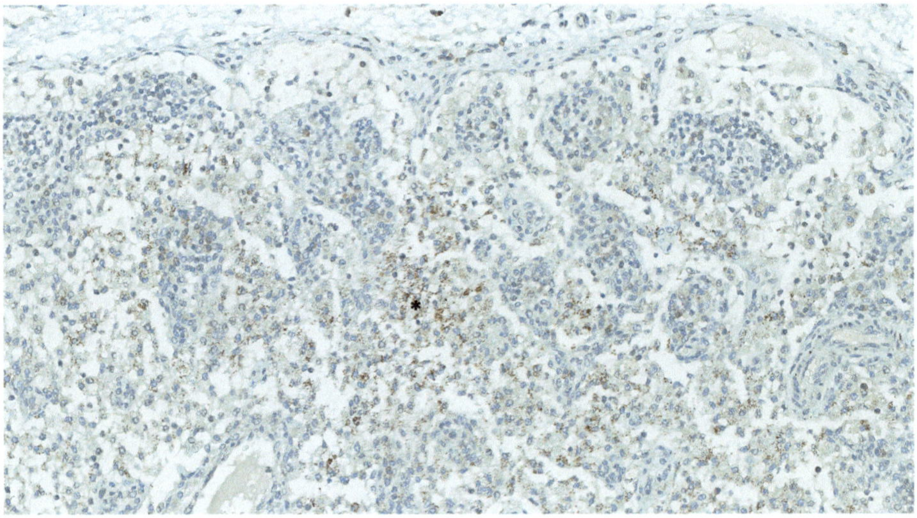

FIGURE 6.19

An immunohistochemical stain for connective tissue growth factor (CTGF) performed on the mesenteric lymph node of a male rhesus macaque necropsied 11 days after irradiation at 12 Gy shows a moderate population of $CTGF^+$ cells (*). Stain performed with 3, 3′-diaminobenzidine chromogen and hematoxylin counterstain, 20 × objective magnification.

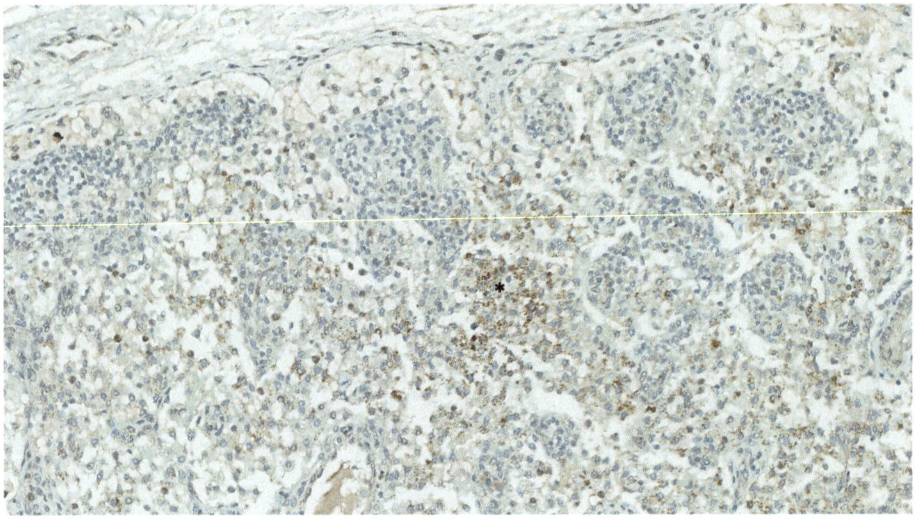

FIGURE 6.20

An immunohistochemical stain for transforming growth factor-beta (TGF-β) performed on the mesenteric lymph node of a male rhesus macaque necropsied 11 days after irradiation at 12 Gy shows a moderate population of TGF-β^+ cells (*). This specimen is from the same animal used to demonstrate connective tissue growth factor IHC staining. Stain performed with 3, 3′-diaminobenzidine chromogen and hematoxylin counterstain, 20 × objective magnification.

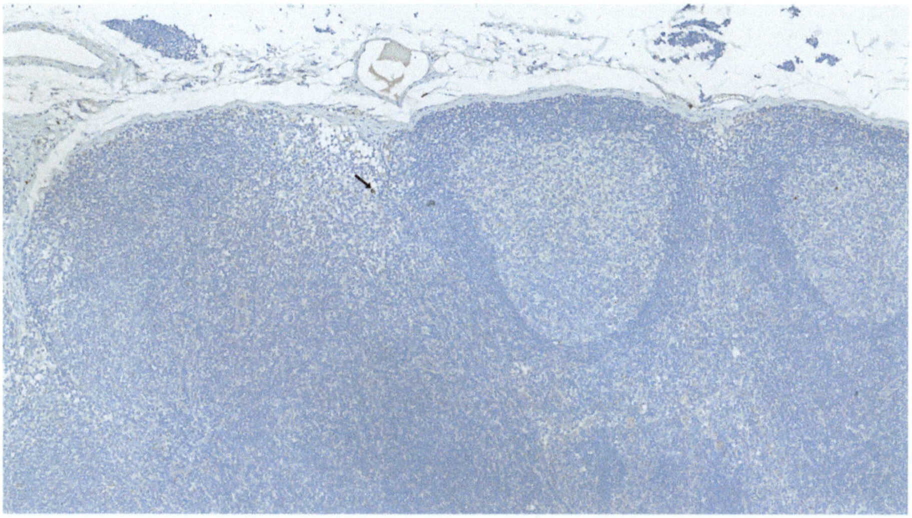

FIGURE 6.21

An immunohistochemical stain for tryptase, a marker for mast cell granules, performed on the mesenteric lymph node of a naive male rhesus macaque shows rare tryptase^{+} mast cells (arrow). Stain performed with 3, 3′-diaminobenzidine chromogen and hematoxylin counterstain, 10 × objective magnification.

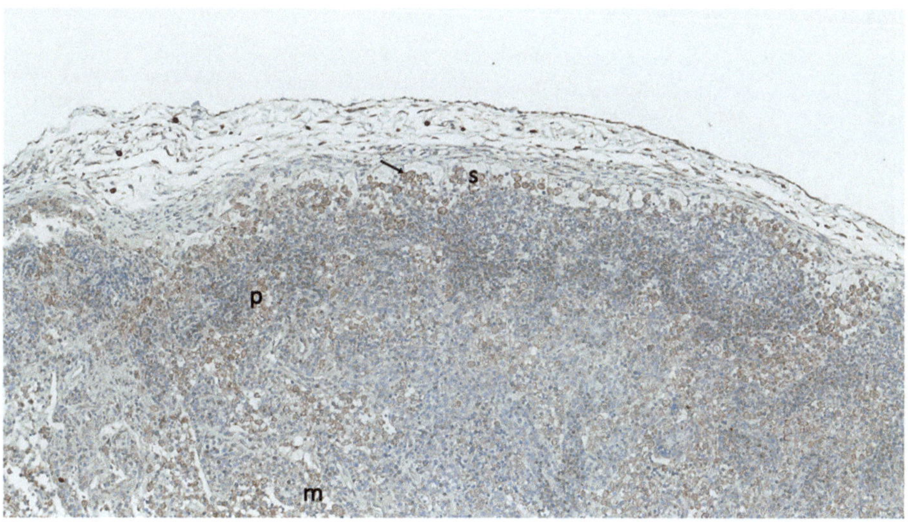

FIGURE 6.22A

An immunohistochemical stain for tryptase, a marker for mast cell granules, performed on the mesenteric lymph node of a male rhesus macaque necropsied 11 days after irradiation at 12 Gy shows a pronounced population of tryptase + cells (arrow) that are located primarily in subcapsular (s) paracortical (p) and medullary (m) sinuses. Stain performed with 3,3′-diaminobenzidine chromogen and hematoxylin counterstain, 10 × objective magnification.

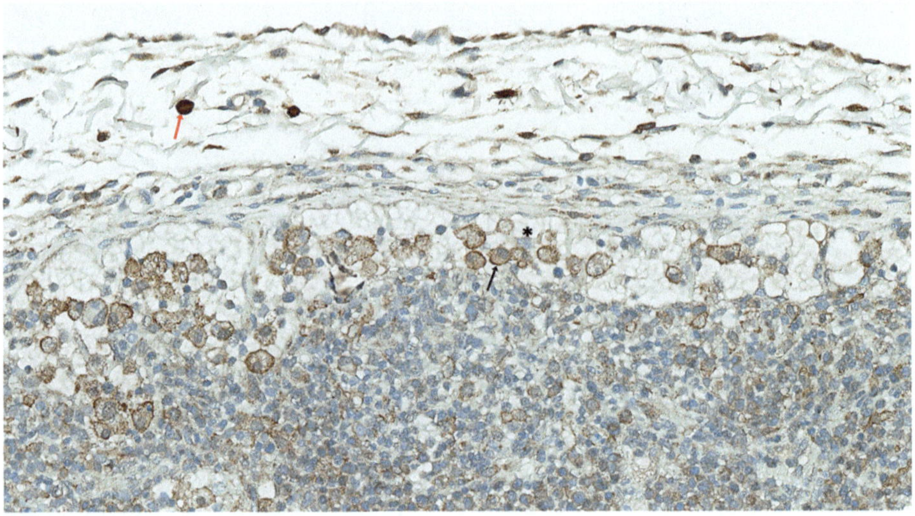

FIGURE 6.22B

Higher magnification of the previous specimen shows a moderate level of tryptase staining on the surface of cells (black arrow) in subcapsular sinuses (*). Tryptase is contained within cytoplasmic granules of mast cells and may be released as a response to irradiation. The origin and significance of the tryptase staining of the large mononuclear cells within mesenteric lymph node sinuses but may have resulted from degranulation of mast cells in the intestinal region drained by the lymph node. The dense tryptase staining of mast cells (red arrow) in the lymph node capsule is shown for comparison. Immunohistochemical stain for tryptase with 3, 3′-diaminobenzidine chromogen and hematoxylin counterstain, 30 × objective magnification.

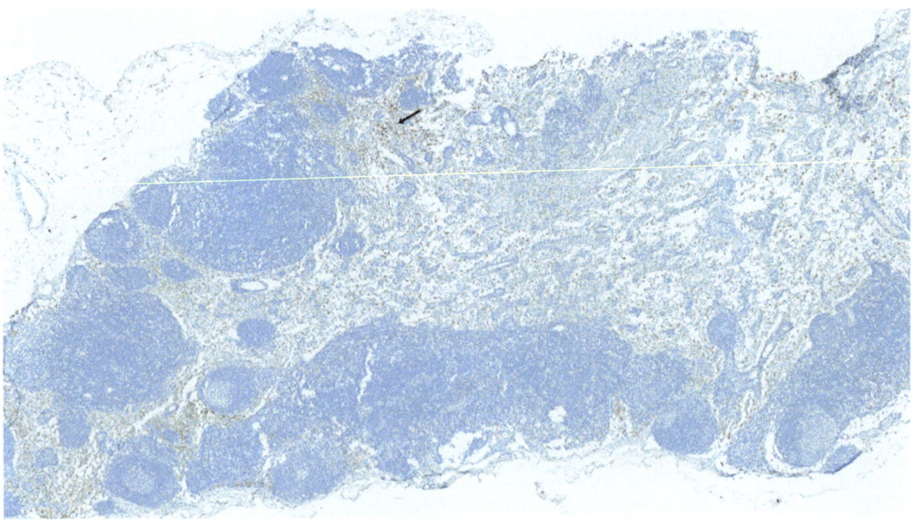

FIGURE 6.23A

Immunohistochemical staining for CD163 reveals only small clusters of brown stained CD163 + cells (arrow) in the mesenteric lymph node of a naive control animal. CD163 IHC staining with 3,3′-diaminobenzidine chromogen and hematoxylin counterstain, 5 × objective magnification.

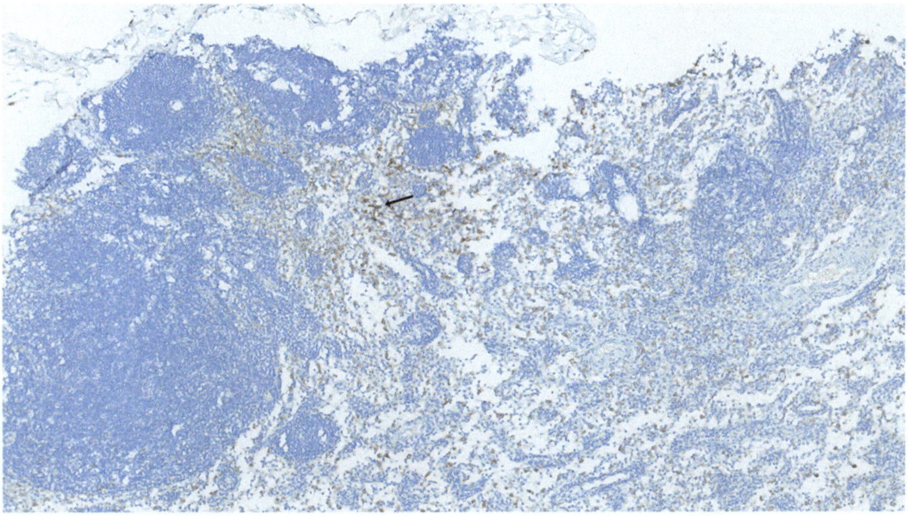

FIGURE 6.23B

Higher magnification of the previous image shows cluster of CD163^{+} cells (arrow) in medullary sinuses of the mesenteric lymph node. CD163 IHC staining with 3,3′-diaminobenzidine chromogen and hematoxylin counterstain, 5 × objective magnification.

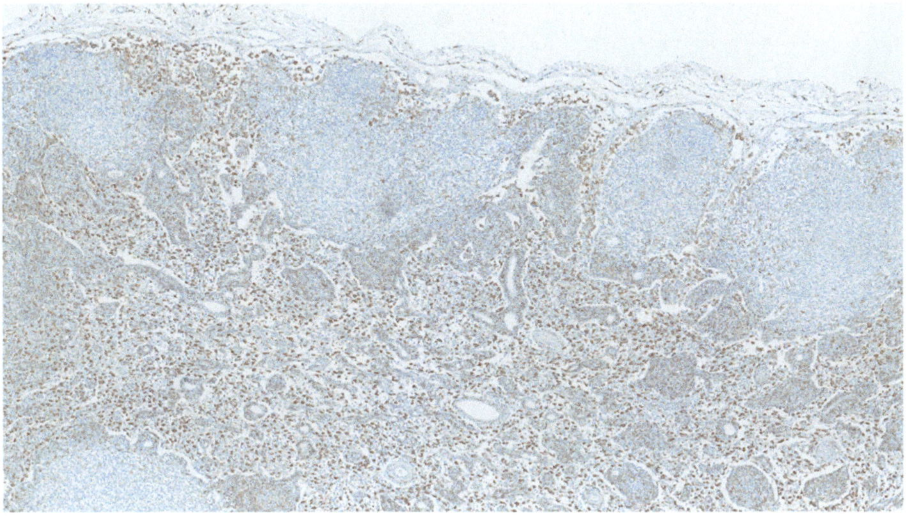

FIGURE 6.24A

Immunohistochemical (IHC) staining for CD163 reveals a massive number of brown stained CD163^{+} cells in the mesenteric lymph node of a rhesus macaque that was necropsied 24 days after irradiation at 12 Gy. CD163 IHC staining with 3,3′-diaminobenzidine chromogen and hematoxylin counterstain, 5 × objective magnification.

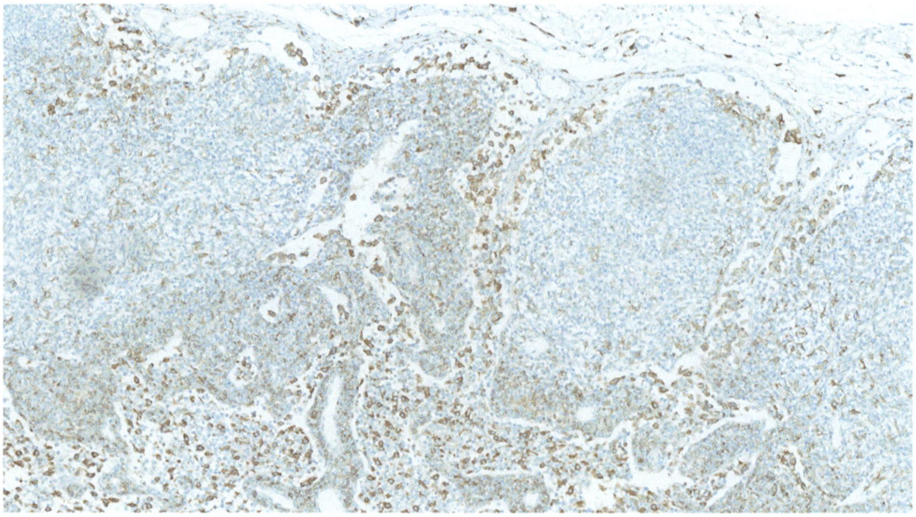

FIGURE 6.24B

Higher magnification of previous image shows numerous $CD163^+$ cells in subcapsular, paracortical, and medullary sinuses. 5 × objective magnification.

FIGURE 6.25

Immunohistochemical staining for CD38 reveals a large number of brown stained $CD38^+$ cells in the mesenteric lymph node of a rhesus macaque necropsied 16 days after irradiation at 12 Gy. Stain performed with 3,3′-diaminobenzidine chromogen and hematoxylin counterstain, 5 × objective magnification.

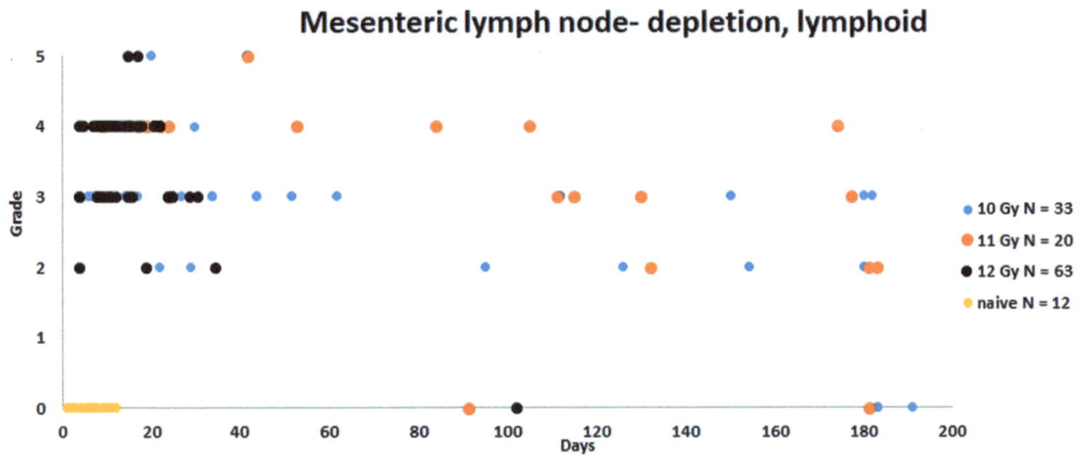

FIGURE 6.26

Scatter plot: Mesenteric lymph node—depletion, lymphoid. Observations based on light microscopic examination of paraffin-embedded specimens stained with hematoxylin and eosin.

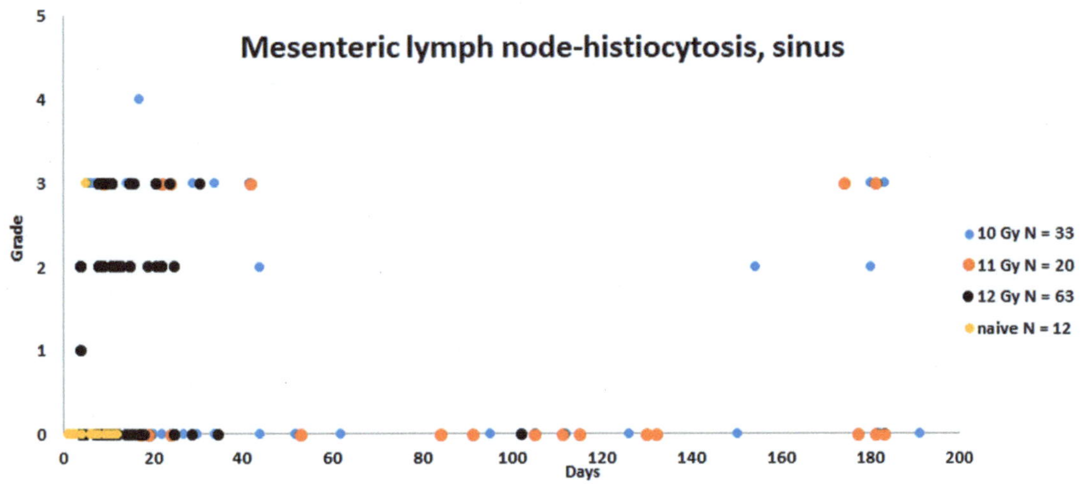

FIGURE 6.27

Scatter plot: Mesenteric lymph node—histiocytosis, sinus. Observations based on light microscopic examination of paraffin-embedded specimens stained with hematoxylin and eosin.

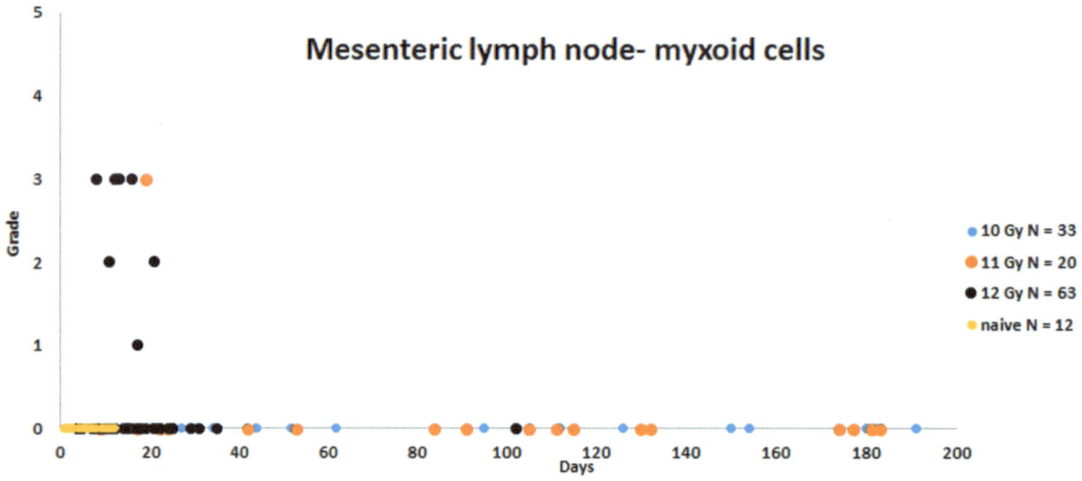

FIGURE 6.28

Scatter plot: Mesenteric lymph node—myxoid cells. Observations based on light microscopic examination of paraffin-embedded specimens stained with hematoxylin and eosin.

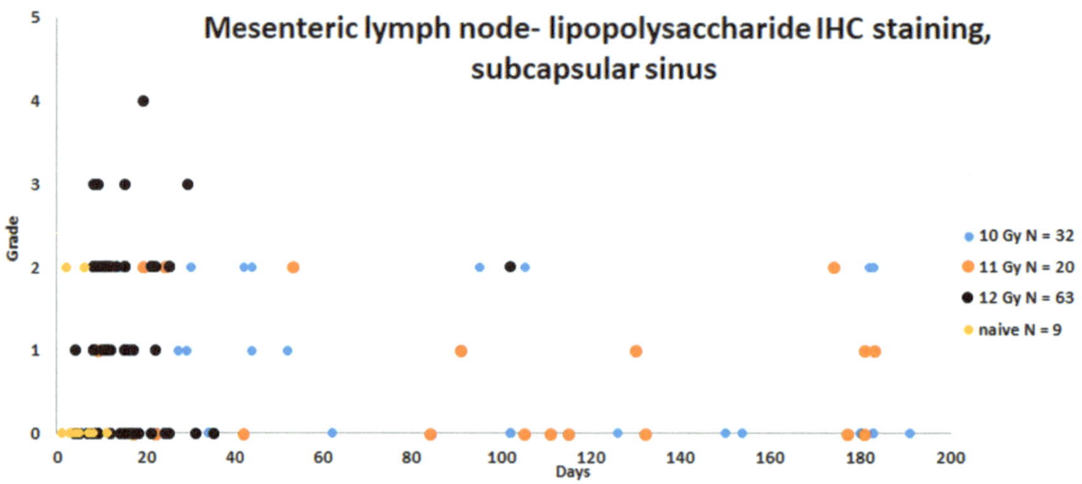

FIGURE 6.29

Scatter plot: Mesenteric lymph node- lipopolysaccharide (LPS) core antigen staining, subcapsular sinus. Observations based on light microscopic examination of paraffin-embedded specimens stained by immunohistochemical procedures for LPS with 3,3′-diaminobenzidine chromogen and hematoxylin counterstain.

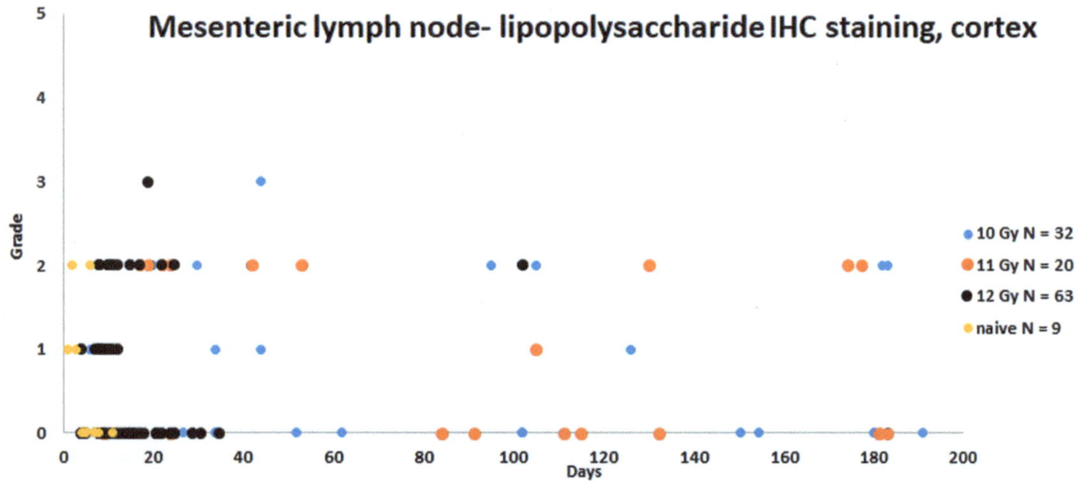

FIGURE 6.30

Scatter plot: Mesenteric lymph node—lipopolysaccharide (LPS) core antigen staining, cortex. Observations based on light microscopic examination of paraffin-embedded specimens stained by immunohistochemical procedures for LPS with 3,3′-diaminobenzidine chromogen and hematoxylin counterstain.

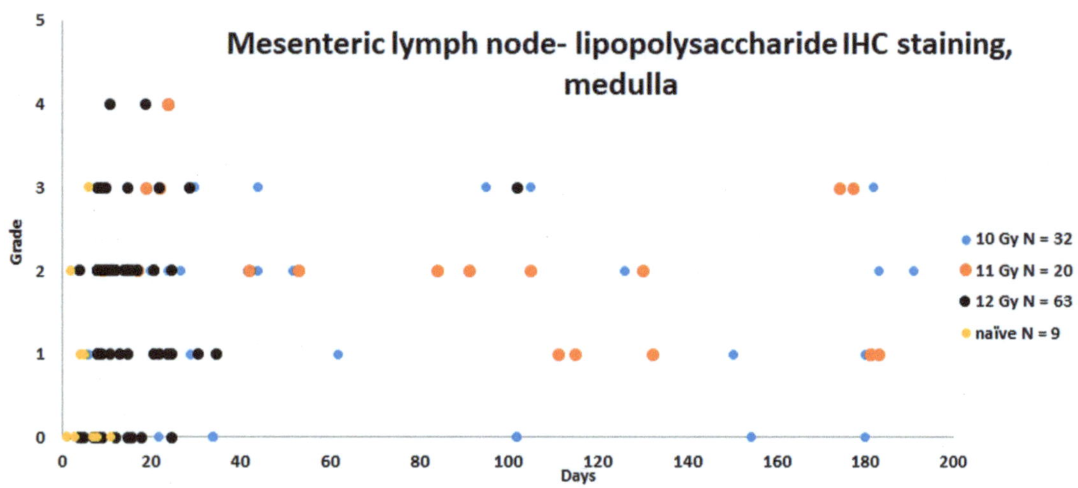

FIGURE 6.31

Scatter plot: Mesenteric lymph node—lipopolysaccharide (LPS) core antigen staining, medulla. Observations based on light microscopic examination of paraffin-embedded specimens stained by immunohistochemical procedures for LPS with 3,3′-diaminobenzidine chromogen and hematoxylin counterstain.

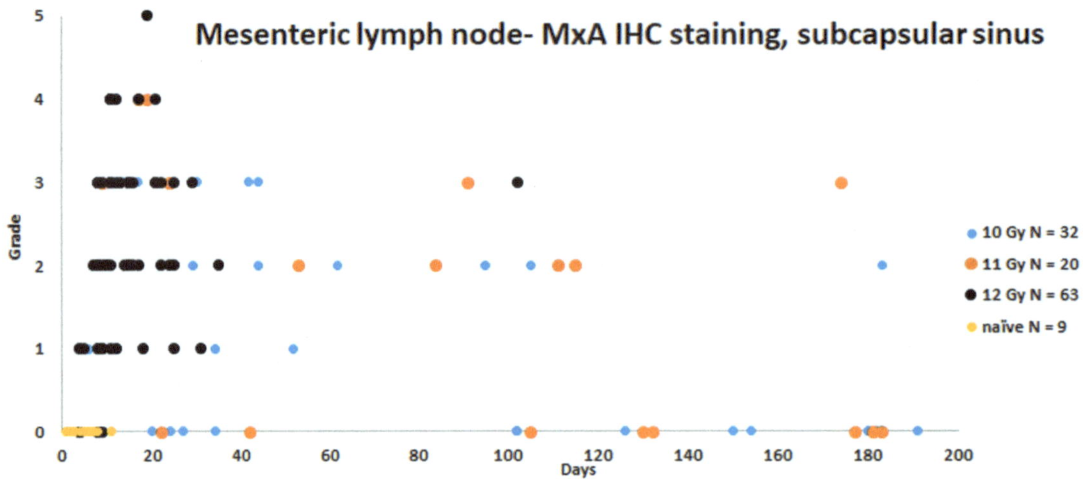

FIGURE 6.32

Scatter plot: Mesenteric lymph node—MxA staining, subcapsular sinus. Observations based on light microscopic examination of paraffin-embedded specimens stained by immunohistochemical procedures for MxA with 3,3′-diaminobenzidine chromogen and hematoxylin counterstain.

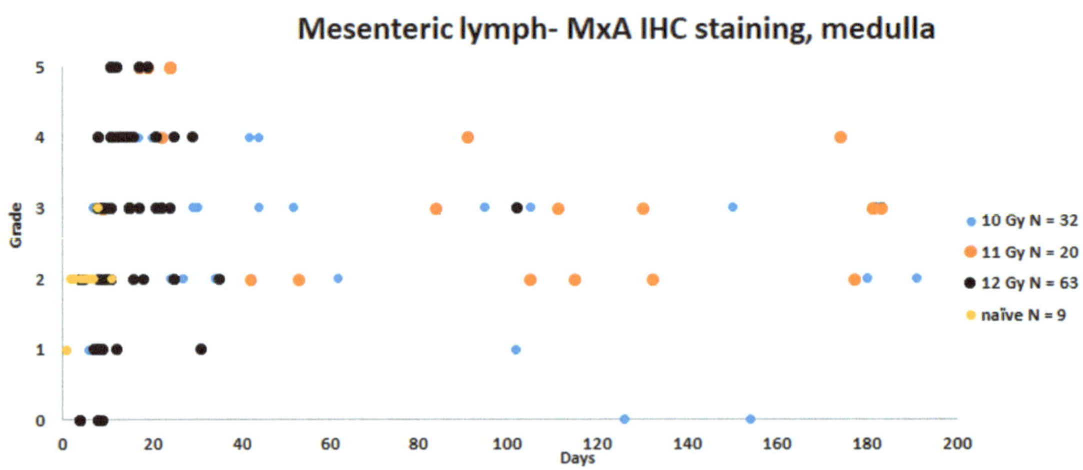

FIGURE 6.33

Scatter plot: Mesenteric lymph node—MxA staining, medulla. Observations based on light microscopic examination of paraffin-embedded specimens stained by immunohistochemical procedures for MxA with 3,3′-diaminobenzidine chromogen and hematoxylin counterstain.

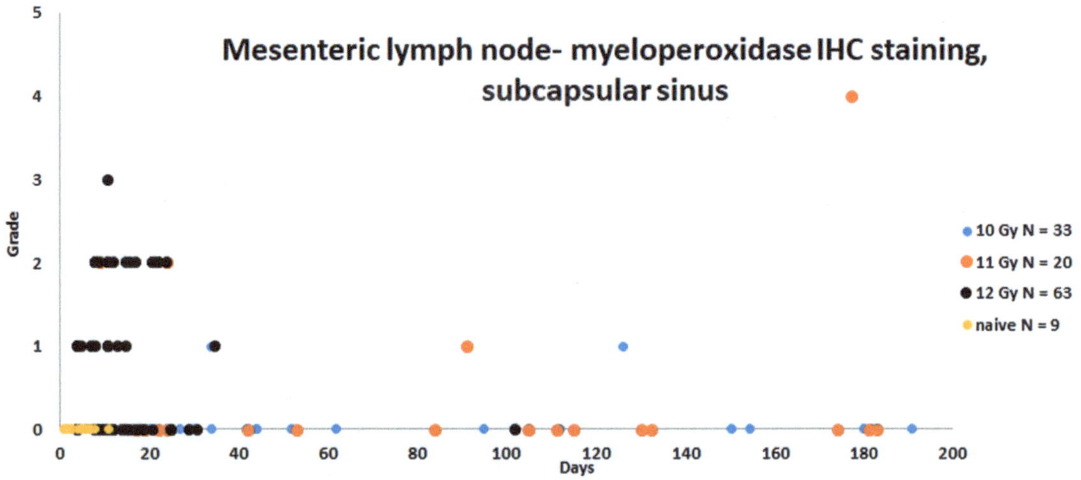

FIGURE 6.34

Scatter plot: Mesenteric lymph node—myeloperoxidase staining, subcapsular sinus. Observations based on light microscopic examination of paraffin-embedded specimens stained by immunohistochemical procedures for myeloperoxidase with 3,3′-diaminobenzidine chromogen and hematoxylin counterstain.

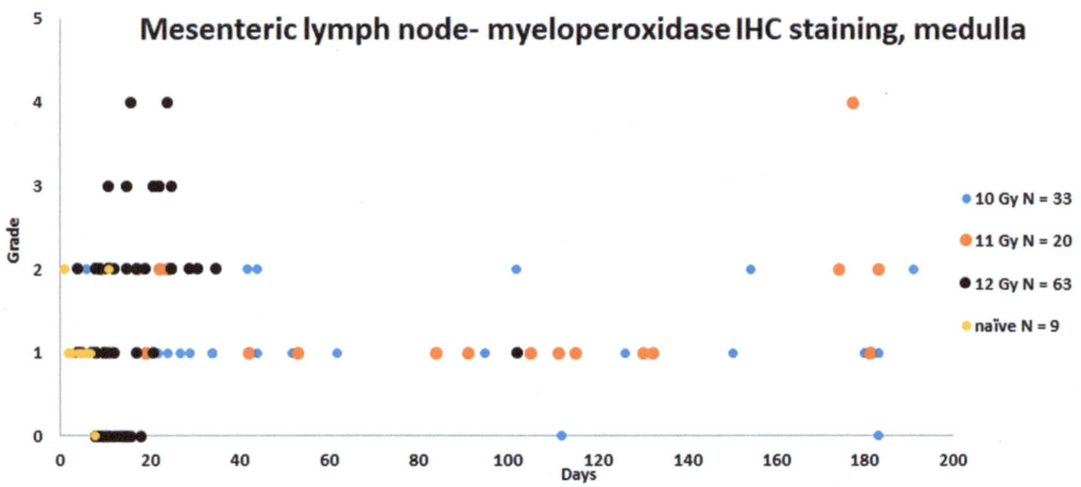

FIGURE 6.35

Scatter plot: Mesenteric lymph node—myeloperoxidase staining, medulla. Observations based on light microscopic examination of paraffin-embedded specimens stained by immunohistochemical procedures for myeloperoxidase with 3,3′-diaminobenzidine chromogen and hematoxylin counterstain.

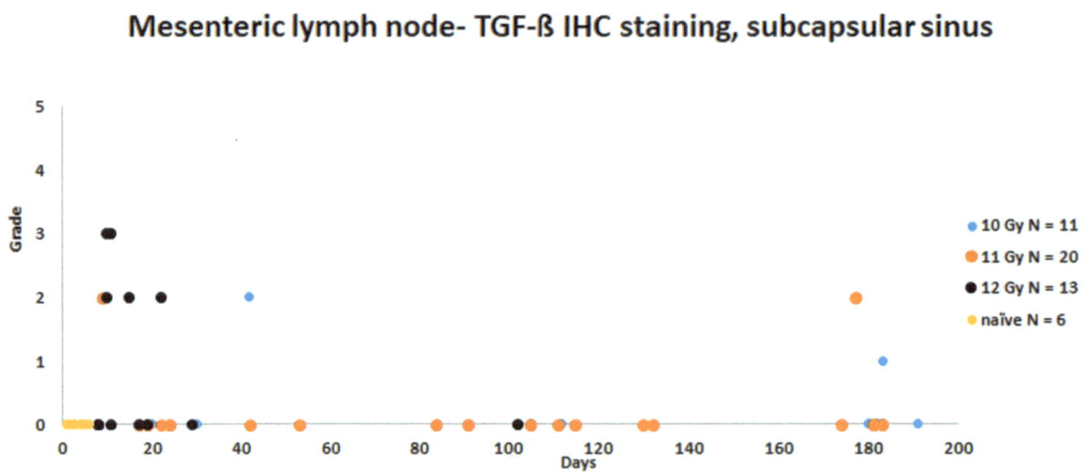

FIGURE 6.36

Scatter plot: Mesenteric lymph node—transforming growth factor-beta (TGF-β) staining, subcapsular sinus. Observations based on light microscopic examination of paraffin-embedded specimens stained by immunohistochemical procedures for TGF-β with 3,3′-diaminobenzidine chromogen and hematoxylin counterstain.

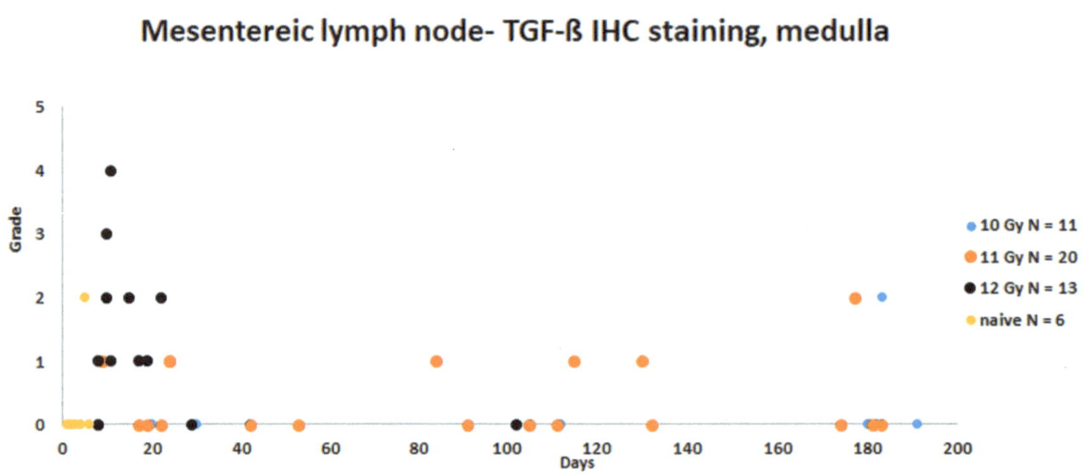

FIGURE 6.37

Scatter plot: Mesenteric lymph node—transforming growth factor-beta (TGF-β) staining, medulla. Observations based on light microscopic examination of paraffin-embedded specimens stained by immunohistochemical procedures for TGF-β with 3,3′-diaminobenzidine chromogen and hematoxylin counterstain.

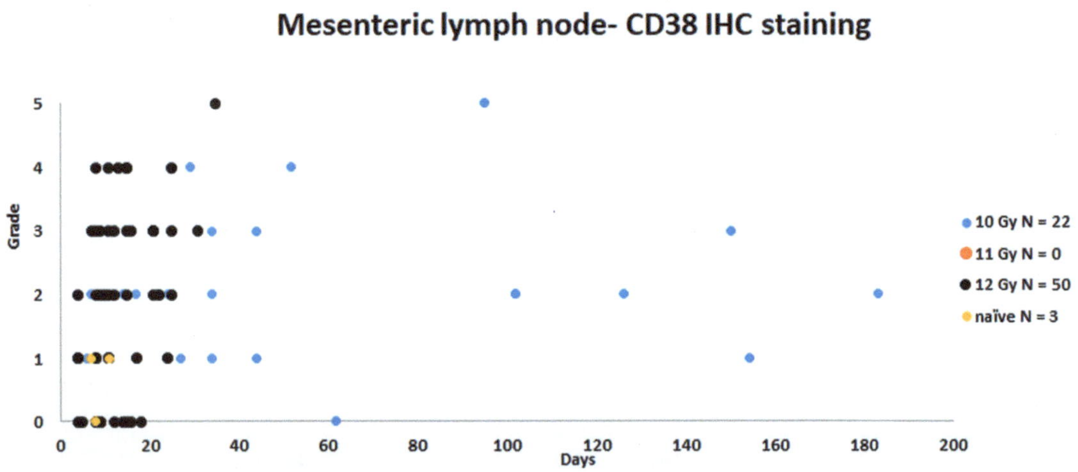

FIGURE 6.38

Scatter plot: Mesenteric lymph node—CD38 staining. Observations based on light microscopic examination of paraffin-embedded specimens stained by immunohistochemical procedures for CD38 with 3,3′-diaminobenzidine chromogen and hematoxylin counterstain.

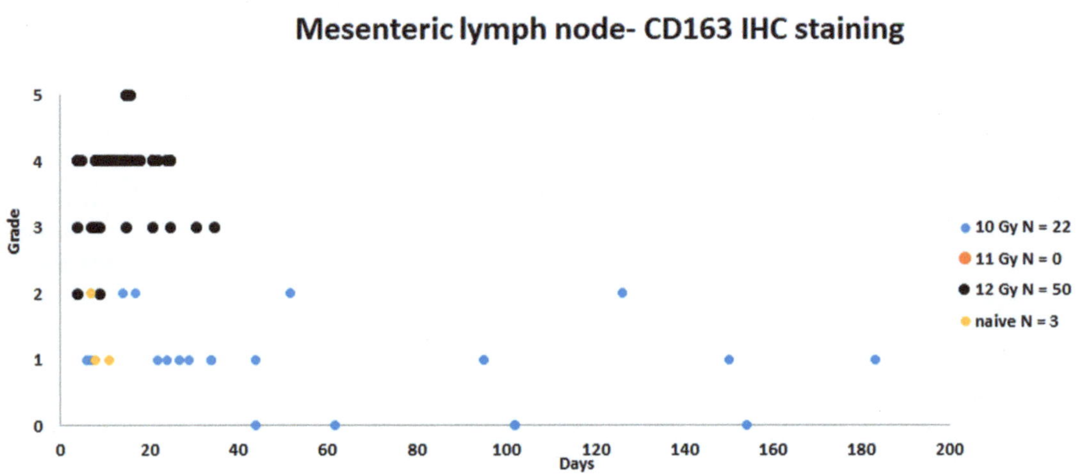

FIGURE 6.39

Scatter plot: Mesenteric lymph node—CD163 staining. Observations based on light microscopic examination of paraffin-embedded specimens stained by immunohistochemical procedures for CD163 with 3,3′-diaminobenzidine chromogen and hematoxylin counterstain.

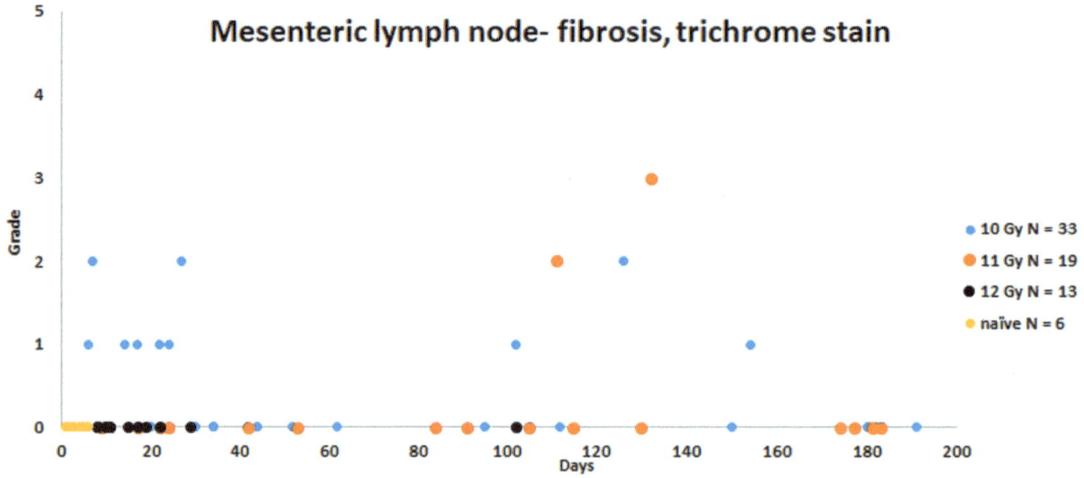

FIGURE 6.40

Scatter plot: Mesenteric lymph node—fibrosis. Observations based on light microscopic examination of paraffin-embedded specimens stained with Masson's trichrome stain.

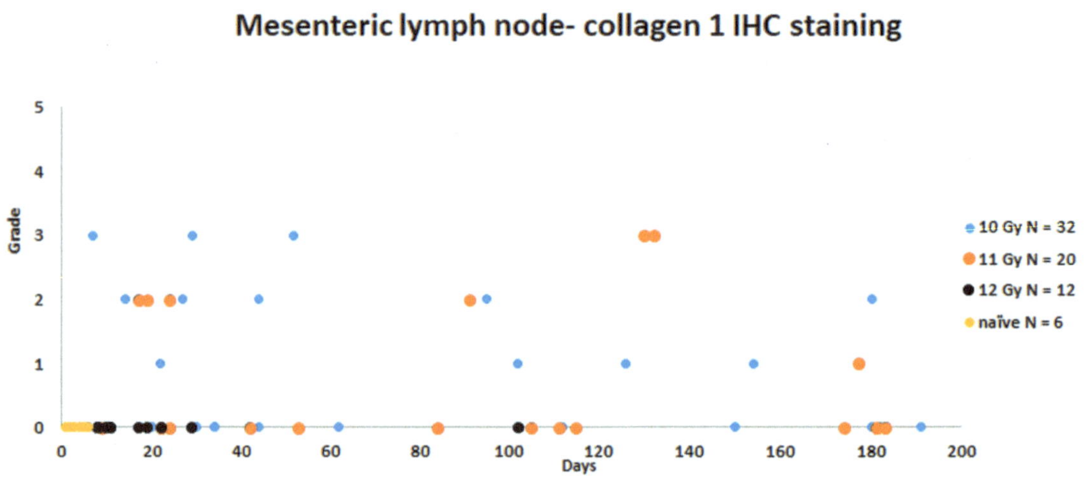

FIGURE 6.41

Scatter plot: Mesenteric lymph node, collagen 1 staining. Observations based on light microscopic examination of paraffin-embedded specimens stained by immunohistochemical procedures for collagen 1 with 3,3′-diaminobenzidine chromogen and hematoxylin counterstain.

Index

Note: Page numbers followed by "*f*" refer to figures.

A

B

C

J

K

L

M

N

P